Integrating Science and Policy

Science in Society Series

Series Editor: Steve Rayner
Institute for Science, Innovation and Society, University of Oxford

Editorial Board: Gary Kass, Anne Kerr, Melissa Leach, Angela Liberatore, Stan Metcalfe, Paul Nightingale, Timothy O'Riordan, Nick Pidgeon, Ortwin Renn, Dan Sarewitz, Andrew Webster, James Wilsdon, Steve Yearley

Animals as Biotechnology
Ethics, Sustainability and Critical Animal Studies
Richard Twine

Business Planning for Turbulent Times
New Methods for Applying Scenarios
Edited by Rafael Ramírez, John W. Selsky and Kees van der Heijden

Debating Climate Change
Understanding Debate and Agreement
Elizabeth L. Malone

Democratizing Technology
Risk, Responsibility and the Regulation of Chemicals
Anne Chapman

Genomics and Society
Legal, Ethical and Social Dimensions
Edited by George Gaskell and Martin W. Bauer

Integrating Science and Policy
Vulnerability and Resilience in Global Environmental Change
Edited by Roger E. Kasperson and Mimi Berberian

Marginalized Reproduction
Ethnicity, Infertility and Reproductive Technologies
Lorraine Culley, Nicky Hudson and Floor van Rooij

Nanotechnology
Risk, Ethics and Law
Edited by Geoffrey Hunt and Michael Mehta

Resolving Messy Policy Problems
Handling Conflict in Environmental, Transport, Health and Ageing Policy
Steven Ney

Unnatural Selection
The Challenges of Engineering Tomorrow's People
Edited by Peter Healey and Steve Rayner

Vaccine Anxieties
Global Science, Child Health and Society
Melissa Leach and James Fairhead

A Web of Prevention
Biological Weapons, Life Sciences and the Governance of Research
Edited by Brian Rappert and Caitrìona McLeish

Integrating Science and Policy

Vulnerability and Resilience in Global Environmental Change

*Edited by Roger E. Kasperson
and Mimi Berberian*

publishing for a sustainable future

London • Washington, DC

First published in 2011 by Earthscan

Copyright © Roger E. Kasperson and Mimi Berberian, 2011

All rights reserved. No part of this publication may be reproduced, stored in a retrieval system, or transmitted, in any form or by any means, electronic, mechanical, photocopying, recording or otherwise, except as expressly permitted by law, without the prior, written permission of the publisher.

Earthscan Ltd, Dunstan House, 14a St Cross Street, London EC1N 8XA, UK
Earthscan LLC, 1616 P Street, NW, Washington, DC 20036, USA

Earthscan publishes in association with the International Institute for Environment and Development

For more information on Earthscan publications, see www.earthscan.co.uk
or write to earthinfo@earthscan.co.uk

ISBN 978-1-84407-605-5 hardback
      978-1-84407-606-2 paperback

Typeset by MapSet Ltd, Gateshead, UK
Cover design by Susanne Harris

A catalogue record for this book is available from the British Library

Library of Congress Cataloging-in-Publication Data

Integrating science and policy : vulnerability and resilience in global environmental change / edited by Roger E. Kasperson and Mimi Berberian.
   p. cm.
 Includes bibliographical references and index.
  ISBN 978-1-84407-605-5 (hardback) — ISBN 978-1-84407-606-2 (pbk.)
 1. Global environmental change. 2 Environmental policy. 3. Environmental sciences.
I. Kasperson, Roger E. II. Berberian, Mimi.
GE149.I525 2010
333.72—dc22

                                                                          2010024665

At Earthscan we strive to minimize our environmental impacts and carbon footprint through reducing waste, recycling and offsetting our CO_2 emissions, including those created through publication of this book. For more details of our environmental policy, see www.earthscan.co.uk.

Printed and bound in the UK by MPG Books,
an ISO 14001 accredited company. The paper used
is FSC certified.

Library
University of Texas
at San Antonio

Contents

List of figures, tables and boxes   *ix*
List of contributors   *xiii*
List of acronyms and abbreviations   *xv*

Part I — Introduction

1. Characterizing the Science/Practice Gap   3
   Roger E. Kasperson

Part II — What Do We Know Now?

2. Knowledge to Practice in the Vulnerability, Adaptation and Resilience Literature: A Propositional Inventory   23
   Rebecca Morgenstern Brenner

3. Integrating Science and Practice for the Mitigation of Natural Disasters: Barriers, Bridges, Propositions   51
   Juergen Weichselgartner

4. Linking Vulnerability, Adaptation and Resilience Science to Practice: Pathways, Players and Partnerships   97
   Coleen Vogel, Susanne C. Moser, Roger E. Kasperson and Geoffrey D. Dabelko

Part III — Growing Political Urgency: Climate Change

5. The US Climate Change Science Program   131
   Zachary Christman

6. Linking Climate Change Science with Policy in California   151
   Guido Franco, Dan Cayan, Amy Luers, Michael Hanemann and Bart Croes

7 Russia's Climate Policy and the Kyoto Ratification Deal:
   Assessing the Science/Practice Interface   169
   Elena Nikitina and Vladimir Kotov

8 Urban and Social Vulnerability to Climate Variability in
   Tijuana, Mexico   187
   Roberto Sánchez-Rodríguez

Part IV — The Science/Practice Gap: Global Perspectives

9 Food Insecurity in South Africa   215
   Scott Drimie and Gina Ziervogel

10 Science and Vulnerability Reduction in Taiwan after the
   1999 Chi-Chi Earthquake   233
   Kuan-Hui Elaine Lin, Huei-Min Tsai and Chang-Yi David Chang

11 Participatory Evaluation of Development Interventions in a
   Vulnerable African Environment   269
   Ton Dietz

12 Science and Indigenous Knowledge in Resource Management in
   the Canadian Arctic   291
   Johanna Wandel, Barry Smit, Tristan Pearce and James Ford

13 Reducing Vulnerability of Rural Communities in the Philippines:
   Modelling Social Links between Science and Policy   307
   Lilibeth Acosta-Michlik and Victoria Espaldon

14 Addressing Vulnerability in the European Programme for Food
   Aid and Food Security: Knowledge Gaps, Obstacles and
   Opportunities Across the Science/Practice Interface   335
   Elia Machado

15 Land in Transition: Coping with Market Forces in Managing
   Rangelands in Mongolia   363
   Togtohyn Chuluun and Dennis Ojima

16 Managing Floods and Scarcity in a Monsoon Climate   381
   Louis Lebel, Po Garden and Phimphakan Lebel

Part V — Where Do We Go from Here?

17 Issues that Need to be Addressed: Assessing Experience 421
    Uno Svedin

18 Directions for Closing the Science/Practice Gap 427
    Roger E. Kasperson

Index *441*

List of Figures, Tables and Boxes

Figures

1.1	Social amplification	7
1.2	Spider web	8
1.3	Boundary organizations in end-to-end systems	10
1.4	Two risk decision models	14
3.1	Great natural disasters and economic losses	52
3.2	Elements of a conceptual framework for analysing the influence of vulnerability assessments	55
3.3	Pipeline knowledge transfer and common failures	75
4.1	Scientific input at various stages of the decision-making process and the nature of science's influence	104
5.1	Structure of the US organizations related to climate change activities	134
5.2	US Climate Change Science Program (CCSP) budget allocated by goal (2009 estimated budget)	137
5.3	CCSP budget allocated by agency (2009 estimated budget, excluding the NASA space budget)	138
5.4	Change in CCSP agency budgets, 2002–2010	139
5.5	Change in allocated resources for US Global Change Research Program (USGCRP)- and US Climate Change Science Program (CCSP)-related activities	139
5.6	Idealized connections for exchange of information and recommendations	141
6.1	Climate science and Californian climate policies: A timeline	153
8.1	A continuum from vulnerability to sustainable development	201
9.1	Integrated Food Security and Nutrition Programme (IFSNP) institutional structure	222
9.2	Proposed Food Insecurity and Vulnerability Information Mapping System (FIVIMS) institutional arrangements	225
10.1	Number of typhoons in Taiwan between 1980 and 2005	241
10.2	Expanding area vulnerable to landslides on the Chen-Yo-Lan River watershed: Typhoons Herb (1996) to Mindulle (2004)	243

10.3 History of system improvement and policy development in disaster reduction in Taiwan — 244
10.4 Schematic diagram illustrating the functional role of the Central Emergency Operation Centre (CEOC) and its interactions with various typhoon disaster-related government agencies through the Assessment Group — 246
10.5 Maps of Taiwan and Shan-an village — 250
10.6 Schematic diagram of adaptive disaster-management governance through a process of cross-scale learning, knowledge co-production and institutional transition — 255
11.1 The research area: Western Pokot in Kenya/Uganda — 271
12.1 Location of Nunavut, Inuvialuit and Ulukhaktok — 296
12.2 Actors in resource management in Inuvialuit — 297
12.3 Actors in Narwhal Resource Management in Nunavut — 299
12.4 Arctic char harvests from Lakes Fish and Red Belly, 1988–2003 — 300
12.5 Length of Arctic char harvested at Fish Lake, 1992, 2004 — 301
12.6 Average age of Arctic char sampled at Fish Lake, 1992, 1994–2004 — 303
13.1 Geographical location of the case study area in the Philippines — 309
13.2 Conventional production and marketing systems in Tanauan City — 311
13.3 Percentage of farmers not adopting adaptation measures, by typology and reasons — 318
13.4 Network for transferring knowledge among farmers, policy agents and science agents — 326
13.5 Idealized network for transferring knowledge among the agents — 328
13.6 Trends in vulnerability components with the current and idealized social network scenarios, 2006–2025 — 329
13.7 Trends in vulnerability with the current and idealized social network scenarios, 2006–2025 — 330
14.1 Selected population food security trends in Nicaragua — 339
14.2 Institutional context of the European Programme of Food Aid and Food Security — 340
14.3 EP-FAFS programme operation modes according to the food security states of the beneficiary countries and type of implementation partner — 343
14.4 The European Programme for Food Aid and Food Security in Nicaragua: Diagram of web actors and modes of operation — 345
15.1 Population dynamics of Mongolia — 366
15.2 Number of herders in Mongolia — 366
15.3 Livestock dynamics in Mongolia — 369
15.4 Herders' household wealth distribution based on privately owned livestock numbers — 370
15.5 Cashmere export of Mongolia — 371
15.6 Meat production and export — 371

15.7 Mongolian cultural landscape incorporates four seasonal pastures: *Otor* (a long distance from the main camp pasture), reserve pastures (used during extreme events such as drought and *zud*), hay-making lands and sacred lands (worshipped by local people due to their religious, aesthetic and biodiversity values) 373
16.1 (a) Monthly variation in rainfall for Chiang Mai City based on monthly mean, 1971–2000; (b) inter-annual variability in total annual rainfall for Chiang Mai, 1951–2006 386
16.2 Some of the actors involved in the emerging knowledge system for the Upper Ping River Basin 391
16.3 The emerging knowledge system around floods, scarcity, services and quality issues in the Upper Ping River Basin 407

Tables

3.1	Case study characteristics	61
7.1	Russian climate policy: Major recent milestones	177
10.1	Casualties of the most serious typhoons in Taiwan, 1995–2005	242
10.2	Examples of setting typhoons/storms as major issues and the resolution strategies following collaborative learning, discussion and communication processes in Shan-an village	253
11.1	A chronology of events, 1979–2002	278
11.2	Perceived positive and negative changes in six capability domains	279
11.3	Development projects by NGOs, including churches	282
11.4	Arid and Semi-Arid Lands (ASAL) projects	283
11.5	Status assessment of development projects in northwest Pokot, according to type of donor	284
11.6	Capability assessment of development projects in northwest Pokot	285
11.7	'Best' and 'worst' projects for four types of project donors; separate assessments by men and women (all research areas combined)	287
13.1	Profiles of the farmer typologies in the case study area	315
13.2	Percentage of interviewed farmers in Tanauan City carrying out the adaptation measures ranked in terms of adoption rate, by farmer typology	317
13.3	Sources of information on farming practices, by farmer typology and in percentage	325
14.1	European Commission fund allocations in food security for 2003 and 2004	341
15.1	Changes in land-use patterns, land-use regulation and land tenure from the pre-*negdel* to post-*negdel* period: *Negdel* refers to pastoral collective	364
15.2	Exported cashmere during 2005 and 2006	369

16.1 Water management-related vulnerabilities in the Upper Ping
River Basin   385
16.2 Examples of sources of resilience to challenges of water scarcity,
floods, altered hydrological services and quality at different levels
in the social–ecological system; at the smallest level we have
also disaggregated by broad zones   390

Boxes

11.1 Overview of development projects with the most positively
perceived impact   286
11.2 Overview of development projects with the most negatively
perceived impact   286

List of Contributors

Lilibeth Acosta-Michlik, University of the Philippines, Philippines; Potsdam Institute of Climate Impact Research, Germany; University of Edinburgh, UK; Université Catholique de Louvain, Belgium.

Rebecca Morgenstern Brenner, Clark University, US.

Dan Cayan, Scripps Institution of Oceanography, US.

Chang-Yi David Chang, National Taiwan University, Taiwan.

Zachary Christman, Rutgers University, US.

Togtohyn Chuluun, Mongolian Development Institute, Mongolia.

Bart Croes, California Air Resources Board, US.

Geoffrey D. Dabelko, Woodrow Wilson International Center for Scholars, US.

Ton Dietz, University of Amsterdam, The Netherlands.

Scott Drimie, International Food Policy Research Institute, South Africa.

Victoria Espaldon, Université Catholique de Louvain, Belgium.

James Ford, McGill University, Canada.

Guido Franco, California Energy Commission, US.

Po Garden, Chiang Mai University, Thailand.

Michael Hanemann, University of California-Berkeley, US.

Roger E. Kasperson, Clark University, US.

Vladimir Kotov, EcoPolicy Research and Consulting, Russia.

Louis Lebel, Chiang Mai University, Thailand.

Phimphakan Lebel, Chiang Mai University, Thailand.

Kuan-Hui Elaine Lin, National Taiwan University, Taiwan.

Amy Luers, Google Foundation, US.

Elia Machado, Clark University, US.

Susanne C. Moser, Susanne Moser Research & Consulting, US.

Elena Nikitina, Russian Academy of Sciences, Russia.

Dennis Ojima, Colorado State University, US.

Tristan Pearce, University of Guelph, Canada.

Roberto Sánchez-Rodríguez, University of California-Riverside, US.

Barry Smit, University of Guelph, Canada.

Uno Svedin, FORMAS, Sweden.

Huei-Min Tsai, National Taiwan Normal University, Taiwan.

Coleen Vogel, University of Witwatersrand, South Africa.

Johanna Wandel, University of Waterloo, Canada.

Juergen Weichselgartner, GKSS Research Centre, Germany.

Gina Ziervogel, University of Cape Town, South Africa.

List of Acronyms and Abbreviations

921 PERC	921 Post-Earthquake Reconstruction Council
AB 32	Assembly Bill 32
AB 4420	Assembly Bill 4420
ACCK	Associated Christian Churches in Kenya
ACH	Acción Contra el Hambre
ACK	Anglican Church of Kenya
ACPI	Accelerated Climate Prediction Initiative
ACTED	Agency for Technical Cooperation and Development
ADB	Asian Development Bank
AEC	Atomic Energy Commission
AIC	Anglican Independent Communion
AIDS	acquired immune deficiency syndrome
ARC	Agricultural Research Council
ASAL	Arid and Semi-Arid Lands programme
BQCMB	Beverly and Qamanirjuaq Caribou Management Board
CA Registry	California Climate Action Registry
CAFTA	Central America Free Trade Agreement
CalEPA	California Environmental Protection Agency
CAP	California Applications Program
CAT	Climate Action Team
CBDM	Community-Based Disaster Management Project
CCF	Christian Children's Fund
CCSP	US Climate Change Science Program
CEC	California Energy Commission
CEOC	Central Emergency Operation Centre
CGIAR	Consultative Group on International Agricultural Research
CICESE	Centro de Investigación y Ensenanza Superior de Ensenada
CIG	Climate Impacts Group
CLIMAS	Climate Assessment for the Southwest project

CO	community organization
COLEF	El Colegio de la Frontera Norte
COPIBO	Commercialization of Organic Produce
CRIC	Centro Regionale di Intervento per la Coperazione
CRS	Catholic Relief Services
CSIR	Council for Scientific and Industrial Research
CSP	*Country Strategy Paper*
CSREES	Cooperative State Research, Education and Extension Service of the US Department of Agriculture
DDPM	Department of Disaster Prevention and Mitigation
DFID	UK Department for International Development
DFO	Department of Fisheries and Oceans (Canada)
DIAND	Department of Indian Affairs and Northern Development
DMDS	Disaster Management Decision Support Project
DoA	South African Department of Agriculture
DOD	US Department of Defense
DOE	US Department of Energy
DOT	US Department of Transportation
DPR Act	Disaster Prevention and Response Act
DPR Council	Disaster Prevention and Response Council
DPR Plan	Disaster Prevention and Response Basic Plan
DPRA Plan	Disaster Prevention and Response Action Plan
DSD	Department of Social Development (South Africa)
DWR	Department of Water Resources
EC	European Commission
ECHO	European Commission Humanitarian Office
EIRB	Environmental Impact Review Board
EISC	Environmental Impacts Screening Committee
EKV	Ecomuseum Kristianstad Vattenrike
ELCK	Evangelical Lutheran Church in Kenya
ENSO	El Niño/Southern Oscillation
EOC	Emergency Operation Centre
EOP	Emergency Operation Plan
EPA	US Environmental Protection Agency
EP-FAFS	European Programme for Food Aid and Food Security (Nicaragua)
EPRI	Electric Power Research Institute
ESA	Ecological Society of America
ET	emission trading
EU	European Union
FANR	Food, Agriculture and Natural Resources Directorate
FANRPAN	Food, Agriculture and Natural Resources Policy Analysis
FAO	United Nations Food and Agriculture Organization
FEWSNET	Famine and Early Warning System Network

FGCK	Full Gospel Church of Kenya
FIVIMS	Food Insecurity and Vulnerability Information Mapping System
FIVIMS-ZA	Food Insecurity and Vulnerability Information Management System (South Africa)
FJMC	Fisheries Joint Management Committee
GDP	gross domestic product
GEC	global environmental change
GHG	greenhouse gas
GIS	geographic information system
GPS	global positioning system
HCWG	Holman Char Working Group
HFC	hydrofluorocarbon
HHS	US Department of Health and Human Services
HIPC	highly indebted poor country
HIV	human immunodeficiency virus
HSRC	Human Sciences Research Council
HSRC/FIVIMS	Human Sciences Research Council, Food Insecurity and Vulnerability Mapping Systems
HTC	Hunters and Trappers Committee
HWMC	Hydrology and Water Management Centre
IAI	Inter-American Institute for Global Change Research
ICRAF	International Centre for Research in Agroforestry
ICRC	International Committee of the Red Cross
IFSNP	Integrated Food Security and Nutrition Programme
IGU	International Geographical Union
IHDP	International Human Dimensions Programme (on Global Environmental Change)
IIASA	International Institute for Applied Systems Analysis
IMF	International Monetary Fund
IPCC	Intergovernmental Panel on Climate Change
IRAM	Institut de Recherche et d'Application des Méthodes de Développement
IRI	International Research Institute for Climate Prediction
IUCN	World Conservation Union (*formerly* International Union for the Conservation of Nature)
IWMI	International Water Management Institute
IWRM	integrated water resources management
JBIC	Japan Bank for International Cooperation
JI	joint implementation
KAG	Kenya Assemblies of God
MADECOR	Mandala Agricultural Development Corporation
MAGFOR	Ministerio Agropecuario y Forestal (Ministry of Agriculture and Forestry)
MDG	Millennium Development Goal

MOAC	Ministry of Agriculture and Co-operatives
MOI	Ministry of Interior
MONRE	Ministry of Natural Resources and Environment
MOU	memorandum of understanding
MP	member of parliament
NAFTA	North American Free Trade Agreement
NAS	US National Academy of Sciences
NASA	National Aeronautics and Space Administration
NCAR	National Center for Atmospheric Research
NCCK	National Council of Churches of Kenya
NCDR	National Science and Technology Centre for Disaster Reduction
NDPPC	National Disaster Prevention and Protection Commission
NGO	non-governmental organization
NIH	National Institutes of Health
NOAA	National Oceanic and Atmospheric Administration
NRC	US National Research Council
NS	North Slopes
NSF	National Science Foundation
NT	Northwest Territories
ODI	Overseas Development Institute
OECD	Organisation for Economic Co-operation and Development
ONEP	Office of Natural Resources and Environmental Policy and Planning
PACLIM	Pacific Climate Workshop
PDO	Pacific Decadal Oscillation
PFC	perfluorated carbon
PIER	Public Interest Energy Research programme
PNAS	*Proceedings of the National Academy of Sciences*
POKATUSA	Western Kenya Pastoralists
PPIC	Public Policy Institute of California
PRODELSA	Programa de Desarrollo Local y Seguridad Alimentaria (Food Security and Local Development Programme)
PTA	Parent–Teacher Association
RBO	river basin organization
RCA/AIC	Reformed Church in America/Africa Inland Church
RESAL	European Food Security Network
RHVP	Regional Hunger and Vulnerability Programme (southern Africa)
RIACSO	Regional Inter-Agency Coordination Support Office
RID	Royal Irrigation Department
RISA	Regional Integrated Sciences and Assessments programme
RSBO	river sub-basin organization
RSPP	Russian Union of Industrialists and Entrepreneurs

RVAC	Regional Vulnerability Assessment Committee
SADC	Southern African Development Community
SADC–FANR	Southern African Development Community–Food, Agriculture and Natural Resources Directorate
SADC–FANR–RVAC	Southern African Development Community–Food, Agriculture and Natural Resources Directorate–Regional Vulnerability Assessment Committee
SADC RISDP	Southern African Development Community's Regional Indicative Social Development Plan
SADC–RVAC	Southern African Development Community–Regional Vulnerability Assessment Committee
SAP	*Synthesis and Assessment Product* report
SARPN	Southern African Regional Poverty Network
SETSAN	Mozambique Technical Secretariat for Food Security and Nutrition
SGPRS	*Strengthened Growth and Poverty Reduction Strategy*
SICA	Sistema de Integración de Centroamericana
SPPI	science/policy/practice interface
SSCI	*Social Science Citation Index*
START	SysTem for Analysis, Research and Training
SWCB	Taiwan Soil and Water Conservation Bureau
TAO	Tambon administration organization
TDRI	Thailand Development Research Institute
TEI	Thailand Environment Institute
UCS	Union of Concerned Scientists
UK	United Kingdom
UN	United Nations
UNDP	United Nations Development Programme
UNESCO	United Nations Educational, Scientific and Cultural Organization
UNESCO–WHC	United Nations Educational, Scientific and Cultural Organization–World Heritage Convention
UNFCCC	United Nations Framework Convention on Climate Change
UNICEF	United Nations Children's Fund
UNRWA	United Nations Relief and Works Agency
UNWFP	United Nations World Food Programme
UNWFP–VAM	United Nations World Food Programme–Vulnerability Analysis and Mapping
US	United States
USAID	US Agency for International Development
USDA	US Department of Agriculture
USGCRP	US Global Change Research Program
USGS	US Geological Survey
VAC	vulnerability assessment committee

VAM	Vulnerability Analysis and Mapping (of the World Food Programme)
VARIP	Vulnerability and Resilience in Practice project
WFO	World Food Organization
WFP	United Nations World Food Programme
WHC	World Heritage Convention
WMAC	Wildlife Management Advisory Committee
WTO	World Trade Organization
WWF	World Wide Fund for Nature
ZimVAC	Zimbabwe Vulnerability Assessment Committee

Part I

Introduction

1
Characterizing the Science/Practice Gap

Roger E. Kasperson

The gap between science and decision-making has long been a gap of major concern. While science itself cannot provide the full knowledge base needed for complex decisions, the science component is important. This book reports on the Vulnerability and Resilience in Practice (VARIP) project that explores how in one developed scientific and assessment area, knowledge has entered into and informed decision-making. The various chapters enquire into the adequacy of science and assessment, but also explore in some detail the interface between science and decision-making. The scope is international and therefore a wide range of environmental threats and political systems is involved.

Is there a problem? How do we know?

It is widely recognized that a serious gap exists between science and expert assessment, on the one hand, and decision-making, on the other. Major scientific reports relevant to decision-makers frequently pass unnoticed and decision-makers remain largely unaware that scientific analyses have been conducted that may be highly relevant to their decision-making areas of responsibility. As this book will detail, decision-makers typically do not read scientific journals or academic books that regularly appear in their fields of interest. But there is substantial variability in this – between different issues of concern and among different countries.

In the US, political culture has long eschewed the role of expertise in public policy and decision-making. Indeed, the political process has often assumed that political control over scientific assessments is not only appropriate but necessary. Sometimes this control occurs in the form of outright censoring or attempting to silence research findings relevant to some environmental hazard, as has occurred, for example, in broad domains of concern, such as climate change. In other cases, it occurs in the structuring of governmental process and organizations, as was done in the US climate change science programme (US National Research Council, 2009a, 2009c), to stultify effective action or by allocating insufficient funding so that failure is virtually ensured.

The Obama administration, early in its advent, recognized this issue and in a speech before the National Academy of Sciences in October 2009, President Obama assured that science would play a more major role in science-related policy-making and that political or ideological interference aimed at changing scientific findings would no longer occur. This issue is generic, of course, and not restricted to the particular issues in US political culture (Jasanoff, 1990).

Where does the problem come from?

The notion that a linear process exists between science and policy still permeates the thinking of many scientists and decision-makers, and so it is scientists operating in the realm of 'curiosity science' who largely determine how problems of science are defined, framed and communicated to potential users. Predictably, this mode of framing the development of science aims at the needs for greater scientific understanding, rather than the needs of practitioners in their everyday decisions. In both the halls of science and even within governmental agencies, the goal is assumed to be the growth of scientific understanding of some issue of environment or science and then 'outreach', often by posting results on a website, to prospective users.

As noted above, it is also commonplace that prospective users and decision-makers do not regularly read scientific journals and academic books (indeed, often characterizing the authors pejoratively as 'academics'), and do not follow scientific development on websites; as a result, running through the motions of 'outreach' rarely accomplishes the task of transmitting scientific results to those who could use them to inform decision-making.

But the problems go well beyond the issues of effective dissemination and outreach. Political decisions, it has been widely noted, extend well beyond issues of science (Jasanoff, 1990; US National Research Council, 2009d). Inevitably, a host of other issues – cost and benefit considerations, impacts upon industry, reduction of risk achieved, stakeholder views, distributional issues – enter into such decisions. If scientific and assessment studies encompass only the issues of science, they have limited value to prospective users who require a much broader range of considerations and knowledge. As a result, scientific results and assessments of risk are often not used because they have been defined and framed to meet the needs of science rather than the needs of those making decisions. To avert these problems, a different kind of scientific and assessment process is needed, one that begins with user needs and a problem-solving orientation (US National Research Council, 2009d). The communication among scientists, assessors and users has greatest success when 'inreach' as well as 'outreach' is involved (US National Research Council, 2009d). Effective linkage between science and practice occurs not in a linear process of producers and consumers, but in a continuing two-way interaction in which learning occurs among both the scientists and assessors, on the one hand, and the practitioners who make decisions, on the other, with a rich pattern of feedbacks the norm for both.

Thus far, however, we have considered only the roles of scientists and practitioners. But a multitude of other issues and actors are involved. To begin, even the notion of what science is relevant to what problem is controversial and contested. So it is a continuing saga that US National Research Council reports regularly critique federal agency efforts on environmental problems and risks for their failure to incorporate knowledge related to human behaviour and social science within the knowledge base that informs high-level political decision-making. For example, a recent evaluation of the US climate change programme noted among its six top priorities for restructuring climate change research:

- Reorganize the programme around integrated scientific/societal issues to facilitate cross-cutting research focused on understanding the interactions among climate, human and environmental systems and on supporting societal responses to climate change (US National Research Council, 2009a, p4).
- A strong underpinning of observations and models is needed, as well as strengthened research across the board – particularly in the human dimensions of global change and in user-driven (applied) research that supports decision-making – and increased involvement of stakeholders (US National Research Council, 2009a, p5).

Such observations are not uncommon. Although it is abundantly clear that science is needed to inform decision-making on the environmental challenges that face society, it is still the case across many societies that science is still perceived (and financially supported) as only the natural, medical and engineering sciences. Therefore, review report after review report continues the drumbeat for a more integrated concept of science. Meanwhile, the understanding of human interactions with environmental systems continues to lag badly behind what is needed in an adequate knowledge base to inform decisions in both the public and private sectors (see, for example, US National Research Council, 2009a, p5). These problems are deep seated and unlikely to change anytime soon. The personnel and expertise in federal agencies and private corporations continue to share a paucity of social and behavioural science expertise involving such fields as psychology, sociology, anthropology and geography. Only economics among the social sciences has won broad recognition and involvement. Until this long-standing deficit of expertise is rectified, the knowledge base for practice will continue to have serious gaps and deficiencies that will limit the entrance of science into decision-making. What the decision-maker desperately needs to know, in short, will continue to be well beyond the scope of traditional science.

A study by Brewer and Stern (2005) took up the question of what the major research needs are for improving environmental decision-making. It recommended:

- studies focused on decision quality, exploring decision-makers' evaluation of decision processes and outcomes, and improving formal tools for structuring decisions;
- research on information needs: characterizing risk and uncertainty, communication processes, decision-support development and decision-support 'experiments' (Brewer and Stern, 2005).

The recent US National Research Council report (2009b) on *Informing Decisions in a Changing Climate* elaborates upon these issues.

Thinking about the gap: The issue of metaphors

Language, as the deconstructivists have made clear over the past several decades, pervasively influences our thinking on problems and decision issues. This is pronounced in the case of linkage between science and practice. Rhetoric and writings are replete with references to the 'bridge', 'superhighway' and 'pipeline' that connect science to practice. But it is abundantly clear that the linkage is anything but linear. First, a wide cast of intermediaries, including corporate officials, federal agency personnel, state and local officials, and non-governmental organization (NGO) leaders interpret and reframe the results of science for a broad host of decision-makers (Moser and Dilling, 2007). They do this, as we have pointed out in our writings on the social amplification of risk (see, in particular, Pidgeon et al, 2003), by emphasizing particular elements of the risk results and reframing the basic message of the inference to be drawn for risk and environmental management. A wide variety of stakeholders and actors compete in this arena for shaping the risk and other 'signals' to decision-makers (Pidgeon et al, 2003), as shown in Figure 1.1. The net effect is to reframe the message and thereby 'amplify' or 'attenuate' the signal to society and decision-makers of the social meaning of the risk event or research.

Accordingly, a metaphor is needed that is more descriptive of the linkage between science and practice. We propose the use of the term 'spider webs' as a more accurate description of what actually occurs in at least the contested cases that become a matter of dispute and controversy. The image of a spider web is, of course, an imperfect image of a complex and often unstable network. Technically, the image we prefer is what is termed a 'communal spider web',

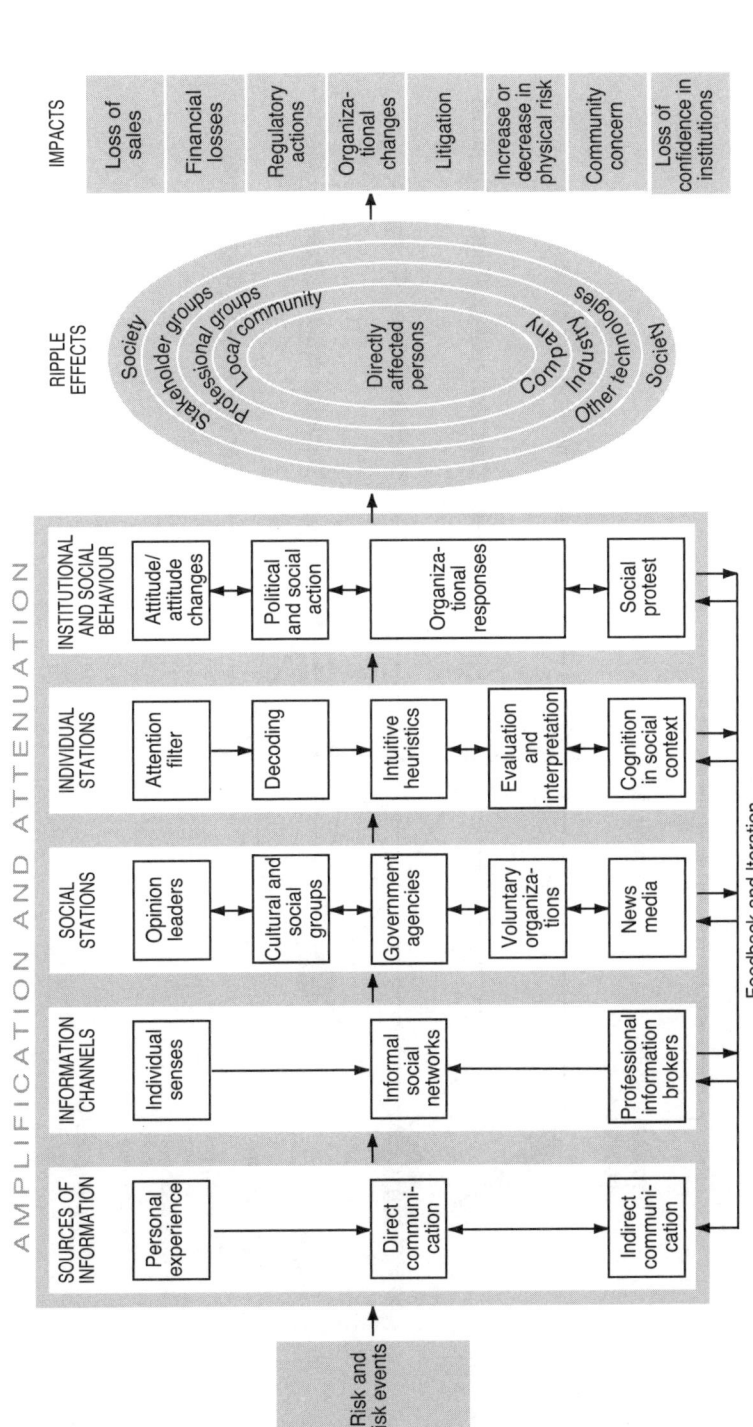

Figure 1.1 *Social amplification*

Source: chapter author

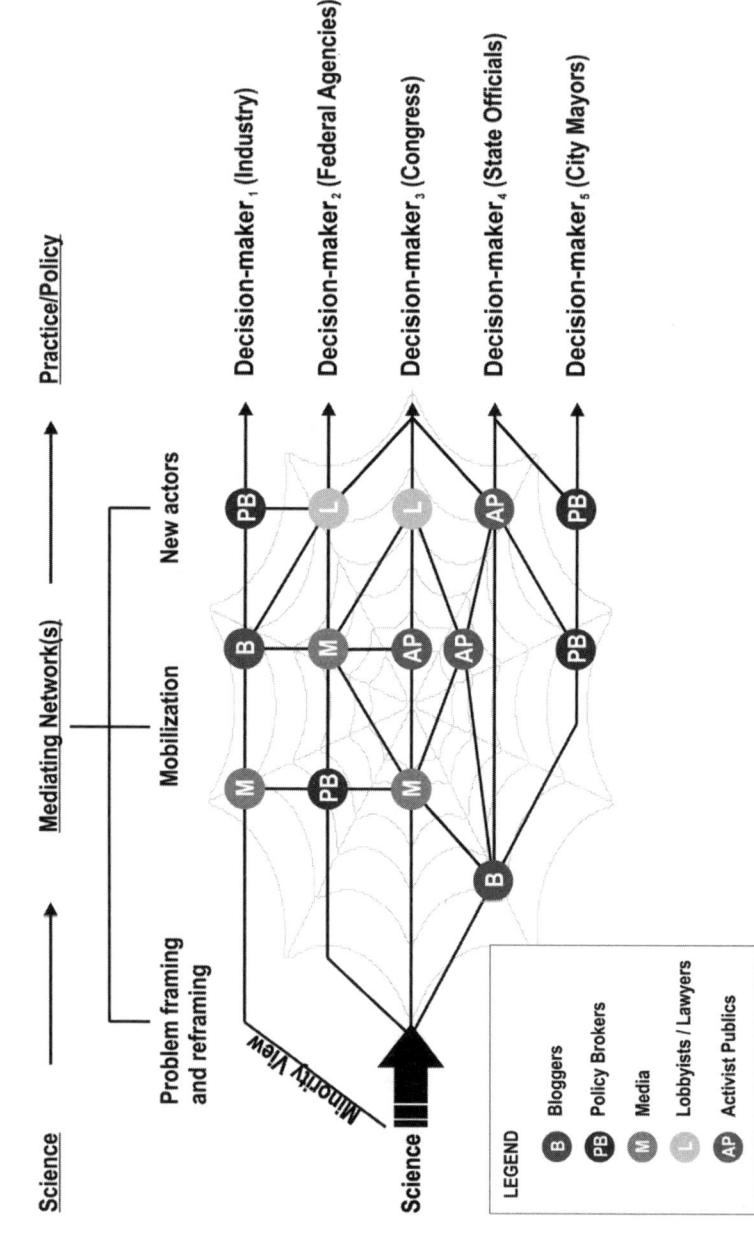

Figure 1.2 Spider web

Source: chapter author

where there are multiple spiders and a dynamic web of linkages. To provide a visual image of such a spider web, Figure 1.2 shows a hypothetical case of a linkage system between sciences and practice. This figure suggests a wide range of intermediaries who may be involved: the media, lobbyists, bloggers, NGOs, industry trade groups and policy brokers. For the purpose of this discussion, science is treated as unitary, which of course it is not. A spectrum of decision-makers is typically involved, ranging over public and private institutions, officials at different government scales, individual resource managers, and individual interest groups and decision-makers.

Of course, there is no one model or image of the linkage system. Here we suggest that there are at least three major types of architecture to these spider webs. The first we term the *simple spider web*, in which there is a high degree of direct linkage between science and the primary decision-makers. The Intergovernmental Panel on Climate Change (IPCC) is a good illustration of this type of architecture. Here, national governments suggest the members who will serve on the panel and participate in the final stage of report drafting in the actual language to be incorporated in the final text. The underlying issue – global climate change – is also an issue that requires involvement and consent by at least the major actors who are involved. The second type of spider web architecture we term *complex but stable spider webs*. These spider webs show a more complex set of actors and less direct linkage between science and end users and decision-makers. An example of this type of architecture is the Nuclear Non-Proliferation Treaty, in which there is a stable negotiating and regulatory institution – the International Atomic Energy Agency – but a complex problem area (ability to make and deliver a nuclear weapon) and shifting national agendas and relationships. The structure of political interactions has some relatively stable power relationships and interests, but long-term technical and political uncertainties. The third type of spider web architecture we term *complex, unstable spider webs*. Here the problem arena is complex, with interests, agendas and actors that are often unstable. Network actors enter and drop out of the web from time to time. A notable example of this spider web architecture is the case of marine fisheries, with diffuse actors facing a rapidly deteriorating risk situation, a weak and unstable regulatory structure and conflicting national priorities.

Running through these different architectures are several important risk management challenges. How to build and maintain social trust across the diverse actors in the spider web is a major need in producing an effective relationship between science and practice (Cvetkovich and Löfstedt, 1999; Hardin, 2006; Siegrist et al, 2007). If the spider web becomes more diffuse and contentious, as amplification and attenuation agents compete for control over the framing of the science results and the associated signals of the associated social meaning to society and decision-makers, concerted action and 'best' practices become more difficult to obtain.

Filling the gap: Are boundary organizations the answer?

One major hypothesis has emerged as to how the gap between science and practice may be narrowed effectively. Organizations that sit astride the domains of research and user communities, it is argued, ensure a greater potential for effective interaction and the production of knowledge more germane to the needs of decision-makers. While the range of empirical studies is thin and largely oriented to unitary decision-makers in the Western context and to international governance regimes based on Western governance notions, several important generalizations, however tentative, have emerged. Boundary organizations are most effective, it is argued, when:

- They are situated at the frontier between science and politics, but have neutral accountability to each.
- Actors from both science and politics and intermediaries in the linkage space between them actively participate in interactions through the linkage domain.
- They provide opportunities and incentives for the creation of shared instruments (e.g. models or research plans) that facilitate collaborations and the pursuit of natural interests.
- They encourage the co-production of information and analysis on behalf of both of the domains of science and practice.

Figure 1.3 nicely summarizes these conceptual arguments.

A series of case studies has explored the role of boundary organizations in a range of empirical situations (Guston, 1999; Agrawala et al, 2001; Cash, 2001; Guston, 2001; Hellstrom and Jacob, 2003; Carr and Wilkinson, 2005).

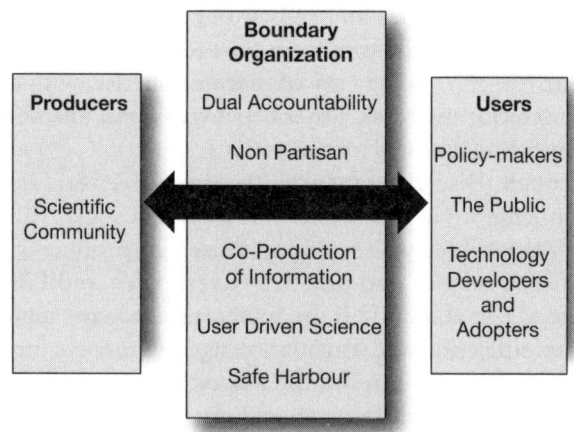

Figure 1.3 *Boundary organizations in end-to-end systems*

Source: adapted from Mitchell et al (2006)

Despite this valuable work, questions remain. As yet, nearly all the studies have addressed only one of our three types of spider web architecture – namely, the simple spider web, where both the research and practice domains tend to be unitary and connections more direct and relatively stable. We still have little empirical evidence drawn from more complex and dynamic spider webs where intermediaries are contentious, vie for power and compete to define framings of the science or risk messages. Organizations may not, however, be the critical issue to begin with. What is most important may be the functions that are achieved and the processes that go forth. A formal structure or organization may well be less important than what they achieve – mainly efforts involving the co-production of knowledge, agreed-upon mutual accountability, mechanisms for jointly produced planning and analytical methods and models, and an arena for ongoing mediation in issues and objectives. Beyond this, we need to see how these processes can be obtained, and whether boundary organizations can be found or created in the more challenging spider web architectures, where actors are more diffuse, decisions-makers widely distributed, intermediaries shifting and in conflict, and the media contributing to amplification and attenuation of the signals to decision-makers and society more generally.

One alternative concept, which emphasizes culture and informal processes rather than structure, is the notion of the *epistemic community*. In his work on the development of a plan to address pollution in the Mediterranean Basin, Peter Haas (1990) charts the evolution of an interactive network of scientists and policy-makers in conducting assessments and formulating an integrative science and risk management policy for the Mediterranean Basin. He calls particular attention to the creation of extensive personal and professional linkages between scientists and practitioners in creating an overall holistic plan for addressing a wide variety of threats to environmental quality in the basin. While organizational mechanisms evolved to facilitate both scientific assessments and policy deliberations, at base they rested upon the emergence of a knowledge system with a shared conception of problems and goals, levels of personal trust among key actors, and continuing adjudication of the tensions between science and policy. This analysis captures many of the elements involved in the notion of a 'boundary organization'; but the focus is clearly upon the emergence of a common knowledge system, shared goals, and ongoing negotiation in analysis and deliberation (see also Haas et al, 1993).

Or do we need a different kind of science?

A major suggestion arising over the last decade is that the problem lies primarily in the area of science – that we need a new kind of science to narrow the gap between science and practice. So, on the eve of the Millennium, two major events occurred. The first was the publication of a landmark US National Research Council (1999) report, *Our Common Journey*. This report highlighted the conclusion that 'tensions exist between broadly based and highly focused research strategies; between integrative problem-driven research and research firmly grounded in particular disciplines; and between the quest for generalizable scientific understanding of sustainability issues and the localized knowledge of environment–society interactions that give rise to those issues and generate the options for dealing with them' (US National Research Council, 1999, p10). The National Research Council identified three priority tasks for overcoming this tension:

1. Develop a research framework that integrates global and local perspectives to shape a 'place-based' understanding between environment and society.
2. Initiate focused research programmes on a small set of understudied questions that are central to a deeper understanding of interactions between science and the environment.
3. Promote better utilization of existing tools and processes for linking knowledge in action in pursuit of a transition to sustainability (US National Research Council, 1999, pp10–11).

A series of actions were undertaken in the wake of this report to address its findings. A major seminar was convened in Sweden to consider the content and prospects of a new sustainability science. Another major seminar followed at the Earth Science Summit in The Netherlands and a series of regional meetings was held throughout the world in the following years. The US National Research Council also established an ongoing round table on sustainability futures which has continued since the 1999 report. This book may be considered a further effort in this direction. With all of these efforts, it is still uncertain, in the absence of any major searching evaluation, how science may or may not have changed in response to these proposals and admonitions.

Yet, the call for change continues. In a further report on how science might best improve to inform better societal decisions on climate change, Brewer and Stern (2005) set forth six major principles for more effective decision support by science, many of which address the notion of sustainability science:

1. Begin with user needs.
2. Give priority to processes over products.
3. Link information producers and users.
4. Build connections across disciplines and organizations.
5. Seek institutional stability.
6. Design for learning.

However general the language, these are clearly concrete measures designed to further the creation of a sustainability science, though not phrased in these terms. Various roads, in short, lead to the notion of a need for science to become more germane to, and integrated with, the needs of practice.

Or do we need a different kind of decision-making?

Another view sees the problem as primarily in the decision-making arena. The rational model approach to decision-making is often equated to 'command-and-control' decisions, following a military model of addressing a problem or threat, conducting scientific assessments delineating the problem and options for dealing with it, and choosing that option which realizes the greatest gain, or utility, in terms of decision-maker goals. This approach can be elaborated, though not fundamentally changed, by the approaches of multi-objective decision-making. Such conceptions rely heavily upon assumptions of a unitary decision-maker, not a distributed decision system, and requisite knowledge to realize decision objectives. But what do we do when uncertainties are very large, decision goals contested, and decision systems distributed and not unitary? This has led to some changed prescriptive views of policy – for example, that at least for certain kinds of risk decisions (such as those with large uncertainties but large irreversible damage) precautionary decision-making should replace rational utility-maximizing approaches (Harremoës et al, 2002).

As a result, for at least certain decision problems for which science is inherently limited in providing the information and analysis needed for the traditional notion of reaching a 'good' rational decision based upon strong scientific analysis, other approaches to decision-making are receiving growing attention. Among these, recent discussions have taken up adaptive management as an effective, and perhaps preferable, alternative to 'command-and-control' management (Holling, 1978; Folke et al, 2005; Renn and Walker, 2008). Here the argument is for a 'go with the flow' kind of decision-making, one based on the notion that knowledge is evolutionary and that learning and incremental decision-making, if well structured, can produce a more intelligent and successful management system over time. Certainly it can be better adapted to situations of emerging knowledge and large uncertainties than 'command-and-control' approaches. Figure 1.4 shows the major differences in how these two decision processes are structured.

But adaptive management may travel the same course as other corrupted concepts, such as 'sustainability'. It is now commonplace to hear all managers assert, of course, that they enthusiastically embrace adaptive management. Since it is assumed that adaptive management means little more than remaining flexible in the face of changing environmental threats or decision contexts, decision-makers all readily claim that they have always embraced adaptive management. After all, who would claim a position of imperviousness to new knowledge, experience or changing social and political contexts? It is further assumed that one size fits all problems – adaptive management suggests little difference to most decision-makers from what they have always done.

A more rigorous embrace of adaptive management, as outlined by serious thinkers such as Holling (1978), Walters (1986) and Lee (1993), suggests something quite different. Decisions are regarded as 'experiments' rather than definitive solutions. It is difficult, as we have learned from high-level radioac-

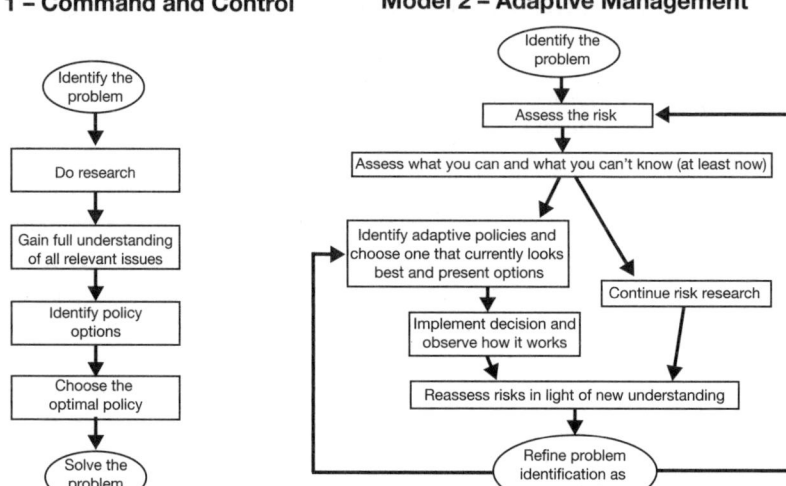

Figure 1.4 *Two risk decision models*

Source: adapted from Morgan (2009)

tive waste storage, to get everything right for 100,000 years when designing a first-of-a-kind facility (US National Research Council, 2002). Much depends upon the presence of effective monitoring systems and strategies to maximize social learning from the errors, missteps and surprises that inevitably occur. Learning can be very difficult to achieve because of the tendency of most management systems for 'path dependency' (Brown et al, 2007). Decision-makers are prone to embark upon a particular management direction or embrace a particular preferred solution; as a result, changes in basic thinking or approach become difficult to achieve (Kingdon, 1995). Consequently, learning tends to be restricted to low-level learning induced by better data, and not basic rethinking of problems and approaches (Lee, 1993).

There are institutional impediments as well, and they are more far-reaching and fundamental than is commonly assumed. In essence, they involve designs and mechanisms not only to encourage flexible management approaches, but more fundamentally to encourage and facilitate social learning from experience. It has been widely noted that managers typically repeat errors rather than learn from them. And in adaptation and reconstruction, basic vulnerabilities are often recreated, as the Hurricane Katrina experience in New Orleans suggests (Kates et al, 2006).

To facilitate social learning from experience and other sources of knowledge, a diversity of inputs to the decision-makers needs to be maximized (Social Learning Group, 2001). Thus, the boundaries of the agency or organization need to be highly permeable and contrary or contested views not only accepted, but actively sought out. The typical approach to critics and opponents of an agency or developer plan is to isolate them as much as possible and to put them at the end of public involvement and communication initiatives. Adaptive management approaches require a major turnaround in these approaches. The strategy needs to engage critics and opponents from day one. It must be understood that valuable things can be learned from such interactions, perhaps about the nature of the risks and related vulnerabilities, but surely about the issues that developers, regulators and other decision-makers will face across the table in public hearings and (increasingly) in the courtroom. More importantly, developers and decision-makers may come not only to understand better the issues of contention, but also to begin early to explore how opposing views may be negotiated, where gaps may exist in the knowledge base assembled for those issues, or whether new monitoring systems may be needed.

But also it is the case that, as technological options move to more mature stages of development, development agencies and corporations will face a broader range of issues associated with vulnerabilities, public concerns, risk communication and public involvement. Accordingly, a wider range of expertise, which as yet may not have been brought into the decision-making organizations, may be needed. It is instructive that review committees of the US National Research Council repeatedly single out the lack of expertise in the social and behavioural sciences as one of the limiting factors in the performance of federal agencies charged with managing environmental and risk questions (see, for example, US National Research Council, 2009c). Given the slow pace of turnover in federal and state governmental agencies, this is likely to remain a significant limiting factor to a greater movement to adaptive management approaches.

Yet another issue relates to the management process itself in 'go with the flow' adaptive management. This process calls for open acknowledgement of uncertainties that may limit the adoption of particular solutions and frank admission of errors that have occurred in past management choices. Adaptive management assumes evolutionary knowledge about environmental changes and risk (Jasanoff, 2004; Lemos and Morehouse, 2005). Reacting to this is an evolutionary development of management strategies in which decisions and solutions are incremental. Therefore, false starts and solutions at one point in time are expected to change and develop. This is in accord with what expertise would suggest. But two points are relevant. First, politically acknowledging uncertainties that are not well understood and errors that have occurred in past management strategies requires a degree of candour and openness that developers and decision-makers rarely have, and political prices are to be expected. Second, negative effects on social trust should be anticipated. After all, perceived competence is one of the major dimensions of social trust, and if doubts arise in major stakeholders and the public as to whether managers possess the needed competence and expertise, then erosion of trust in the managers might well be expected. This may well move decision-making of all kinds to become more contentious and more conflicted, and threaten greater paralysis and delay than otherwise might occur.

Finally, the current rush to embrace adaptive management often proceeds with the assumption that 'one size fits all'. A sound approach to risk management typically begins with an enquiry into the nature of the hazard. 'Learning by experience' and 'going with the flow' are attractive where uncertainties are especially large and the consequences accumulate over time. This is typically the case for hazards such as drought and flood management, or current concerns over climate change. But for hazards where uncertainties may be more limited and consequences abrupt and potentially catastrophic, such as nuclear war or nuclear plant accidents, than a management strategy that proceeds by learning from experience is less appropriate. So distinctions among hazards are essential – what is the nature of the risks and what consequences to society and ecology may be involved. Learning and flexibility need not be lost, but more reliance upon risk analysis may be needed, as informed by both analytical models and experience.

The chapters that follow take up the questions enumerated above to explore the barriers, dynamics and conflicts that beset efforts to narrow the gap between science and decision-making. In these chapters, we recognize that a gap often occurs for good reason and should not disappear. On the other hand, the gap is also clearly limiting how science might more productively inform decision-making. In the final chapter we explore what these studies, as a group, tell us about how navigation in different spider webs might best proceed, so that what is most useful about scientific analysis can be harnessed to improve the ways in which decisions are made about environmental change and risk.

Core questions

In the chapters that follow, we explore the varying interfaces, or architectures, between science and decision-making in a wide variety of settings and national contexts. It should be noted that while vulnerability and resilience are common subjects throughout the book, the specific hazards treated and the types of political system and civil society vary widely. To produce greater coherence and to allow comparative analysis across quite dissimilar cases, we identify a number of core questions that we asked all authors to address wherever possible in their case studies. Of course, no one case study can take up all these issues.

The core questions are as follows:

- How adequate is the knowledge base to support efforts for vulnerability reduction and building resilience and adaptation? How is that knowledge distributed among actors?
- How may the science/practice interface best be structured and characterized? Who are the principal actors? What are their roles and interests? (Note: every case study author was encouraged to present a graphic depicting the structure of the actors and their relationships with boxes and arrows, although we point out in the conclusions that this did not happen in each case.)

- To what extent do the actors make use of the knowledge available to them? How do the actors structure the vulnerability and resilience discussion in the transfer of knowledge? How relevant and pertinent is the knowledge to the needs of decision-makers and other actors?
- What barriers and failures limit the transfer of knowledge and feedbacks in the science/practice interface? Do the barriers and failures occur in the transfer of knowledge principally from science to practice or from practice to science? How important are the intermediaries between science and practice and who are they?
- How does the nature of institutions shape the science/practice interface? To what extent is institutional fragmentation a problem? How permeable are the boundaries of institutions to new information?
- What major conflicts exist among actors and institutions in the interactions between science and practice? To what extent are the conflicts primarily about values or facts? Does social justice enter into the decision-making process?
- What is the role of consensus in the science/practice interface? How is consensus built? How are conflicts resolved? Who are the consensus builders and mediators and what are the major processes that they use?
- What factors contribute most to adaptive capacity? How large is the gap between the capacity to adapt and the adaptation that actually occurs? What causes this gap and how can it be reduced? What new elements of enlarged capacity would contribute most to greater resilience in the face of environmental change over the short run or the longer term?
- To what extent has social learning evolved among the principal actors? To what extent has social learning ameliorated or exacerbated vulnerability? What most limits or facilitates the ability to learn from one's own experience and the experience of others?
- Where is the science/practice interface vulnerable to failure? Where is the science/practice interface most vulnerable to failure to future risks?
- How can the science/practice interface best be improved?

In the last chapter of the book, we return to these questions for a summary of major comparative findings emerging from these various case studies, drawn from different parts of the globe.

Exploring the science/decision-making interface

The first step in this research project was to assess what is already known. This involved greater effort than might be anticipated because although countless case studies have examined issues relevant to vulnerability and resilience in far-flung corners of the globe, few efforts have sought to assemble this knowledge in any systematic way, to draw comparative conclusions, or to test hypotheses in a rigorous manner against empirical evidence. The literature is rich in intuition and speculation, sparse in rigorous analysis. Accordingly, our first step was to mount a propositional analysis (Chapter 2) in which we identified a large number of case studies and then conducted something akin to a meta-analysis, or at least a content analysis, to identify major findings anchored in conceptual thinking and to assess the extent of supporting or conflicting evidence across the cases. This allowed some judgement as to the confidence we believed should be placed in the validity of the proposition. Next, we conducted an analysis (Chapter 3) of some selected scientific assessments, gathering information from both the producers (scientists) of the assessments and potential users. Finally, we asked several analysts with broad experience in vulnerability and resilience work to reflect on the literature and to assess the challenges that pervade the science/decision interface and their implications for attempts to close the gap that currently exists (Chapter 4).

We then turn to one of the global environmental problems of greatest urgency – the growing dilemma of how to respond to climate change. Chapter 5 takes stock of the history of climate change research and action in the US, evaluating the yawning gap between the two since the passage of the US Global Change Research Act of 1990. Since many of the creative initiatives on climate change in the US have occurred at state and local levels, Chapter 6 enquires into science/policy interactions in California, one of the leading states in the US innovating in environmental management, generally, and climate change, in particular. We then contrast the science/decision interface in the US with two other cases – the role of science in Russia in reaching an accord on the ratification of the Kyoto Protocol (Chapter 7) and the handling of vulnerability to climate variability in a Mexico/US border city (Chapter 8).

The chapters then move on to consider a broad range of hazards and interactions between science and decision-making across the globe. Chapter 9 examines an attempt to establish a boundary organization in South Africa around the issue of food security – what worked and what did not. Next, a team of analysts scrutinizes a unique attempt to link science and local knowledge in reducing vulnerabilities to earthquake risks in Taiwan (Chapter 10). Chapter 11 examines the role of 'outside' development and layers in responding to environmental vulnerability in an East African setting. In Chapter 12 a Canadian team of researchers probes into the thorny issue of how to utilize both 'formal' science and indigenous knowledge in a Canadian Arctic case. Shifting from the array of substantive spider web issues, a Philippine case concentrates on methodological issues – how to use agent-based modelling to characterize complex science/decision-making interfaces in Chapter 13. Chapter 14 then turns to an international programme – the European Programme for Food Aid – to identify the barriers that arise in the science/decision-making arena, particularly in implementing policies aimed at reducing vulnerabilities in Central America. Chapter 15 reviews the attempts in Mongolia to expand the use of market forces in managing rangelands, with the role of science weighed against other political considerations. Chapter 16, in turn, puts the knowledge system for managing floods in northern Thailand under the microscope – what is the anatomy of the knowledge system spider web and how did it inform the flooding vulnerabilities of various actors?

The book then takes stock of what has been learned from this review of past work and this impressive array of empirical case studies across many countries. A thoughtful scholar – Uno Svedin from Sweden – offers a perspective on the overall contribution of this work and its implications for efforts to increase the value of science in decision-making (Chapter 17). In Chapter 18, the editor returns to the basic questions set forth in Chapter 1, exploring what has been learned from the various efforts about these questions and what their implications suggest about more effective ways of navigating and traversing the complexity of spider webs so that effective and appropriate uses of science in decision-making can be enhanced.

References

Agrawala, S., Broad, K. and Guston, D. H. (2001) 'Integrating climate forecasts and societal decision making: Challenges to an emergent boundary organization', *Science, Technology and Human Values*, vol 26, no 4, pp454–477

Brewer, G. D. and Stern, P. C. (2005) *Decision-making for the Environment: Social and Behavioral Science Research Priorities*, The National Academies Press, Washington, DC

Brown, M. A., Chandler, J., Lapsa, M. and Sovacool, B. (2007) *Carbon Lock-in: Barriers to Deploying Climate Change Mitigation Technologies*, ORNL/TM-2007/124, Oak Ridge National Laboratory, Oak Ridge, TN

Carr, A. and Wilkinson, R. (2005) 'Beyond participation: Boundary organizations as a new space for farmers and scientists to interact', *Society and Natural Resources*, vol 18, pp255–565

Cash, D. W. (2001) 'In order to aid in diffusing useful and practical information: Agricultural extension and boundary organizations', *PNAS*, vol 100, no 14, pp8086–8091

Cvetkovich, G. and Löfstedt, R. (eds) (1999) *Social Trust and the Management of Risk*, Earthscan, London

Folke, C., Hahn, T., Olsson, P. and Norberg, J. (2005) 'Adaptive governance of social–ecological systems', *Annual Review of Environment and Resources*, vol 30, no 1, pp441–473

Guston, D. H. (1999) 'Stabilizing the boundary between us, politics and science', *Social Studies of Science*, vol 29, no 1, pp87–111

Guston, D. H. (2001) 'Boundary organizations in environmental policy and science: An introduction', *Science, Technology and Human Values*, vol 26, no 4, pp399–408

Haas, P. M. (1990) *Saving the Mediterranean: The Politics of International Environmental Cooperation*, Columbia University Press, New York, NY

Haas, P. M., Keohane, R. O. and Levy, M. A. (1993) *Institutions for the Earth: Sources of Effective Environmental Protection*, MIT Press, Cambridge, MA

Hardin, R. (2006) *Trust*, Polity Press, Cambridge, UK

Harremoës, P., Gee, D., MacGarvin, M., Stirling, A., Keys, J., Wynne, B. and Guedes Vaz, S. (eds) (2002) *The Precautionary Principle in the 20th Century: Late Lessons from Early Warnings*, Earthscan, London

Hellstrom, T. and Jacob, M. (2003) 'Boundary organizations in science: From discourse to construction', *Science and Public Policy*, vol 30, no 4, p235

Holling, C. S. (1978) *Adaptive Environmental Assessment and Management*, John Wiley and Sons, London
Jasanoff, S. (1990) *The Fifth Branch*, Harvard University Press, Cambridge, MA
Jasanoff, S. (2004) *States of Knowledge: The Co-Production of Science and Social Order,* Routledge, New York, NY
Kates, R. W., Colten, C. E., Laska, S. and Leatherman, S. P. (2006) 'Reconstruction of New Orleans after Hurricane Katrina: A research perspective', *PNAS*, vol 103, no 40, pp14653–14660
Kingdon, John W. (1995) *Agendas, Alternatives, and Public Policies,* 2nd edition, Harper, New York, NY
Lee, K. N. (1993) *Compass and Gyroscope: Integrating Science and Politics for the Environment*, Island Press, Washington, DC
Lemos, M. C. and Morehouse, B. J. (2005) 'The co-production of science and policy in integrated climate assessments', *Global Environmental Change*, vol 15, pp57–68
Mitchell, R., Clark, W. C., Cash, D. W. and Dickson, N. M. (eds) (2006) *Global Environmental Assessments: Information and Influence,* The MIT Press, Cambridge, MA
Morgan, M. G., (2009) *Best Practice Approaches for Characterizing, Communicating, and Incorporating Scientific Uncertainty in Decisionmaking*, Synthesis and Assessment Product 5.2, US Climate Change Science Program, Washington, DC
Moser, S. C. and Dilling, L. (eds) (2007) *Creating a Climate for Change: Communicating Climate Change and Facilitating Social Change*, Cambridge University Press, Cambridge, UK
Pidgeon, N., Kasperson, R. E. and Slovic, P. (eds) (2003) *The Social Amplification of Risk*, Cambridge University Press, Cambridge, UK
Renn, O. and Walker, K. (2008) *Global Risk Governance*, Springer, Dordrecht, The Netherlands
Siegrist, M., Earle, T. C. and Gutscher, H. (eds) (2007) *Trust in Cooperative Risk Management*, Earthscan, London
Social Learning Group (2001) *Learning to Manage Global Environmental Risks, Volume 1: A Comparative History of Social Responses to Climate Change, Ozone Depletion, and Acid Rain; Volume 2: A Functional Analysis of Social Responses to Climate Change, Ozone Depletion, and Acid Rain*, MIT Press, Cambridge, MA
US National Research Council (1989) *Improving Risk Communication*, National Academy Press, Washington, DC
US National Research Council (1999) *Our Common Journey: A Transition toward Sustainability*, The National Academies Press, Washington, DC
US National Research Council (2002) *Disposition of High-Level Waste and Spent Nuclear Fuel: The Continuing Societal and Technical Challenges*, The National Academies Press, Washington, DC
US National Research Council (2009a) *Evaluating Progress of the U. S. Climate Change Science Program*, The National Academies Press, Washington, DC
US National Research Council (2009b) *Informing Decisions in a Changing Climate*, The National Academies Press, Washington, DC
US National Research Council (2009c) *Restructuring Federal Climate Research to Meet the Challenge of Climate Change*, The National Academies Press, Washington, DC
US National Research Council (2009d) *Science and Decisions: Advancing Risk Assessment*, The National Academies Press, Washington, DC
Walters, C. (1986) *Adaptive Management of Renewable Resources,* Blackburn Press, Caldwell, NJ

Part II

What Do We Know Now?

2

Knowledge to Practice in the Vulnerability, Adaptation and Resilience Literature: A Propositional Inventory

Rebecca Morgenstern Brenner

Introduction

> *I never perfected an invention that I did not think about in terms of the service it might give others... I find out what the world needs, then I proceed to invent.* (Thomas Edison)

The words of Thomas Edison testify that necessity should inspire invention. How unfortunate, then, that the science community so often does not prioritize practical necessity as a major motivator for its work, or that decision-makers equally as often do not prioritize science-based knowledge when making policy decisions. Research in the Human Dimensions of Global Change community reveals that this science/practice interface is often disconnected and producers and users of knowledge do not always communicate or consider each others' interests. In 2004, in an effort organized by the Union of Concerned Scientists (UCS), over 60 prominent scientists, including Nobel laureates and experts from various disciplines, signed an open letter to President George W. Bush about his alleged 'blatant and unprecedented disregard' for science in his policy decisions. The letter stated that:

> *When scientific knowledge has been found to be in conflict with its political goals, the administration has often manipulated the process through which science enters into its decisions. This has been done by placing people who are professionally unqualified or who have clear conflicts of interest in official posts and on scientific advisory committees; by disbanding existing advisory committees; by censoring and suppressing reports by the government's own scientists; and by simply not seeking independent scientific advice.* (Union of Concerned Scientists, 2004)

This letter from the UCS, while an extreme case, testifies to the need for greater dialogue between scientists and practitioners. The literature in this review suggests that a gap often exists between scientists and practitioners. In order to be experts in their respective domains, scientists (knowledge producers) and practitioners (decision-makers and/or knowledge users) must be intrinsically separate. Scientists and policy-makers have different motives, cultures, language, funding and political agendas (Agrawala et al, 2001; Lemos and Morehouse, 2005). Their specialization in their respective fields requires inherently separate and different goals, focus and training. Yet, both camps are affected by the work and decisions of the other. Both are driven by the current state of affairs in the world and by available economic support.

Globally, the resilience of communities is threatened by change and stochastic events that, in turn, can change the vulnerability of communities. 'Examining globalization in a more analytic fashion may help to illuminate some of the fundamental changes involved, particularly those of special relevance to systemic properties such as resilience, vulnerability, and adaptability' (Young et al, 2006, p306). Globalization brings communities together and alters their vulnerability, resilience and adaptation. As communication increases through globalization, the interactions between scientists and practitioners are increasingly important in efforts to increase resilience, decrease vulnerability, and implement management strategies that promote adaptation. 'Thus, vulnerability research and resilience research have common elements of interest – the shocks and stresses experienced by the social–ecological system, the response of the system, and the capacity for adaptive action' (Adger, 2006, p269). A discussion of the science/practice interface needs an assessment of what research has been done with regards to the fields of vulnerability, adaptation and resilience. In the words of Smit and Wandel (2006, p289):

> *Studies of adaptation to climate change have provided many insights, but to date, have shown only moderate practical effect in reducing vulnerabilities of people to risks associated with climate change. One widely acknowledged lesson is that adaptations are rarely undertaken in response to climate change effects alone, and certainly not to climatic variables that may be of importance to decision makers... There has been considerable*

scholarship in the climate change context on calculating indices of vulnerability and adaptive capacities, and on evaluating hypothetical adaptations, yet the practical applications of this work (in reducing vulnerabilities of people) are not yet readily apparent.

It is also important to note that there is important and valuable empirical work in progress to assess, investigate and make recommendations at the science/practice interface (Jasanoff, 1990; Gieryn, 1995; Social Learning Group, 2001). There are also science reports written directly to the policy-maker (IPCC, 2001; Kitagawa and Yakamoto, 2006). However, this literature review focuses on the vulnerability, adaptation and resilience literature, and more specifically what, if anything, the literature says about the science/practice interface. What are the fields of vulnerability, adaptation and resilience doing to close the gap and to make assessments and investigations more useful?

Frame and focus

Framing the literature review

In this examination, we first need to define vulnerability, adaptation and resilience. As discussed later in this chapter, the foregoing fields are not universally agreed upon. For the purposes of this literature review, the terms will be defined as follows:

- *Vulnerability* 'is the degree to which a system, subsystem, or system component is likely to experience harm due to exposure to a hazard, either a perturbation or stress/stressor' (Turner et al, 2003a, p8074, citing White, 1974).
- C. S. Holling first described *resilience* as 'the persistence of a relationship within a system and ... a measure of the ability of these systems to absorb changes of state variables, driving variables, and parameters, and still persist' (Holling, 1973, p17).
- *Adaptation* refers to 'adjustments in a system's behavior and characteristics that enhance its ability to cope with external stress' (Smit et al, 2000, p225).

The scope of this literature review is limited. Our investigation attempts to uncover references to the science/practice interface, not only among the vulnerability, adaptation and resilience literature, but also from science/practice writings. The intent is to grasp what research has been done, what has been concluded, and what is well established and supported by empirical research in the science/practice interface. In short, we seek to provide useful insights to guide future research to enhance scientists' and practitioners' efforts.

Methodology

A propositional inventory was used to assess and present the empirical evidence generated by the literature. A proposition, as used here, is an analytical statement rooted in concepts or theory from which inferences can be made. For this literature review, we are looking for conceptual statements rooted in empirical evidence. We then use those statements to form cross-cutting propositions about the science/practice interface.

The first step is to identify the key vulnerability, adaptation and resilience publications, then search those works and the works cited for references to the science/practice interface. Relevant literature is then reviewed for assessing empirical evidence. These works are mined in order to find concrete statements that address the interface of scientists and practitioners.

Janssen et al (2006) present a revealing bibliometric analysis of the most cited literature in the fields of vulnerability, adaptation and resilience. Another valuable source in this enquiry included interviewing experts in those fields and consulting peer-review journals. Web-based searches were not as helpful because the terms science/practice often do not appear in the title or abstract, and because key words produced a plethora of literature largely irrelevant to this review.

During the search process, it became apparent that there are two distinct sources of information regarding the science/practice interface – tested empirical evidence and intuitive inferences based on expert experience. Both sources were collected.

A main goal of this literature review is to determine what is now known about the science/practice interface and what needs to be tested and assessed using empirical research. This review determined that many of the research findings about the science/practice interface are based on intuitive inferences rather than specific empirical research. The statements are often not supported by direct empirical evidence, but are based on the experience and insights of experts in the vulnerability, adaptation and resilience fields. Although such statements are valuable, they are not supported by empirical evidence. To separate the empirical research from intuitive statements, this review will be divided into two sections: one that lists propositions based on empirical studies and evidence (see 'Propositional inventory') and another list of propositions based on untested inferences and experts' insights (see 'Inferences or experimental propositions').

Propositional inventory

The propositions and key findings that follow are based on empirical evidence. Each proposition is discussed in detail below.

Inconsistent definitions

The concepts of vulnerability, adaptation and resilience are not consensually defined within the global change literature. Lacking common definitions, it is difficult to apply concepts to policy decisions.

Disciplinary incoherence

Scientific evidence differs across disciplines, leaving decision-makers to determine which discipline's scientific evidence to use. This compromises the legitimacy of, and confidence in, scientific evidence.

Credibility

The impact of science advice is only as strong as its credibility.

Usefulness

The usefulness of knowledge depends upon the saliency, credibility and legitimacy of science.

Tailored assessments

Assessments tailored to the producer and user are used more frequently.

Iterative assessments

Assessments that are iterative generally are more useful than those conducted at one point in time.

Specificity

It is difficult for decision-makers to form generalizations and draw conclusions from vulnerability, adaptation and resilience assessments when the assessments are specific to particular cases.

Boundary organizations

Boundary organizations link science and practice, but simply linking science and practice is not sufficient to generate successful assessments. External factors also contribute to the feasibility and effectiveness of an assessment.

Political economy

Environmental policy decisions are often driven by economic factors and are influenced by political agendas. Thus, the information and results that science provides must work within those constraints.

Advisers

Science advisers (to decision-makers) are often sheltered from the scrutiny of the scientific community, a fact which compromises the dependability of available knowledge.

Communication

Science and practice typically have different languages, which hinders greater collaboration.

Scale

Practice operates on a shorter timeframe and science operates on a longer timescale.

Uncertainty

The uncertain nature of much knowledge generated by science impedes its use by decision-makers in policy decisions.

Variability

Conflicting scientific theories and the variable nature of predictions inhibit the use of science by practitioners.

Adaptive management

Participatory adaptive management practices allow greater flexibility, experimentation and iteration by decision-makers in managing complex social–ecological systems.

Perception

Social constructs and public perceptions often drive policy decisions more than scientific information and results.

Connectivity

Interconnections among scales (such as local to global) enhance the robustness of social–ecological systems.

Assessing the empirical evidence

Inconsistent definitions

The concepts of vulnerability, adaptation and resilience are not consensually defined within the global change literature. Lacking common definitions, it is difficult to apply concepts to policy decisions.

In order to discuss vulnerability, adaptation and resilience, each must be defined. Although definitions abound in the literature, there are none that are universal or generally accepted. Rather, definitions are derived or chosen from multiple sources (Cutter, 1996; Turner et al, 2003a; Janssen and Ostrom, 2006; Janssen et al, 2006; Young et al, 2006). These multiple definitions make analysis difficult for experts and practitioners alike; thus, it is hard for practitioners to utilize the knowledge derived in policy decisions. Without a common definition, it is difficult to apply any conceptual framework:

- *Vulnerability.* The term vulnerability was first discussed in the natural hazards field (Janssen and Ostrom, 2006). Vulnerability is defined in a variety of ways (Burton et al, 1978; Adger, 2006; Gallopín, 2006). 'Vulnerability is the degree to which a system, subsystem, or system component is likely to experience harm due to exposure to a hazard, either a perturbation or stress/stressor' (Turner et al, 2003a, p8074, citing White, 1974). Vulnerability is also 'the degree to which a system is susceptible to, or unable to cope with, adverse effects (of climate change)' (IPCC, 2001, p995). Vulnerability also refers to 'the characteristics of a person or group in terms of their capacity to anticipate, cope with, resist, and recover from the impact of a natural hazard' (Blaikie et al, 1994, p9).
- *Adaptation.* Adaptation was introduced in anthropology during the early 1900s (Janssen and Ostrom, 2006). Beginning in the 1960s, cultural ecologists within geography and ecological anthropology studied the adaptations of humans to ecosystem constraints (Brookfield, 1962; Brookfield, 1964; Netting, 1993; Rappaport, 2000). Culture, for cultural ecologists, served an ecologically adaptive function 'in which human activities are structured that have the effect of reducing or enlarging damage from extreme natural events' (Burton et al, 1978, p39). Adaptation later came to mean 'adjustments in a system's behavior and characteristics that enhance its ability to cope with external stress' (Smit et al, 2000, p225). The IPCC (2001, p643) defines adaptation as:

    > ... *adjustment in ecological, social, or economic systems in response to actual or expected climatic stimuli and their effects or impacts. This term refers to changes in processes, practices, or structures to moderate or offset potential damages or to take advantage of opportunities associated with changes in climate. It involves adjustments to reduce the vulnerability of communities, regions, or activities to climatic change and variability.*

- *Resilience.* Resilience is 'a system's ability to bounce back to a reference state after a disturbance and the capacity of a system to maintain certain structures and functions despite disturbance' (Turner et al, 2003a, p8075). This definition of resilience is inclusive. Much of the ecological community is divided into two main camps: 'ecological resilience' as defined by Holling (1973) and 'engineering resilience' as defined by Pimm (1984). In Holling's classic work of 1973, ecological resilience was seen as 'the persistence of a relationship within a system and is a measure of the ability of these systems to absorb changes of state variables, driving variables, and parameters, and still persist' (Holling, 1973, p17). Engineering resilience (a concept largely discredited by the ecological resilience community) is the time to return to a state of equilibrium following a perturbation (Pimm, 1984; Tilman and Downing, 1994).

Many land managers and agencies adopt engineering resilience as their operative concept. The Resilience Alliance promotes Holling's definition of ecological resilience. The definition by Turner et al (2003a) includes both ecological resilience and engineering resilience; but these two definitions are at odds with each other, thereby dividing the field. This discrepancy in definition is relevant to the science/practice interface since agencies (practitioners) may be using engineering resilience, while academic analysts are likely to use ecological resilience.

Since these three similar fields of science generally do not collaborate, it is even more difficult for practitioners to apply knowledge generated by these disciplines. Gallopín, (2006, p302) discusses the linkages among vulnerability, adaptation and resilience. 'Interdisciplinary research on the Earth System and social–ecological systems at other scales would clearly benefit from having a general, self-consistent set of these basic concepts that could be applied across disciplines.' Also, although the fields of vulnerability and adaptation often cite each other, resilience is disconnected from the other two fields: 'It is remarkable that major publications on the knowledge domain resilience do not cite the other two knowledge domains vulnerability and adaptation, and the other way round' (Janssen and Ostrom, 2006, p249).

Disciplinary incoherence

Scientific evidence differs across disciplines, leaving decision-makers to determine which discipline's scientific evidence to use. This compromises the legitimacy of and confidence in scientific evidence.

Historically, science has often been perceived as strongly based on objective 'facts'. Only recently have analysts recognized that value-laden judgements enter into what is acceptable to science (Ravetz, 1990). The scientific community may disagree on methods and processes of research, thereby compromising universal acceptance of the generated knowledge. As a result, how science conducts research to frame problems or to accept or reject hypotheses is where scientific opinion may differ and create controversy:

> *Controversy is thus not some exceptional phenomenon but some intrinsic element of the scientific process. This contrasts with the demands made on science in terms of decision-making and public policy. Science is put in a difficult, even an impossible, position to provide 'truth' if all it can actually provide are hypotheses of different degrees of acceptability.* (Rüdig, 1993, pp18–19)

Rüdig (1993) provides evidence in a case study of nuclear permitting during the early 1970s that discrepancy in scientific research can arise in the validation of experiments. Opposition between the regulatory Atomic Energy Commission (AEC) and scientists (some involved in the AEC's own assessment process and other experts in the field) brought confusion to the public debate; not all scientists agreed on the methods to assess and contain a core meltdown. The AEC, meanwhile, had multiple roles as the decision-maker (balancing the capital already invested in nuclear projects and the need for additional energy in the electrical grid) and as scientific advisers. Thus, the AEC acted as both the promoter and the regulator of nuclear energy. The AEC ultimately decided to permit the siting of reactors despite the doubts of some scientists. The study by Rüdig shows, first, that scientific opinion and legitimacy often depend upon the acceptance of the methods used, and, second, that the multiple hats which decision-makers wear cause decisions to be made not only on the basis of scientific evidence, but other external factors as well.

Credibility

The impact of science advice is only as strong as its credibility.

If science lacks credibility, practitioners will not use the scientific knowledge. Practitioners need to see a source as credible before applying and relying on the information that the source provides. This requires a high standard of expert opinion (Rüdig, 1993). For example, scientific information posted on the internet may be inaccurate, but may be cited many times (Emanuel, 1998). Furthermore, when information lacks credibility it undermines decision-makers' efforts. The integrity of the science 'depends on its ability to respond effectively to challenges to its integrity as an advisor to the government and as a guardian of science' (Hilgartner, 2000, p148).

The boundary between science and policy limits discussion, thereby concealing some of the ambiguity of the advice; the advice then appears more unified than the technical science actually is. Hilgartner (2000, p149) argues that different levels of involvement by participants and disclosure of information affect the acceptance of knowledge produced by science:

> *Struggles over information control are not something that analysts of science and technology can afford to ignore because they are a prominent feature in knowledge production in precisely those organizational contexts, such as states and corporations, where extremely consequential decisions are routinely made.*

Hilgartner provides evidence for the value of credibility to practitioners, using three controversial reports on diet and health as case studies to show how the advice of the National Academy of Sciences was produced, contested and maintained (Hilgartner, 2000).

Usefulness

The usefulness of knowledge depends upon the saliency, credibility and legitimacy of science.

Scientific information needs to be accepted by practitioners as legitimate and, thus, reliable in order for practice to follow scientific advice (Cash et al, 2003; Parris and Kates, 2003; Adger et al, 2005a). Decision-makers need saliency (relevance), credibility (accuracy) and legitimacy (impartiality), Cash and colleagues (2003) argue, before they can accept, trust and use the knowledge that science provides. The balance of these three characteristics of information determines how useful this information is to decision-makers. Cash et al (2003) present empirical evidence from three investigations consisting of a literature review, a workshop and a comparative case study to better understand the importance of the saliency, credibility and legitimacy of science usefulness of generated knowledge. The Knowledge Systems for Sustainable Development project at Harvard University found that the more diverse the actors who accept information as salient, credible and legitimate, the more the knowledge is utilized. Frequently, increases in the salience, credibility or legitimacy of science come with trade-offs in the other two areas. As Cash et al (2002) point out: 'Efforts to increase credibility, by increasing the isolation of science (maintaining strong boundaries), often have the cost of decreasing salience by removing the decision-maker from the scoping or agenda setting process'. The acceptance of scientifically based technologies needs consensus among the scientific community in order to gain legitimacy.

Tailored assessments

To be useful, assessments must be tailored to the producer and user.

The usefulness of assessments depends upon fit, flexibility and availability of resources. The Climate Assessment for the Southwest (CLIMAS) project (Lemos and Morehouse, 2005) aimed at interdisciplinary collaboration as well as joint conceptions between knowledge producers and users to develop regional assessments and to make the information helpful for decision-making:

> *Despite their focus on application, the reality of regional assessments is that they require a combination of knowledge-driven, applied and interactive science which strikes the delicate balance between what we need to know to understand complex problems and what stakeholders perceive to be their immediate needs for making decisions ... years may need to pass before stakeholders perceive this new knowledge as 'usable' for decision-making.*
> (Lemos and Morehouse 2005, p58)

The CLIMAS project compared several climate studies to develop usable regional assessments. Although the fit, flexibility and availability of resources and the collaboration between the producers and users of knowledge were factors in the success of the assessment, other components mattered as well. 'Among the many potential barriers to use are lack of resources to implement changes, political impediments, professional resistance to change, and cultural resistance' (Lemos and Morehouse 2005, p66). The CLIMAS project demonstrates that successful assessments must be tailored specifically to the producer and user, but must also consider external barriers.

Iterative assessments

In order to generate useful assessments, the assessment process must be iterative.

In addition to tailoring assessments for both the producer and user (as discussed above), the process to generate assessments should be iterative (have multiple trials). It is important to have such a process in order to generate assessments that are useful to a variety of actors. Using an iterative method, which is useful for problems with a large number of variables, will generate assessments that fit multiple uses. 'We argue that the higher the level of fit, interdisciplinary and personal flexibility, and availability of resources, the more likely it will be that iterativity in the relation between science and decision-making will occur' (Lemos and Morehouse 2005, p58).

Specificity

It is difficult for decision-makers to form generalizations and draw conclusions from vulnerability, adaptation and resilience assessments because the assessments are usually specific to particular cases.

Scientists usually craft their research to be as specific as possible. This often, however, makes any insights difficult to generalize or extrapolate from in order to meet the broad needs of practitioners. As Young et al (2006, p311) argue: 'Scholars need to develop empirically supported theories that are the foundation for policy analysis rather than presumptions about the best ways of solving problems derived strictly from idealized models'. Therefore, as Turner and colleagues (2003a) note, vulnerability assessments must be rooted in local case studies. Luers et al (2003, p257) present a framework to make vulnerability assessments more useful by providing a recipe for assessments that puts 'a general definition of vulnerability, the susceptibility of damage, into a mathematical equation', using sensitivity and exposure to assess vulnerability but to consider adaptive capacity as a factor as well.

Luers et al (2003) also present a case study of the Yaqui Valley in northern Mexico that attempts to quantify vulnerability. They find that:

> Vulnerability measures can only accurately relate to specific variables, rather than the generality of a place. Even the simplest system is so complex that it is difficult to account for all the variables, processes and disturbances that characterize it. (Luers et al, 2003, p257)

Thus, it is often difficult for practitioners to use case-based vulnerability assessments for broad-based policy decisions or for cross-cutting generalization. For the same reason, it is difficult for practitioners to conduct vulnerability assessments. Turner and colleagues (2003a) propose guidelines to make vulnerability analysis more usable for decision-makers:

- Human and biophysical vulnerability are linked and should be treated accordingly.
- Beware of one-dimensional vulnerability analyses and be cognizant of varied components and scalar linkages in the coupled systems which increase the range of expected outcomes.
- Do not assume that broadly similar coupled systems have the same vulnerabilities; complex dynamics may cause consequences to vary by system or locale.
- Do not assume that all parts of the coupled system have the same vulnerability; subsystems and components, especially social units, may experience exposure differently, register different impacts, and maintain different response options.
- Although comprehensive vulnerability analysis and place-based variations in the coupled systems and processes affecting them favour multiple approaches, vulnerability assessments should follow a common general methodological framework.
- Critical response opportunities are contingent upon the coupled system or place in question; thus, general guidelines for response options should be malleable.
- Conscious efforts must be made to create institutional structures that link vulnerability analyses to decision-making (Turner et al, 2003a, p8078).

Boundary organizations

Boundary organizations link science and practice; but simply linking science and practice is not enough to generate successful assessments – external factors also contribute to the feasibility of an assessment.

Boundary organizations attempt to identify the line between science and practice and then seek to blur that line (Jasanoff, 1990; Gieryn, 1995; Guston, 2001). Blurring the boundaries between science and politics can, arguably, lead to more productive policy making (Guston, 2001). For boundary organizations to be effective, science and practice must understand and adopt some of the norms and practices of the other.

Boundary management is a method to resolve conflicts between scientists and practitioners and encourage successful dialogue (Cash et al, 2003). Boundary organizations attempt to arbitrate situations using mediators as participants, and seek to make all actors accountable (Guston, 2001). The efforts that scientists put into knowledge generation and that practitioners put into action implementation are linked (Jäger et al, 2001). Using case studies of the recent historical analysis of knowledge generation and implementation, Jäger and colleagues (2001) find that the linkages between intensity of knowledge and action implementation affect environmental risk management generally.

Agrawala et al (2001) identify the International Research Institute for Climate Prediction (IRI) as an example of a boundary organization whose focus is to create an approach to climate change that incorporates both scientists and policy-makers. IRI participants identified that they needed to conduct science research with the end user in mind, known as the 'end-to-end' method. Despite efforts to consider users when generating science research, the poorest and most vulnerable populations did not have the means to act on decision-making and the vulnerable had a difficult time with the unpredictability of climate change predictions (Agrawala et al, 2001). Thus, merely engaging knowledge producers and users is not enough, Agrawala and colleagues would argue, to generate successful assessments.

Scientists and practitioners address knowledge production differently. Science builds knowledge by falsifying information, whereas practitioners build evidence from past experience (Popper, 1972; Gieryn, 1995). Jasanoff (1987) distinguishes between science and policy as rival forces with different interests and motivations. Regulatory agencies must consider legal and economic constraints and use scientific rationale to support their positions. Although boundary work attempts to bring science and practice together, other external forces (such as economic factors) remain as a barrier to successful integration of science and practice (Jasanoff, 1987; Agrawala et al, 2001).

Political economy

Environmental policy decisions are often driven by economic reasons and are influenced by political agendas; thus the information that science provides must work within those constraints.

The importance of political economy in determining people's differential vulnerability to natural hazards and global environmental change is widely acknowledged (e.g. Watts, 1983; Blaikie et al, 1994; Adger, 1999; Adger, 2006). People's vulnerability and resilience in the face of natural hazards and disasters, including their capacity to adapt, is typically strongly shaped by, and sometimes products of, political economic forces (e.g. Watts, 1983; Blaikie et al, 1994). Science, from this perspective, often serves the interests of powerful political leaders and global capital to the disadvantage of local people, as abundant examples of misguided sustainable development projects testify (see Chapter 11 in this volume; see also Peet and Watts, 2004; Robbins, 2004). Thus, any science aimed at reducing vulnerability must provide insights that are locally relevant given political/economic realities (Schrecker, 2001; Franklin and Agee, 2003).

Furthermore, political actors determine which knowledge is applied to political decisions. Scientists in government positions are constrained and influenced by political structure and agendas (Franklin and Agee, 2003). As Schrecker (2001, p34) notes: 'Understanding the place of science in environmental policy requires reference to both characteristics of an industrialized market economy and to the organizational setting within which scientists work'. Independent academics are also subject to the agendas of funders, as research funding is often a significant factor for tenure and promotion. Environmental policy decisions are often driven by political economy (economic reasons influenced by political agendas); thus the information that science provides must work within the political economy. Decision-makers must consider economic factors along with scientific knowledge when debating a political issue. This clutters the scientific basis of the argument with other agendas (Guston, 2001). Using fisheries management as an example, Schrecker (2001) describes the influence of funding compromising the position of science.

Advisers

Science advisers (to decision-makers) are sheltered from the scrutiny of the scientific community, a fact which compromises the dependability of available knowledge.

Science advisers to decision-makers are often not subjected to rigorous peer review; thus policy-makers do not have access to the best available information to use for their decision-making. This compromises the legitimacy and effectiveness of the science advisers' knowledge (Jasanoff, 1990; Hilgartner, 2000). Jasanoff (1987, p196) contends that 'the process of peer review, devised by scientists to validate each other's discoveries, reinforces the position of science as an autonomous social institution requiring no external control'. If valuable peer review is circumvented, the credibility of good science is compromised and the legitimacy of scientific knowledge becomes questionable. Hilgartner (2000) presents three case studies on the National Academy of Sciences (NAS) proceedings on diet and health to show how information was masked by science advisers to the decision-makers and, in turn, the integrity of the information provided was compromised.

Scientific facts define environmental problems. Science may influence the regulatory standards that define regulatory decisions regarding environmental quality, and decision-making is based on scientific assessments and cost-benefit analysis (Brickman et al, 1985; Jasanoff, 1990; Fischer, 2000; Hilgartner, 2000). However, contrary evidence shows that scientists are not involved enough in decision-making. Although regulatory standards should rely on science-based knowledge, in actuality other influences, such as economic pressure from political actors, often shape regulatory standards. Rüdig's (1993) discussion of nuclear permitting (as discussed previously in this chapter) is an example of where not all actors have access to all of the information, which warps decision-making. Ultimately, the goal is for the science/practice interface to have the knowledge to make an informed decision. Some decision-makers may lack the expertise to weigh the information or all of the uncensored data and, thus, may not have access to all the knowledge that science has to offer.

Communication

Science and practice have different languages, which hinders collaboration. In order for science and practice to be able to work collaboratively, both must be able to understand the others' language and jargon (Jasanoff, 1987; Adger et al, 2005b; Metzger et al, 2005; Rudel et al, 2005). Practitioners lack the training or skills to interpret scientific data, which prevents an even playing field for controversial dialogues (Jasanoff, 1987).

Scale

Practice operates on a shorter timeframe and science operates on a longer timescale. Social and ecological systems are difficult to predict and decision-makers need definitive information on which to base decisions. Explorations in science often use models and predictions as tools to formulate hypotheses. These predictions are frequently not the definitive information that decision-makers need upon which to base policy decisions. Scientists need large data sets with long timescales to develop effective models (Ravetz, 1990; Fischer, 2000; Yorque et al, 2002; Rudel et al, 2005). Science expects practice to rely on predictive models. But in order for predictive models to be useful to decision-makers, scientists must first show that the models are accurate, through time and experience. Decision-makers need to meet deadlines that scientists are not restricted to (Ravetz, 1990; Fischer, 2000; Yorque et al, 2002; Rudel et al, 2005). For example, Cane and colleagues published El Niño/Southern Oscillation (ENSO) models that were not used by decision-makers until their predictions proved accurate at the next ENSO event (Cane et al, 1986; Agrawala et al, 2001).

Uncertainty

The uncertainty of knowledge generated by science is often too great for decision-makers to base policy decisions upon it. In order to apply predictive models (as discussed above), practitioners must rely on uncertain outcomes and well-educated guesses to base critical decisions:

> *Efforts to predict events or phenomena with complex, diffuse and regional impacts ... have rarely contributed to the resolution of policy debates and have often contributed to political gridlocks. This experience in part reflects the intrinsic scientific challenges of prediction, but it also derives from the complex scientific and policy context within which the predictive research takes place.* (Yorque et al, 2002, pp427, 429, citing Sarewitz and Pielke, 2000)

Science often cannot offer reliable enough predictions of the future for political actors based on the accumulation of evidence (Popper, 1972). For example, the findings of carbon emissions in developing a strategy for meeting reduced greenhouse emissions are uncertain; but science still anticipates action to meet regulation as defined by the United Nations Framework Convention on Climate Change (UNFCCC) and the Kyoto Protocol (Bolin and Kheshgi, 2001). According to the 2001 Intergovernmental Panel on Climate Change report, for example, the 'current knowledge of adaptation and adaptive capacity is insufficient for reliable prediction of adaptations; it also is insufficient for rigorous evaluation of planned adaptation options, measures, and policies of governments' (IPCC, 2001, p880).

Ecosystems do not follow predictable patterns. The unrealistic expectation that ecosystems follow predictable patterns following disturbance are based on outdated notions of resilience (Holling, 1973). As Holling and colleagues subsequently explain, 'mediation among stakeholders is irrelevant if it is based on ignorance of the integrated character of nature and people. The results may be momentarily satisfying to the participants but ultimately reveal themselves as based upon unrealistic expectations about the behavior of natural systems and the behavior of people' (Holling et al, 2002, pp7–8). Predictable outcomes and controllable variables are rarely the case in real-world situations (Holling and Meffe, 1996).

Human response to environmental catastrophes must be measured in terms of the hazard of the event. The magnitude, frequency and duration of the event, speed of onset, special dispersion and temporal spacing all shape human responses (Burton et al, 1978). The natural world is riddled with uncertainties that can undermine the effectiveness of decision-makers' responses. Unfortunately, the uncertainty and unpredictability of science and prediction of stochastic natural hazards often prevent premeditated preparatory action by political actors. Instead, they respond to situations after the fact. Furthermore, in addition to scientific uncertainty, competing political objectives often complicate policy decisions (Yorque et al, 2002). The magnitude and frequency of the event may also affect the social amplification of the risk and, therefore, political response (Pidgeon et al, 2003).

Variability

Conflicting scientific theories and the variable nature of predictions create political gridlock. Expert science advice may create political gridlock and taint the integrity of science. Yorque et al (2002) contend that science advice in the form of predictions has rarely contributed to the resolution of policy debates and instead has often resulted in political gridlock. Gunderson et al (2002) attribute this to experts thinking that they can solve social–ecological–economic problems, often with conflicting theories. Holling and colleagues (2002, p7) argue that 'the complex issues connected with the notion of sustainable development are not just ecological problems, or economic, or social ones, but a combination of all three'.

Adaptive management

Participatory adaptive management practices permit both decision-makers and scientists to be malleable when investigating and managing complex social–ecological systems. Work during the 1990s on resilience (e.g. Gunderson et al, 1995; Holling and Meffe, 1996; Gunderson, 1999) provides both empirical case studies and conceptual/theoretical work on the integrated nature of social–ecological systems and their dynamics. Many of these case studies show how efforts to negate the uncertainty and unpredictability in social–ecological systems result in unexpected resource management outcomes and policy crises (Holling and Meffe, 1996). Gunderson (1999) uses resilience to explain how broad global processes present themselves locally, using the example of ENSO fluctuations. This work represents an unusual attention to the science/practice interface in the resilience literature in its call for participatory adaptive management and new ways of scientific thinking.

Gunderson and Holling (2002) discuss Holling's (1973) new take on resilience and how it can lead to better-integrated science and improved participatory resource management. This work offers considerable empirical evidence for the concept of ecological resilience and for social–ecological system linkages (Holling, 1973; Gunderson and Holling, 2002). In fact, this work is uncommonly strong on empiricism, yet also explicitly addresses the science/practice interface.

Perception

Social constructs and perception often drive policy decisions. Public perception drives policy decisions and the direction of political action. Whether or not a crisis is perceived determines the level of reaction. People's perceptions differ when evaluating the significance of a hazard; individual financial positions and lifestyle choices determine the significance of a hazardous event on one's life (Burton et al, 1978). For example, water management policy decisions, such as in the Florida Everglades, are subject to perceptions that drive policy decisions and the direction of science (Gunderson et al, 2002). Gunderson (1999) provides evidence using the 1940s Florida Everglades land-use plan, which designated three distinct land-use areas altering the chemistry of the water and soil. He notes: 'In the freshwater Everglades, the external climate variation creates a set of disturbances (dry period, freezes, or fire) that intersect with a slowly accumulating phosphorous (as one mechanism for the switch in stability domain). If these shifts in stability domains are viewed as a crisis, then understanding how and why people choose to react is key to managing for resilience' (Gunderson, 1999, p7).

Connectivity

Interconnections among scales (such as local to global) enhance the robustness of social–ecological systems. The connectivity of distinct local and global events (as they relate to vulnerability, adaptation and resilience) determines the effectiveness of the responses to unpredictable events. This connectivity is important to explore for both science and policy in order for there to be changes in the response to unpredictable events in the future:

> *The existence of many interconnections may enhance the robustness or resilience of large-scale social/ecological systems by diluting and distributing the impact of strong changes in individual elements upon other elements of the system.* (Young et al, 2006, p309)

More research is needed in the interconnectedness of the macro-scale in globalization as it relates to vulnerability, adaptation and resilience. The cross-scale (size, time and local to global) linkages are also important for both science and practice. In order to achieve sustainability, research must include global and local processes across different scales, such as local management issues to global climate change (Stern et al, 1992; Cash and Moser, 2000; Kates et al, 2001).

As a result, the politics of global environmental change, state-level environmental policy and local vulnerability often tend to widen the gap between the production of knowledge and its practical implementation at local scales. Then again, the anticipation of the consequences of global change is often more effective at the local scale. This has led some (e.g. Naess et al, 2006) to call for a true exchange of data and insights between local actors and scientists, rather than the often suggested interdisciplinary integration of science.

Local knowledge is neither science nor practice, but a separate contributor to the science/practice interface to affect the coupled social–ecological system. Both local knowledge and science, as generated research, contribute to policy decisions and influence practice. Local knowledge provides additional site-specific information to augment science and challenges science with self-oriented interests. While some societies welcome expert science, others may reject the intrusion. Gadgil and colleagues (2003) present evidence from three case studies where experts in science and local knowledge were brought together adaptively to manage ecological changes. Then the combined (scientist and local) expert knowledge was considered in management.

Blann and colleagues (2003) used a resource crisis in Minnesota as a case study for practitioner collaboration. They found that the 'new' practitioners, as facilitators of learning and change, need different skill sets, including the ability to articulate vision and metaphor for double-loop learning and to create a safe, open and respectful platform for dialogue, learning and relationship-building, and experimentation (Blann et al, 2003). Thus, the practitioner can then apply collaborated resource knowledge with the goal to transfer adaptive capacity across scales and to re-establish the social–ecological system.

Inferences or experimental propositions

The statements that follow are made by individuals who have conducted empirical work, and their intuitive accounts are based on experience. In this chapter, we try to distinguish between insights and evidence. Insights are valuable, but they have not been tested through an empirical study; rather, they are based on researcher experience. Although there is not direct empirical evidence to support these statements, they are made by knowledgeable experts in the fields of vulnerability, adaptation and/or resilience. This chapter section presents these insights of researchers into the science/practice interface:

- Science cannot provide answers with enough precision to be effective for decision-makers. Science often raises more questions than it answers (Fischer, 2000).
- The science/practice interface is weak and needs to be strengthened. Both science and policy need to communicate better and engage in dialogue (Jasanoff et al, 1997; Lubchenco, 1998; Adger, 1999; Cash et al, 2003; Adger et al, 2005b; Jury and Vaux, 2005).
- Social scientists need to be more involved in policy decisions and assessments. They have not generally been engaged in investigations to inform institutions or policy-makers (Lubchenco, 1998; Jury and Vaux, 2005; Selin and Pierskalla, 2005).
- Scientific evidence differs among interpretations and disciplines, leaving users to determine which scientific evidence to use. This compromises the legitimacy of, and confidence in, scientific evidence. (Ravetz, 1990; Fischer, 2000; Hilgartner, 2000; Gibson, 2003; Bammer, 2005).
- Science and practice are separate entities with separate interests, but are interdependent and solutions are intertwined and interdisciplinary. Scientific solutions to global problems rely on a multidisciplinary approach and do not fall under a single discipline (Otway and Ravetz, 1984; Ravetz, 1990; Stern et al, 1992; Jasanoff et al, 1997; Gunderson, 1999; Jäger et al, 2001; Bammer, 2005; Jury and Vaux, 2005; Lemos and Morehouse, 2005; Palmer et al, 2004; Young et al, 2006).
- Science and practice have different norms for evaluating credibility and developing indicators (Cash et al, 2003; Parris and Kates, 2003; Adger et al, 2005b) and so it is difficult to evaluate success.
- Policy has a deeper focus on economic concerns and science's focus is on scientific concepts and processes. These distinct approaches offer different perspectives and priorities (Franklin and Agee, 2003; Jury and Vaux, 2005; Rudel et al, 2005). Politics are often driven by economics, not necessarily research, while science and practice have different incentives.
- Scientific facts often define environmental problems. Science determines the regulatory standards that define regulatory decisions regarding environmental quality (Brickman et al, 1985; Jasanoff, 1990; Fischer, 2000; Hilgartner, 2000; Dessai et al, 2004).

- Decision-making is often determined by a cost-benefit analysis and science (knowledge) is generated on a basis of fact and under the subjective lens of the discipline in which the research is conducted (Ravetz, 1990; Adger et al, 2005a).
- Decision-makers at some point will likely be incorrect. Every action, even a decision for no action, poses consequences. Conversely, scientists do not make value-laden decisions – the facts generated either support or do not support a hypothesis, and findings are statistically significant or not. However, the methodology can influence the outcome of the study (Schrecker, 2001).
- An open decision-making process causes science to deconstruct technical arguments for the political environment (Jasanoff 1990; Hilgartner, 2000).
- The achievements of science to enhance human society have caused many environmental problems (such as air pollution and water contamination) and practice relies on science to find the solution to the same problems that science caused (Ravetz, 1990; Fischer, 2000).
- Society relies on and expects science to lead improvements based on society's goals. This is typically conveyed in the allocation of funds and resources (Hays, 1987; Lubchenco, 1998; Fischer, 2000; Turner, 2002; Bammer, 2005).
- Public interest in an issue often drives management or government to take action with science (IPCC, 2001; Palmer et al, 2004).
- Environmental decision-making is often closed to public scrutiny, where the science cannot be contested or assessed by interested parties. Thus, the science that drives the political agenda, not the best available science, may be used to support decisions (Schrecker, 2001).
- Some management policies that are developed do not consider ecosystem variability. Different ecosystems cannot be managed by the same policy (Franklin and Agee, 2003).
- Existing knowledge is difficult to locate comprehensively. Both researchers and policy-makers must expend considerable time and effort to identify and obtain existing knowledge and contacts (Bammer, 2005).
- To make scientific knowledge comprehensible to decision-makers, classification systems can be generated (and are already available) so that practice can understand the differences and needs of different ecosystems when forming broad policies (Franklin and Agee, 2003).

Discussion

The propositions identified in this literature review are interconnected empirically based propositions regarding gaps in the science/practice interface. After reviewing the literature, it is clear that the vulnerability, adaptation and resilience communities need to better connect knowledge produced with knowledge users.

What can be done is collaboration; but is this feasible with competing timeframes and agendas? Knowledge producers and users in the field of vulnerability, adaptation and resilience need to develop an action plan together, iteratively. Knowledge needs to be generated by both science and practice working together (Clark and Dickson, 2003; Lemos and Morehouse, 2005). To connect science and policy, the social science community needs to develop an action plan to enhance social science participation in scientific review, policy-making and agenda setting (Selin and Pierskalla, 2005).

Unfortunately, although the literature repeatedly states that there is a need to connect decision-makers and scientists more closely, there is not much action to this end. Valuable boundary work is under way, but not necessarily by the vulnerability, adaptation and resilience scientists. The nature of vulnerability, adaptation or resilience makes it difficult to generalize or provide a recipe for these types of iterative assessments.

Science and practice communities need a forum for both parties to engage in a conversation to discuss similar interests, agendas, dialogues and goals. A forum for scientists and practitioners to engage in a conversation concerning mutual agendas and goals would promote a problem-solving dialogue about similar interests of both communities. Scientists and practitioners need a forum for discussion and additional development of communication tools (Adger et al, 2005b). Yorque and colleagues (2002) suggest co-designed workshops with practitioners that can promote adaptive learning. Websites such as Conservation Ecology (www.consecol.org) and the Vulnerability Network (www.vulnerabilitynet.org) provide forums for practitioners and scientists to discuss and exchange knowledge. The internet is valuable since most decision-makers and scientists have access to it, regardless of location, language or economic status.

Conclusions

Although the empirical data found by this literature review are uneven, common themes are nonetheless apparent. The science community and the practice community operate in different ways and have different priorities, pressures and timelines. In order to bridge the gap, both communities need to reach out and communicate to maximize each other's resources. Without the practice community, scientists may gain knowledge; but the practical application is missing. Without science-based knowledge, decision-makers do not have the tools or knowledge to implement changes to increase resilience, decrease vulnerability, and adopt management strategies that promote adaptation.

There is valuable boundary work investigating the science/practice interface (Jasanoff, 1987, 1990; Gieryn, 1995; Hilgartner, 2000; Agrawala et al, 2001; Guston, 2001; Jäger et al, 2001; Cash et al, 2003) and valuable vulnerability, adaptation and resilience explorations (Holling, 1973; Luers et al, 2003;

Turner et al, 2003a, 2003b; Adger, 2006; Folke, 2006; Smit and Wandel, 2006). However, these two communities are working independently. The vulnerability, adaptation and resilience literature discusses the science/practice interface and the need to close the gap, bridge the communities, collaborate on research and provide forums for dialogue; but few are actually doing this. In fact, a review by Janssen and colleagues (2006) discovered that among scholarly networks in vulnerability, adaptation and resilience, the resilience publications less often cite works from the other two fields. The literature from the vulnerability, adaptation and resilience communities is missing an action plan to connect scientists and practitioners.

As a result of this literature review, the following actions are proposed:

- Generate roles in both the science and practice communities for cross-boundary experts – people with a science background and skilled in decision-making. These boundary experts can act as a communication link, or translators, to transfer better knowledge between science and practice.
- Locate funding resources to hold annual collaborative workshops. Currently, the science community and decision-making community lack a forum to collaborate regularly or to meet face to face, so there is little vested interest to build a permanent relationship.
- Use resources such as the internet to develop specific discussion forums as well as news briefings for developments in science and in practice. Both communities are dynamic, and keeping both abreast of changes will help the communities to better understand the needs of the other. The webpage should be brief, concise and organized by issue area and specific categories.
- Develop a biannually updated medium with contact information. Knowing whom to contact with questions will get more questions asked and more relevant answers.

Acknowledgements

Very special thanks to Jake.

References

Adger, W. N. (1999) 'Social vulnerability to climate change and extremes in coastal Vietnam', *World Development*, vol 27, no 2, pp249–269

Adger, W. N. (2006) 'Vulnerability', *Global Environmental Change*, vol 16, no 3, pp268–281

Adger, W. N., Arnell, N. W. and Tompkins, E. L. (2005a) 'Successful adaptation to climate change across scales', *Global Environmental Change*, vol 15, no 2, pp77–86

Adger, W. N., Brown, K. and Hulme, M. (2005b) 'Redefining global environmental change: Editorial', *Global Environmental Change*, vol 15, no 1, pp1–4

Agrawala, S., Broad, K. and Guston, D. H. (2001) 'Integrating climate forecasts and societal decision making: Challenges to an emergent boundary organization', *Science, Technology and Human Values*, vol 26, no 4, pp454–477

Bammer, G. (2005) 'Integration and implementation sciences: Building a new specialization', *Ecology and Society*, vol 10, no 2, p6, www.ecologyandsociety.org/vol10/iss2/art6/

Berkes, F. (1999) *Sacred Ecology: Traditional Ecological Knowledge and Management Systems*, Taylor and Francis, Philadelphia and London

Berkes F. and Folke, C. (1998) 'Linking social and ecological systems for resilience and sustainability', in F. Berkes and C. Folke (eds) *Linking Social and Ecological Systems: Management Practice and Social Mechanisms for Building Resistance*, Cambridge University Press, Cambridge, UK

Berkes, F. and Folke, C. (2002) 'Back to the future: Ecosystem dynamics and local knowledge', in L. H. Gunderson and C. S. Holling (eds) *Panarchy: Understanding Transformation in Human and Natural Systems* Island Press, Washington, DC, pp121–146

Blaikie, P., Cannon, T., Davis, I. and Wisner, B. (1994) *At Risk: Natural Hazards, People's Vulnerability, and Disasters*, Routledge, London

Blann, K., Light S. and Musumeci, J. A. (2003) 'Facing the adaptive challenge: Practitioners' insights from negotiating resource crisis in Minnesota', in F. Berkes, J. Colding and C. Folke (eds) *Navigating Social/Ecological Systems: Building Resilience for Complexity and Change*, Cambridge University Press, Cambridge, UK, pp210–240

Bolin B. and Kheshgi, H. S. (2001) 'On strategies for reducing greenhouse gas emissions', *Proceedings of the National Academy of Sciences*, vol 98, no 9, pp4850–4854

Brickman, R., Jasanoff, S. and Ilgen, T. (1985) *Controlling Chemicals: The Politics of Regulation in Europe and the United States*, Cornell University Press, Ithaca, NY

Brookfield, H. C. (1962) 'Local study and comparative method: An example from Central New Guinea', *Annals of the Association of American Geographers*, vol 52, no 3, pp242–254

Brookfield, H. C. (1964) 'Questions on the human frontiers of geography', *Economic Geography*, vol 40, no 4, pp283–303

Burton, I., Kates, R. W. and White, G. F. (1978) *The Environment as Hazard*, Oxford University Press, New York, NY

Cane, M. A., Zebiak, S. E. and Dolan, S. C. (1986) 'Experimental forecasting of El Niño', *Nature*, vol 321, pp827–832

Cash, D. W. and Moser, S. C. (2000) 'Linking global and local scales: Designing dynamic assessment and management processes', *Global Environmental Change*, vol 10, no 2, pp109–120

Cash, D. W., Clark, W., Alcock, F., Dickson, N., Eckley, N. and Jäger, J. (2002) *Salience, Credibility, Legitimacy and Boundaries*, Faculty Working Paper RWP02-046, John F. Kennedy School of Government, Harvard University, Cambridge, MA

Cash, D. W., Clark, W. C., Alcock, F., Dickson, N. M., Eckley, N., Guston, D., Jäger, J. and Mitchell, R. (2003) 'Knowledge systems for sustainable development', *Proceedings of the National Academy of Sciences of the United States of America*, vol 100, no 14, pp8086–8091

Clark, W. C. and Dickson, N. M. (2003) 'Sustainability science: The emerging research program', *Proceedings of the National Academy of Sciences*, vol 100, no 14, pp8086–8091

Cutter, S. (1996) 'Vulnerability to environmental hazards', *Progress in Human Geography*, vol 20, no 4, pp529–539

Dessai, S., Adger, W. N., Hulme, M., Turnpenny, J. R., Köhler, J. and Warren, R. (2004) 'Defining and experiencing dangerous climate change', *Climatic Change*, vol 64, pp11–25
Emanuel, K. A. (1998) 'The power of a hurricane: An example of reckless driving on the information superhighway', *Weather*, vol 54, pp107–108
Fischer, F. (2000) *Citizens, Experts, and the Environment: The Politics of Local Knowledge*, Duke University Press, Durham, NC
Folke, C. (2006) 'Resilience: The emergence of a perspective for social–ecological systems analyses', *Global Environmental Change*, vol 16, no 3, pp253–267
Franklin, J. F. and Agee, J. K. (2003) 'Forging a science-based national fire policy', *Issues in Science and Technology*, Fall, pp1–8
Gadgil, M., Olsson, P., Berkes, F. and Folke, C. (2003) 'Exploring the role of local ecological knowledge in ecosystem management: Three case studies', in F. Berkes, J. Colding and C. Folke (eds) *Navigating Social–Ecological Systems: Building Resilience for Complexity and Change*, Cambridge University Press, Cambridge, UK, pp189–209
Gallopín, G. C. (2006). 'Linkages among vulnerability, resilience, and adaptive capacity', *Global Environmental Change*, vol 16, no 3, pp293–303
Gibson, B. (2003) *From Transfer to Transformation: Rethinking the Relationship between Research and Policy*, PhD thesis, National Centre for Epidemiology and Population Health, Australian National University, Canberra
Gieryn, T. F. (1995) 'Boundaries of science', in S. Jasanoff, G. E. Markle, J. C. Peterson and T. Pinch (eds) *Handbook of Science and Technology Studies*, Sage, Thousand Oaks, CA, pp393–443
Gunderson, L. H. (1999) 'Resilience, flexibility and adaptative management: Antidotes for spurious certitude?', *Conservation Ecology*, vol 3, no 1, p7, www.consecol.org/vol3/iss1/art7/, last accessed 07/24/08
Gunderson, L. H. (2000) 'Ecological resilience: In theory and application', *Annual Review of Ecology and Systematics*, vol 31, pp425–439
Gunderson, L. H. and Holling, C. S. (eds) (2002) *Panarchy: Understanding Transformations in Human and Natural Systems*, Island Press, Washington, DC
Gunderson, L. H., Holling, C. S. and Light, S. (1995) *Barriers and Bridges to Renewal of Ecosystems and Institutions*, Columbia University Press, New York, NY
Gunderson, L. H., Holling, C. S. and Peterson, G. D. (2002) 'Surprises and sustainability: Cycles of renewal in the Everglades', in L. H. Gunderson and C. S. Holling (eds) *Panarchy: Understanding Transformation in Human and Natural Systems*, Island Press, Washington, DC, pp315–332
Guston, D. H. (2001) 'Boundary organizations in environmental policy and science: An introduction', *Science, Technology, and Human Values*, vol 26, no 4, pp399–408
Hays, S. (1987) *Beauty, Health, and Permanence: Environmental Politics in the United States 1955–1985*, Cambridge University Press, Cambridge, UK
Hilgartner, S. (2000) *Science on Stage: Expert Advice as Public Drama*, Stanford University Press, Stanford, CA
Holling, C. S. (1973) 'Resilience and stability of ecological systems', *Annual Review of Ecology and Systematics*, vol 4, pp1–23
Holling, C. S. and Meffe, G. K. (1996) 'Command and control and the pathology of natural resource management', *Conservation Biology*, vol 10, no 2, pp328–337
Holling, C. S., Gunderson, L. H. and Ludwig, D. (2002) 'In quest of a theory of adaptive change', in L. H. Gunderson and C. S. Holling (eds) *Panarchy:*

Understanding Transformations in Human and Natural Systems, Island Press, Washington, DC, pp3–24

IPCC (Intergovernmental Panel on Climate Change) (2001) *Climate Change 2001: Impacts, Adaptation and Vulnerability*, Cambridge University Press, Cambridge, UK

Jäger, J., van Eijndhoven, J. and Clark, W. C. (2001) 'Knowledge and action: An analysis of linkages among management functions for global environmental risks', in The Social Learning Group (ed) *Learning to Manage Global Environmental Risks: A Functional Analysis of Social Responses to Climate Change, Ozone Depletion, and Acid Rain – Volume 2: A Functional Analysis of Social Responses to Climate Change, Ozone Depletion, and Acid Rain*, MIT Press, Cambridge, MA, pp165–180

Janssen, M. A. (2006) 'Resilience, vulnerability, and adaptation: A cross-cutting theme of the International Human Dimensions Programme on Global Environmental Change', *Global Environmental Change*, vol 16, no 3, pp237–239

Janssen, M. A. and Ostrom, E. (2006) 'Resilience, vulnerability, and adaptation: A cross-cutting theme of the International Human Dimensions Programme on Global Environmental Change. Editorial', *Global Environmental Change*, vol 16, no 3, pp237–239

Janssen, M. A., Schoon, M. L., Ke, W. and Börner, K. (2006) 'Scholarly networks on resilience, vulnerability and adaptation within the human dimensions of global environmental change', *Global Environmental Change*, vol 16, no 3, pp240–252

Jasanoff. S. (1987) 'Contested boundaries in policy-relevant science', *Social Studies of Science*, vol 17, no 2, pp195–230

Jasanoff, S. (1990) *The Fifth Branch: Science Advisers as Policymakers*, Harvard University Press, Cambridge, MA

Jasanoff, S., Colwell, R., Dresselhaus, M. S., Goldman, R. D., Greenwood, M. R. C., Huang, A. S., Lester, W., Levin, S. A., Linn, M. C., Lubchenco, J., Novacek, M. J., Roosevelt, A. C., Taylor, J. E. and Wexler, N. (1997) 'Conversations with the community: AAAS at the millennium', *Science*, vol 278, pp2066–2067

Jury, W. A. and Vaux Jr., H. (2005) 'The role of science in solving the world's emerging water problems', *Proceedings of the National Academy of Sciences*, vol 102, no 44, pp15715–15720

Kates, R., Clark, W., Corell, R., Hall, J., Jaeger, C., Lowe, I., McCarthy, J., Schellnhuber, H., Bolin, B., Dickson, N., Faucheux, S., Gallopín, G., Grubler, A., Huntley, B., Jäger, J., Jodha, N., Kasperson, R., Mabogunje, A., Matson, P., Mooney, H., Moore III, B., O'Riordan, T. and Svedin, U. (2001) 'Sustainability science', *Science*, vol 292, April 27, pp641–642

Kitagawa, M. and Yakamoto, R. (2006) *Science on Sustainability: A View from Japan*, Summary Report, E-Square Inc, Tokyo

Lemos, M. C. and Morehouse, B. J. (2005) 'The co-production of science and policy in integrated climate assessments', *Global Environmental Change*, vol 15, no 1, pp57–68

Lubchenco, J. (1998) 'Entering the century of the environment: A new social contract for science', *Science*, vol 279, pp491–497

Luers, A. L., Lobel, D. B., Sklar, L. S., Addams, C. L. and Matson, P. A. (2003) 'A method for quantifying vulnerability, applied to the agriculture system of the Yaqui Valley, Mexico', *Global Environmental Change*, vol 13, no 4, pp255–267

Metzger, M. J., Leemans, R. and Schröter, D. (2005) A multi-disciplinary multi-scale framework for assessing vulnerabilities to global change', *International Journal of Applied Earth Observation and Geoinformation*, vol 7, pp253–267

Moser, S. (2005) 'Impact assessments and policy responses to sea-level rise in three US states: An exploration of human-dimension uncertainties', *Global Environmental Change*, vol 15, pp353–369

Naess, L. O., Norland, I. T., Lafferty, W. M., and Aall, C. (2006) 'Data processing linking vulnerability assessment to adaptation decision-making on climate change in Norway', *Global Environmental Change*, vol 16, no 2, pp221–233

Netting, R. N. (1993) *Smallholders, Householders: Farm Families and the Ecology of Intensive, Sustainable Agriculture*, Stanford University Press, Stanford, CA

Otway H. and Ravetz, J. R. (1984) 'On the regulation of technology 3: Examining the linear model', *Futures*, vol 16, pp217–232

Palmer, M., Bernhardt, E., Chornesky, E., Collins, S., Dobson, A., Duke, C., Gold, B., Jacobson, R., Kingsland, S., Kranz, R., Mappin, M., Martinez, M. L., Micheli, F., Morse, J., Pace, M., Pascual, M., Palumbi, S., Reichman, O. J., Simons, A., Townsend, A. and Turner, M. (2004) 'Ecology for a crowded planet', *Science*, vol 304, pp1251–1252

Parris, T. M. and Kates, R. W. (2003) 'Characterizing a sustainability transition: Goals, targets, trends, and driving forces', *Proceedings of the National Academy of Sciences*, vol 100, no 14, pp8068–8073

Peet, R. and Watts, M. (eds) (2004) *Liberation Ecologies*, Routledge, London

Pidgeon, N., Kasperson, R. E. and Slovic, P. (eds) (2003) *The Social Amplification of Risk*, Cambridge University Press, Cambridge

Pimm, S. L. (1984) 'The complexity and stability of ecosystems', *Nature*, vol 307, pp322–326

Popper, K. R. (1972) *Objective Knowledge: An Evolutionary Approach*, Clarendon, Oxford

Rappaport, R. (2000) *Pigs for the Ancestors: Ritual in the Ecology of a New Guinea People*, Waveland Press, Prospect Heights, IL (originally 1967)

Ravetz, J. R. (1990) *The Merger of Knowledge with Power: Essays in Critical Science*, Mansell Publishing Limited, London

Robbins, P. (2004) *Political Ecology: A Critical Introduction*, Blackwell, Malden, MA

Rudel, T. K., Coomes, O. T., Moran, E., Achard, F., Angelsen, A., Xu, J. and Lambin, E. (2005) 'Forest transitions: Towards a global understanding of land use change', *Global Environmental Change*, vol 15, pp23–31

Rüdig, W. (1993) 'Sources of technological controversy', in A. Barker and B. G. Peters (eds) *The Politics of Expert Advice: Creating, Using and Manipulating Scientific Knowledge for Public Policy*, University of Pittsburgh, Pittsburgh, PA, pp17–32

Sarewitz, D. and Pielke, R. A., Jr. (2000) 'Predictions in sciences and policy', in D. Sarewitz, R. A. Pielke Jr., and R. A. Byerly Jr. (eds) *Prediction, Decision Making and the Future of Nature*, Island Press, Washington, DC, pp11–22

Schrecker, T. (2001) 'Using science in environmental policy: Can Canada do better?', in E. A. Parson (ed) *Governing the Environment: Persistent Challenges, Uncertain Innovations*, University of Toronto Press, Toronto, pp31–72

Selin, S. and Pierskalla, C. (2005) 'The next step: Strengthening the social science voice in environmental governance', *Society and Natural Resources*, vol 18, pp933–936

Smit, B. and Wandel, J. (2006) 'Adaptation, adaptive capacity and vulnerability', *Global Environmental Change*, vol 16, no 3, pp282–292

Smit, B., Burton, I., Klein, R. J. T. and Wandel, J. (2000) 'An anatomy of adaptation to climate change and variability', *Climatic Change*, vol 45, no 1, pp223–251

Social Learning Group (ed) (2001) *Learning to Manage Global Environmental Risks, Volume 2: A Functional Analysis of Social Responses to Climate Change, Ozone Depletion, and Acid Rain*, MIT Press, Cambridge, MA

Stern, P. C., Young, O. R. and Druckman, D. (eds) (1992) *Global Environmental Change: Understanding the Human Dimension*, National Academy Press, Washington, DC

Tilman, D. and Downing, J. A. (1994) 'Biodiversity and stability in grasslands', *Nature*, vol 367, pp363–365

Turner, B. L. II (2002) 'Contested identities: Human-environment geography and disciplinary implications in a restructuring academy', *Annals of the Association of American Geographers*, vol 92, no 1, pp52–74

Turner, B. L. II, Kasperson, R. E., Matson, P. A., McCarthy, J. J., Corell, R. W., Christensen, L., Eckley, N., Kasperson, J. X., Luers, A., Martello, M. L., Polsky, C., Pulsipher, A. and Schiller, A. (2003a) 'A framework for vulnerability analysis in sustainability science', *Proceedings of the National Academy of Sciences*, vol 100, pp8074–8079

Turner, B. L. II, Matson, P. A., McCarthy, J. J., Corell, R. W., Christensen, L., Eckley, N., Hovelsrud-Broda, G. K., Kasperson, J. X., Kasperson, R. E., Luers, A., Martello, M. L., Mathiesen, S., Naylor, R., Polsky, C., Pulsipher, A., Schiller, A., Selin, H. and Tyler, N. (2003b) 'Illustrating the coupled human-environment system for vulnerability analysis: Three case studies', *Proceedings of the National Academy of Sciences*, vol 100, pp8080–8085

Union of Concerned Scientists (2004) '2004 scientist statement on restoring scientific integrity to federal policy making', 18 February, available at www.ucsusa.org/scientific_integrity/abuses_of_science/scientists-sign-on-statement.html

Watts, M. (1983) 'On the poverty of theory: Natural hazards research in context', in K. Hewitt (ed) *Interpretations of Calamity: From the Viewpoint of Human Ecology*, Allen & Unwin, Boston, MA

White, G. F. (ed) (1974) *Natural Hazards: Local, National, Global*, Oxford University Press, New York, NY

Yorque, R., Walker, B., Holling, C. S., Gunderson, L. H., Folke, C., Carpenter, S. R. and Brock, W. A. (2002) 'Toward an integrative synthesis', in L. H. Gunderson and C. S. Holling (eds) *Panarchy: Understanding Transformation in Human and Natural Systems*, Island Press, Washington, DC, pp419–438

Young, O. R., Berkhout, F., Gallopín, G. C., Janssen, M. A., Ostrom, E. and van der Leeuw, S. (2006) 'The globalization of socio-ecological systems: An agenda for scientific research', *Global Environmental Change*, vol 16, no 3, pp304–316

3

Integrating Science and Practice for the Mitigation of Natural Disasters: Barriers, Bridges, Propositions

Juergen Weichselgartner

Introduction

Coupled human/environment systems are undergoing rapid changes and are adapting to changing conditions. This makes the understanding of response mechanisms – and, hence, the state of vulnerability and resilience – one of the most important issues for society, in general, and for science, in particular. Research in the fields of natural hazards, ecology and global environmental change demonstrates that potential losses are determined not by exposure to hazards alone but also reside in the vulnerability and resilience of the society (or system) experiencing such hazards (Burton et al, 1978; Kasperson et al, 1995; Gunderson and Holling, 2002). This recognition has led to questions regarding the understanding of impacts upon and responses by the affected society. During the last decade, an abundant literature has addressed the complex set of processes, factors, causes and agents involved in the analysis of vulnerability and resilience (Dikau and Weichselgartner, 2005; Janssen et al, 2006). Hence, new insights have been gained and researchers have developed different conceptual frameworks for analysis, as well as techniques and tools for assessment (Turner et al, 2003; Birkmann, 2006).

Today, natural hazard management includes many scientists and practitioners from various fields: special research programmes and institutes, numerous journals, advanced technology, private companies and non-governmental organizations (NGOs). In short, it includes a huge variety of knowledge systems and resources. At the same time, great natural disasters set new records in 2005, and the trend towards higher and higher losses continues (Munich Re Group, 2006) (see Figure 3.1). This paradox of concurrent increases in economic loss and in disaster-related research, precisely expressed by White et al (2001) as 'knowing better and losing even more', raises questions about the efficiency of approaches and tools used in hazard assessment and/or disaster mitigation. This, in turn, raises the possibility that progress is being blocked by fundamental conceptual barriers, in addition to profound changes in environmental and social processes, neither of which are adequately being addressed (Weichselgartner and Sendzimir, 2004). A recent prominent example is Hurricane Katrina, after which it was argued that the broad sequence of decisions reflected a long-term pattern of societal response to hazard events: reducing consequences to relatively frequent events while increasing vulnerability to very large and rare events (Kates et al, 2006).

Since an immense enlargement of the literature in the domains of natural hazards, vulnerability and resilience, and practical disaster mitigation efforts has not reversed the upward trend in losses, the use of knowledge in hazard management comes to the foreground. How does hazard-related research-based knowledge relate to the apparent growing toll of losses? Is human knowledge and understanding of the causes of the losses inadequate despite the increasing research effort, or is it that existing knowledge is not applied or not used in an effective fashion?

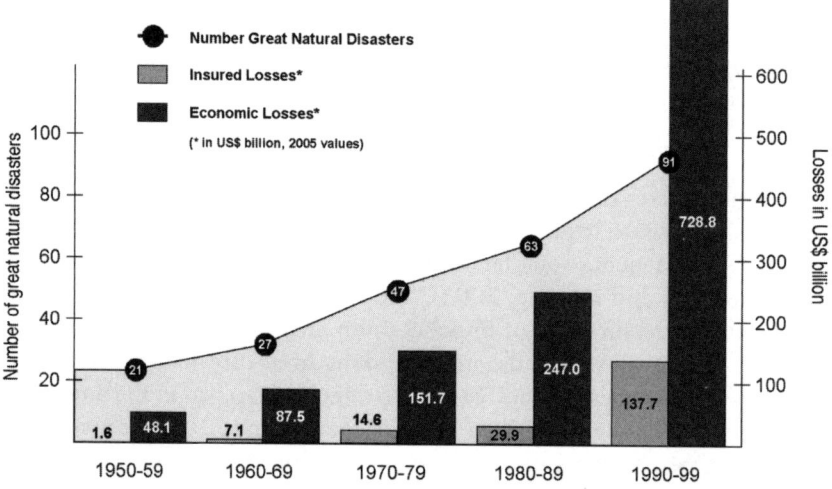

Figure 3.1 *Great natural disasters and economic losses*

Source: Munich Re Group (2006)

Obviously, there are gaps between what is known about natural hazards and disaster mitigation, on the one hand, and how research findings are translated into policies and programmes, on the other (Mileti, 1999; Weichselgartner, 2006). White et al (2001) examined four possible explanations for the situation in which more is lost while more is known:

1 Knowledge continues to be flawed by areas of ignorance.
2 Knowledge is available but not used effectively.
3 Knowledge is used effectively but takes a long time to have effect.
4 Knowledge is used effectively in some respects but is overwhelmed by increases in vulnerability and in population, wealth and poverty.

After reviewing progress in coping with natural hazards, White et al (2001) offered two concluding observations in this context. The first is that better appraisal is needed of the actual results at community and other levels, applying the best available knowledge in the best possible way. The second is that there is a need to build upon past achievements in creating more understanding of natural hazards by better integration of the information within the wider efforts directed at sustainable development. Complementing these findings, Weichselgartner and Obersteiner (2002) argue that the continuing increase of losses from natural disasters is less a consequence of insufficient knowledge than of unsatisfactory transformation of existing knowledge into practical applications. Hence, 'internal' factors of disaster-related science and policy are also responsible for the inability to stem or reverse the upward trend in disaster damage.

Given that less effort has been made to improve the existing gap between scientists (the knowledge producers) and practitioners (the knowledge users), this study focuses on the influence of scientific assessments on decision-making in the practical disaster mitigation arena and the barriers that inhibit the involvement of users in the design of assessments (i.e. the co-production of knowledge). Hazard management, specifically the domains of vulnerability and resilience, is the context but not the focus of the study.

Study objectives

By comparatively analysing 20 scientific assessments from the research domains of vulnerability and resilience, the study addresses the question 'How appropriate and influential is research-based knowledge to support decision-making for vulnerability reduction and building resilience?' in order to provide some empirical evidence for successes and failures. The aim is to survey the knowledge base in order to identify gaps and bridges in the science/policy/practice interface (SPPI) and to support efforts for reducing vulnerability and building resilience. Analysing the influence of vulnerability/resilience assessments requires a certain understanding of social–ecological systems, of the immediate activities taking place on local levels (proximate causes), and of the underlying driving forces and fundamental societal processes (root causes). The same applies with regard to knowledge systems and the modes of knowledge transfer. Much about what makes some assessments more influential than others seems to be associated with the *process* by which they are developed, rather than just the *product* itself. According to a study on global environmental assessments (Mitchell et al, 2006), assessments are better conceptualized as social processes rather than published products. An assessment's formal output – scientific article, report, model or forecast – is only one visible indicator of a larger social process. Thus, focusing exclusively on the content, framing or components of an assessment can never fully discover the real source of its influence (or non-influence).

Ideally, the analysis of the influence of vulnerability assessments should take into account all determinants of this larger social process, ranging from vulnerability-determining factors to the assessment process and its impacts upon disaster management practice. As illustrated in the conceptual framework elaborated upon in Figure 3.2, an analysis of vulnerability assessments and their influence is a challenging endeavour, requiring extensive resources. However, this said, the author's intention is neither to specify factors for conducting vulnerability analyses, nor to critique or promote specific frameworks or methods. Rather, his premise is that analysing case study results can bring some light to the clouded interpretation of barriers and bridges in the SPPI and thus support collaboration between scientists and practitioners in the field of disaster mitigation.

Given the availability of time and the limited workforce for this study, the effort focused on identifying potential linkages between specific vulnerability and assessment determinants, and failure and success in bridging the SPPI. The vulnerability framework developed by Clark University and the Stockholm Environment Institute (Turner et al, 2003) serves as a basis for the determinants of vulnerability. In determining factors of knowledge transfer and influence, the study builds on research undertaken by the Science, Environment and Development Group at Harvard University's Center for International Development (Mitchell et al, 2006; see also www.ksg.harvard.edu/sed). Of particular interest for this study is Harvard University's finding that saliency, credibility and legitimacy are critical attributes of an assessment around which audiences make judgements, and which determine whether they will change their thoughts, decisions and behaviour.

Having outlined the background and objectives, the next section gives a brief rationale concerning the data collection process. Then the main characteristics and qualities of each case study analysed are described. Results of the questionnaire survey follow. By cross-correlating different case study characteristics, patterns in vulnerability assessments that explain successes and failures in knowledge transfer are discussed. Based on the case study publications and the empirical findings of the questionnaire survey, we then examine barriers and bridges in the disaster-related SPPI. The final section summarizes the findings and concludes with suggestions that may be usefully considered by

Figure 3.2 Elements of a conceptual framework for analysing the influence of vulnerability assessments

Source: chapter author

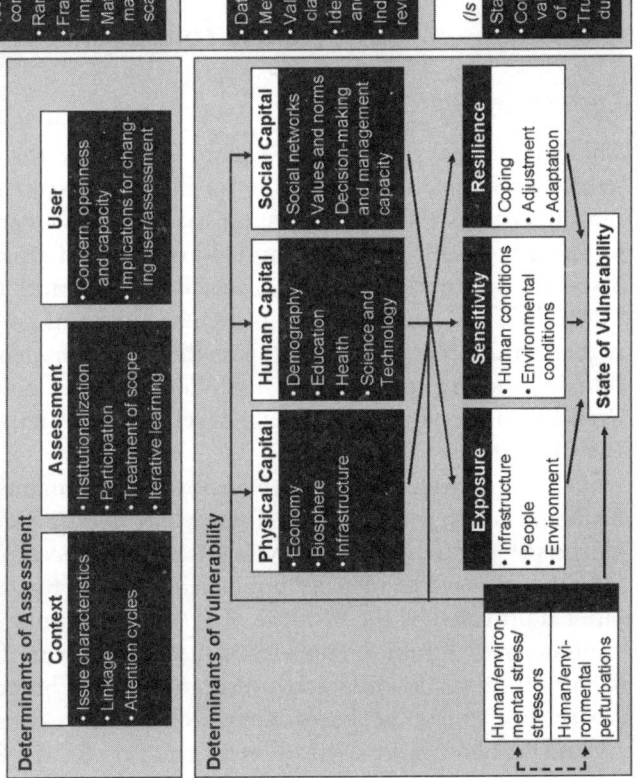

those concerned with improving linkages between research-based knowledge and action to achieve – or, at least, move towards – the reduction of natural disaster impacts.

Note that this study should be considered as preliminary and exploratory, certainly not as exhaustive and definitive. On the basis of a small sample, the investigation attempts to grasp how scientific assessments are carried out and to uncover what limits the co-production of knowledge. As a result of time constraints, both the literature review and questionnaire sample were restricted. A more comprehensive study would be needed to analyse and determine the full range and quality of linkages between specific vulnerability and assessment determinants, and failure and success in bridging the SPPI.

Data collection

Case study search

The overall goal of the study was to investigate the interface between the science developed in vulnerability/resilience assessments and the actions put into practice by those people in positions to implement such research-based knowledge in mitigating natural disasters. Consequently, the selection of assessments to study was made on the basis of breadth rather than depth so as not to bias the study from a particular subset of assessments, whether by region, hazard type or publication method. More in-depth analyses of vulnerability studies are provided by the Vulnerability and Resilience in Practice (VARIP) partnership team, which developed and analysed the case studies in Parts III and IV of this volume.

Most research results in the knowledge domains of vulnerability and resilience are published in journals. The *Hollis Catalog* of the Harvard University Libraries, *Google Scholar* and the *Social Science Citation Index* (SSCI) were used to identify published material for this meta-analysis. The latter is provided by the Institute of Scientific Information; the *Hollis Catalog* contains over 9 million records for more than 15 million publications in all kind of formats. Identification of literature was based on a keyword-based search across these databases. Keywords that were combined with vulnerability and resilience were used to retrieve papers for the knowledge domains of natural hazards, technological hazards, assessment, evaluation, coping, response, case study and disaster. Seminal papers referred to frequently by key scholars who publish on vulnerability and resilience were included to complement the keyword-based search.

The acquired dataset had a number of potential shortcomings that introduced bias into particular streams of research. Most significantly, the non-English literature was largely excluded due to the database and keywords used for the literature search. Since the database covered mostly journal papers, particular book chapters and reports might have been missed as they are not included in the SSCI. Moreover, the concepts of vulnerability and resilience have developed over time, and have been used in various ways. As a result, relevant documents are sometimes published in a 'non-scientific' form and/or medium or for internal use only. Such documents were not retrieved, nor were ones that did not contain or use the keywords. Additionally, interesting and valuable studies were excluded because their assessment was carried out in locations where identification and contact of potential knowledge users was considered difficult (e.g. extreme rural areas and regions with restricted information available on the internet). The limitations induced by these factors were partly balanced by including certain documents in place of others to broaden the set of case studies.

A second issue was time constraint. The time available to conduct the investigation limited the sample size of case studies and the number of knowledge users included in the questionnaire survey. Although realizing that a more comprehensive sample would have improved the quality and informational value of the study, the author views the dataset as usable to provide some empirical evidence on the characteristics of scientific assessments and their influence on decision-makers, thereby adding value to the VARIP core project.

Case study selection

To provide appropriate scope to this research exploration, a number of limiting decisions were made. Each document that was retrieved by a keyword-based search was checked manually for its suitability for the study. Criteria for the initial selection were publication date, hazard type and assessment mode and scale. The decision to include or exclude a paper was based on the information provided in the title, abstract and date of publication. A first selection was made by the date when the document was published. Assuming that newer publications contain more valuable information for knowledge users, priority was given to documents over the last eight years. This period allowed the analysis of whether there are differences in influence between recently published papers and documents already published some years ago. A second criterion was that the paper should refer to natural hazards. Studies that focused exclusively on ecological vulnerability or resilience (e.g. animals and plants) or on infrastructure (e.g. buildings and facilities) were excluded. For the remaining publications in the area of natural hazards, a third criterion was that a concrete assessment be carried out. This study focuses on the influence of research-based knowledge on practice, and this necessitated that the case studies examined had some specific or practical application. Thus, publications that were purely analytical or theoretical were not included. Fourth, it was expected that assessments on local and regional levels would provide more valuable information concerning the identification of successes and failures in the SPPI than national or global assessments. With the exception of Dennis Mileti's *Disasters by Design* (1999), national and global assessments were excluded.

The final selection was designed to obtain a set of case studies with a broad range across geographic location, hazard type and recognition of the publication. Recognition was measured by means of the SSCI, recording the number of times that a given report was cited; wider public impact was evaluated by the number of *Google* hits each publication title earned. Using *Google* English, measurements took place between 19 and 21 May 2006. Taking into account the time available for conducting the study, 20 case studies – 17 journal articles, one report, one book chapter and one book – were chosen for meeting the criteria of breadth, time and source of publication and the potential for relevance to practitioners (see Appendix 1). Case studies were located in Africa, North America, South America, Asia, Europe and Australia. For each publication, an analysis sheet was prepared in which information was saved on the author(s), address, title, study location, hazard type, assessment scale, data used, actors involved, research funds, publisher information, SSCI and *Google* hits, main findings, suggestions given, causes of vulnerability/resilience identified, and – if mentioned – barriers and bridges in the SPPI.

To evaluate the influence of the case studies, two questionnaires were designed: one for case study producers and one for potential users of the case study (see Appendices 3 and 4). Both questionnaires start with a question regarding the use of information sources and are followed by more specific questions concerning the assessment. The questionnaire for the case study authors was designed to ask questions regarding the research process, the intended audience, dissemination of findings and the science/practice interface. The questions were followed by multiple choice answers in order to quantify better the responses; but space was left for elaboration and comments, and several questions were open ended. The questionnaire for potential users included questions about the usefulness, scope and real or future impact of the case study publication. As with the producer questionnaire, a combination of multiple choice and open-ended questions was used, with an emphasis on probing into the underlying bases of response.

The authors of each case study were contacted and invited to participate in the study. After establishing contact, the questionnaires were emailed and the authors returned the questionnaires via email as well, with follow-up occurring over the telephone or email. Almost universally, the case study producers contacted were eager to participate and readily responded with completed questionnaires. In total, 64 people authored the 20 case studies selected. Forty-seven authors were contacted and 40 of them participated in the study (85.1 per cent response rate). The rest could not be contacted because authors had died or contact details could not be identified. One author was excluded due to his participation in the VARIP project.

Identifying and contacting potential users of each assessment were more difficult. A case study user was defined as a person in a position to put the ideas and knowledge in each report into practice, whether in the context of a government, an NGO or a private company. It was not necessary that a potential user had heard of the report, as this would have severely limited the pool of practitioners. In fact, it was a central question of the study to see how wide the impact of a given assessment was, and to determine whether there might be a problem of propagation of knowledge from one sphere to another. A few potential users were identified through the assessments themselves. For example, if a government institution was involved in funding or formulating a report, that institution could likely be considered a potential user. In most cases, however, users were identified by researching the region for which the study was performed and identifying key policy-makers in areas of land management, natural hazard planning, emergency planning and response, and other areas specific to each assessment. For each assessment, a strong effort was made to identify a variety of potential users from different levels of government, non-governmental institutions and other areas.

Once identified, potential users were contacted first by telephone, if the number was available. They were asked to participate in the study, and if they agreed, an email was sent with a cover letter containing details of the study, together with the relevant scientific publication and the questionnaire developed for potential users. When reached by telephone, most potential users were interested in the study and willing to participate. However, it was difficult to contact a significant number of potential users by telephone because their number could not be found, or because secretaries and answering machines put up an impassable blockade. In some cases, people did not receive the documents because of spam filters. Additionally, as users were located all around the globe, the time change was occasionally a (surmountable) obstacle. In one case – the case study in the Philippines – even with the support of the author, no users replied. Thus, that case was dismissed in the cross-correlation calculations.

After emailing the questionnaires, the response rate from potential users was disappointingly but not unexpectedly low (42.3 per cent). Reminder emails and follow-up telephone calls were sometimes performed up to ten times for a single potential user. In some cases, understandable delays occurred as potential users were caught up in dealing with current natural disasters in their part of the world. After much follow-up, 52 completed questionnaires were received. However, although expected, it is worth noting that those potential users who agreed to participate and followed through with reading the relevant assessment and filling out the questionnaire were those, for the most part, who were already the most interested in the intersection of science and practice, and usually were those for whom the identified assessment was the most relevant. In general, the more often potential practitioners were accustomed to dealing with the scientific world, the more eager they were to participate in the study. In total, 40 case study producers and 52 potential users participated in the survey (see Appendix 2).

Case study analysis

Characteristics of publications

In total, 20 case studies were analysed, which were authored by 64 people, 42 of whom were male and 22 female (see Table 3.1). The number of authors of a publication ranges from 1 to 11 authors. Seven publications were single authored and 13 were multi-authored. Among multi-authored publications, four had two authors, three had three authors, and the other six had more than four authors. Publication dates of the case studies range from 1998 to 2005, with one unpublished book chapter that was later published in 2007. This publication was excluded in the measurement of recognition. Regarding wider public impact, as measured by the number of *Google* hits that each publication earned, results ranged from no hits to 10,800 for Mileti's book. Concerning scientific recognition, SSCI varies from 0 to 64. On the basis of these two factors, publications were grouped into three recognition classes, with four assessments in the highly recognized group, six in the medium and nine in the low-recognition class.

Case study assessments were carried out for locations in the following countries: Australia, Canada, East Caribbean Islands (including Antigua and Barbuda, Bahamas, Barbados, Dominica, Grenada, Jamaica, St Kitts and Nevis, St Lucia, St Vincent and the Grenadines, and Trinidad and Tobago), Egypt, Germany, India, Mexico, Pakistan, Peru, the Philippines, the US and Vietnam. With 12 case studies, North and South America are represented above the average. Four assessments were made in Asia, followed by Oceania with two, and Africa and Europe with one each. Consequently, case studies in developed countries had a light coverage. Using the World Bank's country classification, which is based on gross national income per capita in 2005, 12 case studies are in high-income countries, five in middle-income countries (both lower and upper middle), and three in low-income countries.

Regarding hazard type, 12 case studies refer to a single hazard and 8 to multiple hazards. The following hazard areas are assessed: climate change (impacts include changes in varying rainfall regimes and soil moisture budgets, potential increases in the frequency and intensity of hurricanes, and changes in regional and local sea levels and patterns of wave action), earthquake, flash flood, flood, globalization (i.e. liberalization of agricultural trade and supporting policies), market fluctuation, sea-level rise (impacts include inundation, erosion, saltwater intrusion, increased soil salinity, changes of coastal ecosystem and loss of productivity), storm, storm surge, tsunami and wildfire. Assessments with composite indices include the following additional hazard sources: chemical release, drought, hail, hurricane, snowstorms, temperature extremes and tornado. Hence, landslide and volcano hazards are missing, resulting in a bias towards atmospheric hazards. Altogether, the case study sample includes 14 local assessments (city, community, district level), five regional (state, province, county level) and one national assessment (US).

Table 3.1 Case study characteristics

No.	Date	Journal/Publisher	No. of Authors	Hazard Type	Location of Assessment	Country's Income Status	Scale of Assessment	Recognition of Assessment	Integration of Assessment
1	1999	World Development	1	Climate change, storm	Vietnam	Low	Local	High	Medium
2	2005	Mitigation and Adaptation Strategies for Global Change	2	Tropical cyclones, storm surges	Australia	High	Local	Medium	Low
3	1999	Disasters	2	Flood	Canada	High	Local	Medium	Low
4	2002	Environmental Hazards	2	Earthquake	US	High	Local	Low	Low
5	1998	Mitigation and Adaptation Strategies for Global Change	10	Storm, flood	US	High	Local	Medium	Medium
6	2005	Environmental Hazards	1	Fire	US	High	Local	Low	Medium
7	2000	Annals of the Association of American Geographers	3	Multiple	US	High	Regional	Medium	Medium
8	1999	Environmental Monitoring and Assessment	4	Sea-level rise	Egypt	Medium	Local	Medium	Low
9	2004	Cities	1	Fire	Australia	High	Local	Low	Medium
10	2002	Natural Hazards Review	1	Earthquake	Peru	Medium	Regional	Low	Low
11	2004	Marine Policy	3	Sea-level rise	East Caribbean Islands	Medium	Regional	Low	Medium
12	2003	Global Environmental Change	5	Climate change, market fluctuations	Mexico	Medium	Local	High	Medium
13	2003	Environmental Hazards	1	Flood	Pakistan	Low	Local	Low	Medium
14	2004	Global Environmental Change	11	Climate change, globalization	India	Low	Local	High	High
15	2002	Natural Hazards Review	1	Multiple	US	High	Regional	Low	Medium
16	2004	Coastal Management	2	Earthquake, tsunami	US	High	Local	Low	High
17	2002	Mitigation and Adaptation Strategies for Global Change	1	Flood	Philippines	Medium	Local	Low	Low
18	2003	German Committee for Disaster Reduction	9	Flood	Germany	High	Regional	Medium	High
19	1999	Joseph Henry Press	1	Multiple	US	High	National	High	High
20	2007	Elsevier	3	Sea-level rise, flood, surges	US	High	Regional	N/A	Medium

Another characteristic analysed is the assessment's capability to integrate multiple-scale, science/practice and social/physical aspects. When grouped according to the degree of integration, four case studies are considered to highly integrate these factors (at least two out of the three factors have been fully integrated), ten assessments did it partly (at least one out of the three factors have been fully integrated and the others partly), and six case studies showed little integration (factors have been only partly integrated or not at all). Sixteen out of the 20 publications acknowledge funding sources.

Qualities of vulnerability and resilience

The knowledge domains of vulnerability and resilience, by virtue of their character which brings together the physical environment, society and the manmade infrastructure, inevitably pose difficulties for discipline-based research, in general, and single-authored knowledge production, in particular. The case study review regarding causes, drivers, characteristics and consequences of vulnerability documents that vulnerability is:

- multidimensional (economic, social, political, institutional, cultural, spatial, temporal and historical, among others);
- socially divergent (individually, between and among social groups);
- dynamic (causes and conditions vary in time);
- interactive (causes and parameters modify each other); and
- scale dependent.

In contrast, much of what is important about the character of vulnerability and resilience tends to be lost by the generalizing, de-contextualizing and reductionist tendencies of single-discipline research, single-hazard approaches, single-scale focus, and concentration on either social or physical aspects.

Single-discipline and single-author knowledge are, by necessity, reductionist in nature and capture only some of the causes, conditions and impacts of vulnerability and resilience. Particularly in the risk-related domain, the epistemological divide between engineering and the natural sciences, on one side, and social sciences, on the other, obstructs a more comprehensive picture (Jasanoff, 1993). Likewise, the interdisciplinary knowledge that exists is often generalized rather than specific, and de-contextualized rather than locally embedded. Although often claiming to have universal applicability, single-discipline and single-author assessments still require exterior inputs from other disciplines in order to properly address the dynamic and multilayered nature of vulnerability and resilience. Additional input is needed from non-research-based knowledge in order to tackle practical issues of the suitability and feasibility of concrete vulnerability reduction and disaster mitigation measures.

Today, there is broad agreement that more integrative assessments are needed. However, less consensus exists on *what* needs to be integrated and *how* that integration should be accomplished. Suggestions range from the integration of scope, research methods and scale to disciplines and stakeholder involvement. During recent years, two interrelated trends of integration have received particular attention: first, an increasing interest in structuring assessments to better integrate science and practice; and, second, an increasing effort to identify, assess and integrate social and geophysical linkages across multiple scales. As a result of these characteristics, an adequate assessment requires integration across all these dimensions of vulnerability and resilience. Although convincing arguments for the importance of scale in global change research exist (e.g. Wilbanks and Kates, 1999), only one fifth of the case studies analysed attempted to integrate practical elements and to consider socio-economic and geophysical aspects across spatial scales.

Existing or potentially vulnerable populations are often institutionally and economically invisible; but their participation in vulnerability assessments is crucial if these assessments are to be useful for decision-makers. Although an increasing number of vulnerability assessments aim to be explicitly stakeholder driven, the subject of social vulnerability (i.e. vulnerable populations) is rather a study object that is assessed than an equal stakeholder to be integrated within the knowledge production process. Interestingly, for the case study that thoroughly integrated the people into the assessment (interviews, participant observation and group discussions to assess the differential impacts of floods among the street children, the urban poor and residents of wealthy neighbourhoods in Metro Manila), none of the contacted users returned a filled-in questionnaire.

Another feature of vulnerability and resilience is the lack of a universal definition: 'Since different intellectual traditions use the terms in different, sometimes incompatible, ways, they emerge as strongly related but unclear in the precise nature of their relationships' (Gallopín, 2006, p293). If care is not used, the author concludes, the field of human dimensions research can become epistemologically very messy. The case study analysis supports this concern. All assessments operate with their own characterization of key terms. While individual use of the vulnerability and resilience concepts is scientifically problematic, it is not clear whether or not the absence of common definitions poses an obstacle for decision-makers to apply the concepts. On the contrary, a diversity of definitions and approaches might even be necessary in order to address the full complexity of the vulnerability and resilience concepts. However, universal standards would certainly facilitate and support the practical application.

Questionnaire survey

Knowledge producers

Almost half of the authors (47.5 per cent) formulated their research questions for the assessment without input from internal or external colleagues. Out of those researchers who included other people in the design of the research, more than three-quarters (76.2 per cent) consulted internal and external colleagues. The remaining 23.8 per cent integrated only internal ones. Regarding frequently used sources for obtaining information, almost all assessment producers regularly use scientific sources (92.5 per cent), followed by governmental (57.5 per cent) and non-governmental sources (42.5 per cent). About one third of the producers frequently drew on internal information sources, personal communication and the media. With regard to rarely or never used information sources, the media ranked top (35 per cent), followed by communication (27.5 per cent) and internal sources (25 per cent). When asked about the most influential and critical information sources for the assessment, half of the producers referred to surveys, fieldwork and communication, 47.5 per cent to statistics, census and maps and 45 per cent to scientific literature. Only about one tenth of the producers considered modelling and non-scientific resources as relevant.

Two-thirds of the producers cited policy-makers and practitioners as the intended audience for the assessment. More than half (57.5 per cent) considered science as the primary audience and only one quarter regarded non-governmental, public and private organizations as potential users. If the purpose of the assessment would have been an action plan for practitioner use, 35 per cent of the producers would have changed their assessment to include the application of theory and clear recommendations. One quarter would have made changes regarding data and method, and 22.5 per cent would have changed the focus, audience and/or the medium. Only 17.5 per cent considered the involvement of stakeholders and 10 per cent the alteration of style as a way of increasing the practical use of the assessment. Another 12.5 per cent of the assessment producers would not have made any changes.

Almost two-thirds (62.5 per cent) of the producers said that the research was disseminated in the form of a report, 50 per cent mentioned scientific papers and 40 per cent circulated it through symposia and meetings. Only 20 per cent used the media, internet, public material or lectures to disseminate their study. When asked for their opinion on whether the assessment addressed the needs of users, 37.5 per cent of the producers answered 'no' and 62.5 per cent 'yes'. Only 17.5 and 12.5 per cent of the producers, respectively, believed that their assessment was relevant for science and the general public, private sector, media or NGOs. Altogether, half of the responders thought that the assessment was relevant for decision-makers in local governments. Almost the same amount (47.5 per cent) considered the assessment as relevant for agencies and practitioners.

Regarding the influence of the study, 45 per cent of the producers were not aware of any impact, whereas 55 per cent believed that it had caused changes. Out of those who were aware of impacts, more than two-thirds (77.3 per cent) affirmed that it caused behavioural changes (i.e. concrete actions were taken as a result of the study, such as new laws), and 36.4 per cent judged that it led to changes in attitude (i.e. awareness and belief). Almost one third (31.8 per cent) were aware of impacts upon science and 22.7 per cent thought that their study caused procedural changes (i.e. methodology and processes were altered). Almost half of the producers (47.5 per cent) believed that their study had an impact upon science but not upon practice. More than one third (35 per cent) considered their study to have influenced the reader's belief and thus would have an impact in the future. Finally, 27.5 per cent believed that it had only limited impact upon both science and practice, and 7.5 per cent were not able to evaluate the influence.

With regards to conflicts among actors and institutions in the SPPI, almost half (47.5 per cent) of assessment producers believed that differences in objectives, needs, scope and priorities were the main sources of conflict. For 32.5 per cent of the producers, divergent institutional settings and standards were the causes for inadequate interactions; for 27.5 per cent, differences in language, a lack of understanding and mistrust were responsible. One fifth saw no conflicts or conflicts that could not be classified into a specific group. When asked how knowledge transfer at the SPPI could be improved, more than half (57.5 per cent) responded that the best way would be through more collaboration, trust and/or outreach. Fifteen per cent of the producers noted that efforts in training, education and/or language could enhance knowledge transfer; 12.5 per cent pointed out that some kind of 'brokers' or intermediary bodies could lead to improvements. Only 7.5 per cent proposed reward application as a possible means to facilitate knowledge transfer between science and practice.

Knowledge users

Governmental (71.2 per cent) and internal (65.4 per cent) sources were the most frequently used information sources for the knowledge users interviewed. While almost all assessment producers regularly used scientific sources (92.5 per cent), only half of the decision-makers in policy and practice did so. With regards to rarely or never used information sources, the media ranked at the top (32.7 per cent), followed by non-governmental (19.2 per cent) and scientific sources (17.3 per cent). When asked about the most influential and critical information sources for their work, half of the potential users responded that media and governmental sources were most important, followed by non-scientific literature (42.3 per cent) and scientific sources (38.5 per cent). Unsurprisingly, the internet (7.8 per cent) and fieldwork and surveys (19.2 per cent) were the least influential information sources for decision-makers.

Regarding awareness, over half (51.9 per cent) of the potential users did not know anything about the specific assessment on which we queried them. A few had heard that something was going on (9.5 per cent); 13.5 per cent knew about the assessment but did not know the assessment report; and 5.8 per cent knew the report but hadn't read it. Only 13.5 per cent were aware of both the assessment and the report and even less (5.8 per cent) were involved in the assessment. When asked about the influence of the assessment report, many users replied that they did or would talk about it within their organization

(69.2 per cent); that they did or would consider the findings in the future (48.1 per cent); and that they did or would use the report to convince other people (42.3 per cent). About one quarter answered that the assessment would influence their beliefs (26.9 per cent). Almost the same number of users replied that the report's findings did not or would not influence their actions (25 per cent).

With regards to the relevance of the assessment, almost half of the users (46.2 per cent) stated that the report addressed some of their needs. For one tenth, the report was not relevant at all; nearly one quarter (23.1 per cent) mentioned that it didn't address their needs, but they saw its potential strength. Almost one third (32.7 per cent) considered the assessment's findings as relevant, compared to only 1.9 per cent who did not. Forty-four per cent of the users who answered the question 'What should the assessment have addressed to be more relevant and useful?' stated that changes in scope (different, broader, and/or more specific) would have improved the saliency of the assessment, followed by 32 per cent who suggested clearer and/or better recommendations.

In contrast, both the perceived credibility and legitimacy of the assessments are very high. The potential users who considered the assessments as accurate and technically sound were 78 per cent (compared to 21.2 per cent who did not), and 82.7 per cent believed that they were respectful of stakeholders and unbiased (compared to 17.3 per cent who did not). The main reasons given for considering an assessment as not credible were missing or ignoring certain aspects of the issue (36.4 per cent), a somehow too-qualitative and/or theoretical approach, and a weak method and/or insufficient data (both 27.3 per cent). According to the potential users interviewed, the credibility of those assessments could have been improved by in-area research and/or applying them to a different location (35 per cent); improving data, information sources and methodology (25 per cent); and involving stakeholders and having a better linkage to actions (10 per cent). Reasons for a perceived lack of legitimacy in the assessment were excluded stakeholders and a biased view that ignored certain aspects (both 44 per cent). Consequently, 57.1 per cent of the users recommended improving data and/or information sources, followed by integrating stakeholders and local knowledge (28.6 per cent). For some users an integration of governmental policy and a better understanding of the 'real world' would improve the legitimacy of assessments (14.3 per cent).

Characteristics and correlations

This section points to some observations that come into view when grouping the returned questionnaires into specific classes. In particular, cross-covariance between factors illuminates some interesting patterns in the SPPI of vulnerability and resilience. For instance, by grouping the answers from assessment producers by gender and number of authors, some characteristics attract attention. Notably, single-authored publications mainly drew on the scientific literature (71.4 compared to 39.4 per cent of multi-authored publications) and used significantly fewer data (14.3 compared to 54.5 per cent of multi-authored publications) and models as information sources. Moreover, the regular use of non-governmental information sources was three times higher for authors of multi-authored assessments than for single-authored ones. The latter also disseminated their findings to a higher degree only through the publication analysed (28.6 compared to 3 per cent of multi-authored publications).

When asked if they were aware of any impacts of the assessments, 71.4 per cent of the authors of single-authored papers said that they were not aware of impacts, compared to 39.4 per cent of authors of multi-authored publications. Nevertheless, 100 per cent of those who were aware of impacts believed that the assessment caused behavioural changes. In contrast, 60.6 per cent of the authors of multi-authored publications were aware of impacts and only one quarter of them believed that they had caused behavioural changes. Of the authors of individual papers, 42.9 per cent evaluated the influence of their assessment as limited both in science and practice, and only 14.3 per cent of them believed that it had had an impact on science, but not on practice. Only one quarter of the authors of jointly written publications evaluated the influence of their assessment as limited, but more than half (54.5 per cent) believed that it had influenced science and not practice.

No significant differences existed between male and female authors. Divergences are apparent only when regarding the conflicts between science and practice, and the suggestions to improve them. Only 36 per cent of the male authors but 66.7 per cent of the females saw existing conflicts in divergent objectives, needs, scope and priorities between scientists and practitioners. When asked how the knowledge transfer between the two arenas could be improved, both male and female authors suggested mostly through collaboration, trust and outreach. However, male authors mentioned a variety of improvement measures, ranging from more application of science, better training and education to reward the application, and use of intermediary bodies. Interestingly, female authors did not consider the application of science and rewards as possible options to improve the SPPI.

Considerable differences were also apparent when grouping the case studies according to their recognition. In particular, the number of people involved in formulating the research questions seems to have various correlations. The number of cases in which only the author(s) prepared the research was significantly higher for low-recognized publications (77.8 compared to 54.5 per cent for highly recognized publications), for assessments with little integration (66.7 compared to 46.7 per cent for highly integrated assessments), and for countries with a low income status (77.8 compared to 33.3 per cent for developed countries). It is assumed that the involvement of external people leads to more integrated assessments, resulting in higher recognition. Authors in developing countries seem to have more difficulties in incorporating other scientists and stakeholders within their research.

Less surprising is the finding that authors of highly recognized (72.2 per cent) and single-authored (85.7 per cent) publications largely regarded scientists as their main audience. Great divergence, however, exists regarding the question of whether or not the assessment addressed the need of users. While 45.5 per cent of the authors of highly recognized documents believed that their assessment was relevant for science, none of the authors of poorly recognized publications believed so. On the other hand, 77.8 per cent of the latter considered their assessment as relevant for agencies and practitioners, compared with only 9.1 per cent of the authors of well-recognized publications.

These findings lead to the conclusion that the producers of highly recognized case studies estimate the assessment's scientific quality and relevance as appropriate, whereas single authors tend to overestimate the relevance. Only 14.3 per cent of the latter believed that the assessment was not relevant (as compared to 42.4 per cent of multi-authored assessments), and 85.7 per cent of those who believed that it was relevant considered agencies and practitioners as the most important users (compared to 39.4 per cent of multi-authored assessments). This is supported by the fact that authors of highly integrated assessments considered different audiences as potential users for their findings, whereas most of the authors of poorly integrated studies believed that their assessment addressed the need of agencies and practitioners (77.8 per cent) and local governments (88.9 per cent). None thought that it addressed the need of science – neither the public nor the private sector.

Authors of publications that were highly recognized used statistical, graphical and census data to a higher degree (63.6 per cent) than the average (47.5 per cent). No huge differences exist regarding the use of information sources, although authors of well-known papers employed media sources less often than their colleagues of less recognized case studies. Divergent opinions are visible when asked for changes to improve the practical use of the publication. Almost half of the authors of highly recognized studies would have changed the method and/or data basis, but only one tenth of the authors of lesser-known publications would have made such changes. In contrast, the latter would have improved their case studies mostly through the application of theory and clear recommendations (44.4 compared to 27.3 per cent). Moreover, these authors disseminated their findings significantly less often by means of reports than authors of well-known case studies (44.4 compared to 81.8 per cent).

Another difference exists when asked how the authors evaluated the influence of their publication. Most of the authors of well-recognized case studies (81.8 per cent) answered that it had an impact upon science, but not upon practice; only 9.1 per cent believed that it had only limited impact upon both science and practice. Almost half of the authors of less-recognized publications, by contrast, claimed influence on both science and practice. Furthermore, only 22.2 per cent of them had the opinion that their assessment had scientific impact, but no impact upon practice. Between both groups no significant differences exist regarding conflicts and improvements of knowledge transfer at the SPPI.

By asking the question 'Is the assessment structured in a way to integrate science and practice, social and geophysical factors, and multiple scales?', case studies were grouped into three classes: high, medium and low. Concerning this characteristic, as already mentioned, authors of less integrated assessments mostly formulated their research questions without input from internal or external sources. When asked about the intended audience for the assessment, a surprisingly high number of authors of both highly and poorly integrated assessments named policy-makers and practitioners as their primary target group (80 and 78.8 per cent, respectively). While highly integrated assessments are doubtlessly of higher relevance for policy and practice, one possible explanation for the high response of poorly integrated case studies might be the fact that such assessments are very place and case specific and, thus, local decision-makers are considered to be the primary users.

Furthermore, authors of assessments with a high degree of integration used a variety of information sources, whereas less integrated assessments used primarily one specific source. The highest discrepancy exists regarding the regular use of governmental information sources: 60 per cent of the former authors but only 22.2 per cent of the latter used them regularly. For more than half of the authors of highly integrated assessments, conflicts at the SPPI are due to different objectives, needs, scope and priorities, as well as institutional settings and standards. By contrast, authors of less integrated assessments hardly cite these factors, but rather point mainly to differences in understanding, language and mistrust. Another apparent divergence was already mentioned above. Concerning the assessment's utility for users, almost all authors of poorly integrated assessments believed that their study addressed the need of agencies, practitioners and local governments. No single author considered it as practical for science and the public and private sector.

With regards to single- and multi-hazard assessments, no surprising differences are apparent between the two groups. Authors of multiple-hazard assessments used governmental information sources more regularly and media sources more rarely than authors of single-hazard assessments. Not astonishing, also, is the fact that the latter would make fewer changes in data and methods employed in order to make their study more usable for practitioners than their colleagues of multi-hazard assessments. Evidently, data are more accessible and well-established methods are more frequently in use for single-hazard assessments. However, more authors of multi-hazard assessments disseminated their findings through reports, media and the internet. Moreover, they considered agencies and practitioners as primary users for their study to a lesser degree than did authors of single-hazard assessments. Only 16.7 per cent of the latter, on the other hand, are aware of rational or attitudinal changes that their assessment caused, compared with 60 per cent of the authors of multi-hazard assessments. Considering the fact that single-hazard assessments are quite common today, this finding is also not surprising.

When comparing assessments in developing countries with those undertaken in developed countries, the number of cases in which only the author(s) prepared the research is significantly higher for assessments in countries with a low income status (77.8 per cent compared to 33.3 per cent). Authors of assessments undertaken in developing countries more often used internal information sources (66.7 per cent compared with 42.9 and 25 per cent of countries with middle- and high-income status) as well as surveys and fieldwork as suppliers of information (77.8 per cent compared to 28.6 and 45.8 per cent of countries with middle- and high-income status). Moreover, they didn't disseminate their findings in the form of public material, offprints and lectures (0 per cent compared to 28.6 and 25 per cent of countries with middle- and high-income status). One explanation could be the limited availability of or access to [public] data and fewer financial resources to produce public and offprint material. The fact that these authors would primarily make changes in data collection and methods in order to improve the practical use of their study supports this assumption. Given that some of the authors of an assessment undertaken in a developing country are from developed countries, the differences mentioned might be even more significant when considering only authors from developing countries.

Also interesting is the finding that one third of the authors of case studies in low-income countries who believe that their assessment addresses the need of users think that it addresses the need of science, but to a lesser degree the need of agencies, practitioners and local governments. By contrast, only 12.5 per cent of the authors of publications with assessments in developed countries considered that their study addressed the need of science. The majority holds the view that it is practical for local governments (58.3 per cent) and agencies and practitioners (66.7 per cent). When looking at the impacts of the assessment, no author of case studies in developing countries was aware of procedural changes as a consequence of the assessment (compared to 35.7 per cent of authors of case studies in developed countries), but 60 per cent were aware of changes in awareness (compared to 21.4 per cent). Moreover, a vast majority of authors believed that the assessments influenced science but not practice (77.8 compared to 37.5 per cent). Also interesting is the fact that one third suggested intermediary bodies and brokers to improve knowledge transfer at the SPPI, whereas only 4.2 per cent of the authors of case studies in developed countries did so.

When grouping the answers of case study producers according to the date of the publication, only a few differences are apparent. While the authors' ranking of regularly used information sources is the same, the use of non-scientific information sources generally increased. Authors of 'newer' publications frequently used more internal, governmental and especially non-governmental sources for their work. They also considered to a higher degree scientific literature and data as most influential and critical for scientific assessments and disseminated their findings more often through reports as compared with authors of 'older' publications. Not surprisingly, the latter were more aware of scientific impacts that their assessment had caused, whereas authors of recently published case studies were more aware of changes in perception due to their research.

Interesting observations also come into view when grouping the questionnaires from knowledge users into specific classes. For instance, there were differences in the use of information sources with respect to the gender of the potential user. Male decision-makers to a higher degree frequently used internal sources (70.7 compared with 45.5 per cent of female users) and scientific sources (53.7 compared with 36.4 per cent). By contrast, female users in policy and practice frequently used governmental information sources (81.8 compared with 68.3 per cent of male users), personal communication (81.8 compared with 39 per cent), media sources (72.6 compared with 36.6 per cent) and non-governmental sources (54.5 compared with 29.3 per cent). In general, some male users did not employ specific information sources at all, whereas female users mostly used a variety of sources. Consequently, slight differences exist regarding the information considered to be most important. While 41.5 per cent of the male decision-makers affirmed that scientific sources were most critical to their work, only 27.3 per cent of the female users believed likewise. Moreover, 73.2 per cent of the men stated that they did or would talk about the assessment within their organization (compared with 54.5 per cent of the women), but only 36.6 per cent said that they did or would use the report for their work or to refer to it to convince other people (compared with 63.6 per cent of female users). A slightly higher number of women considered the findings of the assessments as relevant (45.5 compared with 29.3 per cent of male users). More interesting is the fact that significant gender differences existed concerning the credibility of assessments. Among female users, 66.7 per cent did not believe in assessment findings because of weak methodology and/or insufficient data (compared with 0 per cent of the men), whereas for 42.9 per cent of the male users, the use of approaches that were too qualitative and/or theoretical were the main reason for a lack of trust in the assessments (compared with only 14.3 per cent of the women who listed the same reason).

Likewise, divergent views existed between decision-makers working in the domain of disaster management (e.g. emergency planning, fire department) and those whose work was not primarily disaster related (e.g. development planning, health, city council). For example, the former used to a lesser degree internal information sources (54.8 compared with 81 per cent) and considered scientific information as more critical to their work (45.2 compared with 28.6 per cent) than decision-makers in the non-disaster domain. It also seems that the assessments were more influential to users in the disaster arena. When asked about the influence the assessment did or would have, decision-makers working in disaster-related policy and practice affirmed to a higher degree each level of influence, ranging from 'talking about' to 'influencing my beliefs' and 'considering the finding in the future'. The assumption is underlined by the fact that almost half (48.4 per cent) of the users whose work is related to disaster management issues considered the assessments' findings as relevant, as compared with only 9.5 per cent of decision-makers working in the non-disaster arena. On the contrary, the assessments addressed to a larger extent the needs of the users not working on disaster management (66.7 compared with 32.3 per cent).

Also interesting were the different views regarding credibility and legitimacy. Although both user groups considered the assessments generally as true and legitimate, divergent opinions existed with regard to the reasons for not considering an assessment as true and legitimate and the recommendations to increase both factors. Decision-makers in disaster policy considered an assessment as having low credibility if certain aspects were ignored or missed (60 compared with 20 per cent of decision-makers in the non-disaster field), and suggested to a higher degree that credibility be improved by providing clear recommendations (18.2 compared with 0 per cent). On the other hand, users in the non-disaster arena stated more often that an assessment's credibility would be higher if the authors had chosen more in-area research and/or a different location. Differences with regard to legitimacy were even higher. While for 80 per cent of the disaster-related decision-makers the exclusion of stakeholders was the reason for considering an assessment as illegitimate (compared with 0 per cent of the users in the non-disaster arena), the use of weak theory, methodology and recommendations was the main cause cited by those decision-makers who were not working directly in disaster management (compared with 0 per cent of the users in the disaster arena).

A little surprising is the finding that in developing countries twice as many decision-makers considered scientific information as most critical to their work and used it slightly more often than the government and the media as information sources than users from developed countries. Moreover, decision-makers were to a lesser degree aware of both the assessment and report: 71.4 per cent were not aware at all (as compared to 51.4 per cent of users in developed countries); none of the users in the developing countries had read the report before nor were involved in the assessment (in developed countries, 16.2 per cent of the users had read the report and 8.1 per cent were involved). More interesting is the fact that almost the same number of users from developed countries stated that the assessment did or would influence their beliefs (29.7 per cent) but would not influence their actions (35.1 per cent); however, none of the users from developing countries concurred. The latter almost exclusively responded that they would talk about the assessment within their organization (71.4 per cent) and would use it for their work and for convincing other people (57.1 per cent). Furthermore, differences existed when an assessment was not considered to be salient. While 42.1 per cent of the users from developed countries suggested clearer recommendations to improve the saliency of the report, none of the users from developing countries did so. By contrast, half of them suggested the use of 'better' methodologies, which none of the users from middle-income countries and only 5.3 per cent from high-income countries recommended. Also, when considering legitimacy, it seems that decision-makers in developed countries were more concerned about stakeholder involvement than their colleagues in developing countries.

When grouping the answers of case study users according to the recognition of the assessment report, it becomes evident that highly recognized papers have not only a higher scientific impact (as measured by the SSCI), but are also more influential. Of the potential users of well-recognized documents, 22.2 per cent were not aware of the assessment and report compared with 60.9 per cent of the users of little-known papers. One third of the former knew about both the assessment and report before they had been contacted, but only 8.7 per cent of the potential users of lowly recognized papers knew about it. Moreover, all of the users of highly recognized case studies affirmed that they would talk about the assessment's findings within their organization, compared with 'only' 60.9 per cent of the decision-makers who evaluated less recognized reports. Likewise, users of highly integrated assessments stated to a higher degree that they would talk about them within their organization. However, these assessments did or would not influence the actions of their organization for almost half of the potential users (42.9 per cent). Only 13 and 26.7 per cent of the users of middle- and lowly integrated assessments testified that they would have no influence on their behaviour. Interestingly, when asked about saliency, 42.9 per cent of the users also confirmed that the findings of the well-integrated assessments are relevant (compared with only one quarter for hardly integrated assessments) and no user stated that the assessment did not address needs (compared with 20 per cent).

Of particular interest is the saliency of an assessment. When asked 'Does the assessment report address your needs?', middle-recognized assessments received the highest amount of 'No, not all' answers, followed by low-integrated and single-hazard assessments. The highest amount of 'Yes, and the findings are relevant' answers given by decision-makers were received by highly recognized and highly integrated papers. More surprising, however, is the fact that the findings of single-authored papers are considered as more relevant than multiple-authored ones. Similarly, the findings of single-hazard assessments are considered as slightly more relevant than the ones of multiple-hazard assessments. A possible explanation is that most of the single-authored papers refer to a single-hazard assessment carried out for a specific local area, resulting in clearer recommendations for the potential decision-makers and, thus, in a higher saliency (taking into account that the Mileti book is edited by a single author, but the assessment involved many authors). Less significant is the publication date, although users were slightly less aware of more recently published assessments (60 per cent were not aware at all) than they were concerning older reports (46.9 per cent). This is another indicator that knowledge is transferred via the pipeline model (for more detail, see Chapters 1 and 4 of this volume).

Interesting findings also appear when we compare the answers of producers and users to the same assessment. For instance, assessments that (according to the answers of potential users given in the questionnaire) did not or would not influence the actions of the decision-makers were in many cases not very well-integrated assessments. Moreover, the authors of such assessments estimated the influence of their report quite properly. Most knowledge producers affirmed that their assessment did not address the need of users and did or would have – at maximum – an impact upon science but not upon practice. On the other hand, authors of well-recognized and highly integrated assessments were correctly pessimistic about the influence of their assessments on practice. Although aware that their report was scientifically highly recognized, most of them estimated properly the limited influence on practical decision-making. This led to the assumption that the case study producers were fairly well acquainted with the needs of decision-makers in policy and practice. Hence, knowledge producers could estimate more or less accurately the influence of their scientific work and were cognizant of the limiting factors that prevented a higher impact upon policy and practice. Moreover, there was a significant relationship between credibility and legitimacy in the user's view. If a potential decision-maker considered an assessment as not credible, in most cases the same user also regarded it as not legitimate. A negative credibility and legitimacy, however, did not necessarily lead to a low saliency and vice versa.

Conclusions

Barriers and bridges

What are the factors that hinder the co-production and transfer of knowledge across the boundaries of science, policy and practice? Generally speaking, many factors that aggravated greater coherence among and between these arenas were attributed to the 'socially absent' character of most natural hazards, the 'structural' character of both academic and public organizations, and the 'public good' character of natural disaster management (see Chapter 1 in this volume; see also Weichselgartner and Obersteiner, 2002). Considered more precisely, a large variety of influential factors exist to limit the co-production of knowledge, comprising such diverse aspects as different needs and objectives, institutional reward systems and incentives for collaboration, language and cultural differences, and divergent standards of credibility and legitimacy. Such barriers contributed to and resulted in failures that typically occur when knowledge is transferred through the traditional pipeline mode in which scientists set the research agenda, do the research and then transfer the results to potential users, assuming that they will diffuse automatically through the practice community (see Figure 3.3).

A first group of factors is functional, including divergent objectives, needs, scope and priorities. Responders of the questionnaire survey pointed out that many practical problems are not relevant for or not known to scientists. As one producer pointed out, researchers 'often work on some sort of obscure, trivial issue that doesn't impact practical decision-making'. In other words, researchers do not necessarily pick research questions that make a difference in the lives of those studied, and then take action to implement those research findings. On the contrary, scientific research grants increasingly encourage

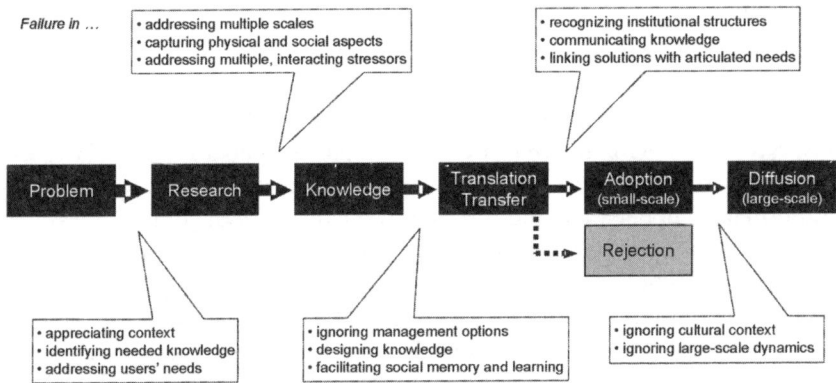

Figure 3.3 *Pipeline knowledge transfer and common failures*

Source: chapter author

huge research teams working on extensive, multifaceted questions that don't translate well to practitioners. Moreover, in this project practical disaster risk assessment was often limited by the available data or the budget for data collection. In contrast, scientific research frequently focuses on methodologies that might not be accurate in the absence of detailed and accurate input data. While science is based on facts, decision-making – especially in traditional risk management – is more cost-benefit oriented, determined by risk-benefit analyses, and often ends up in the domain of finance departments. However, the benefits of mitigation – in the best case the absence of disasters – are hardly measurable in monetary terms. A well-known but still apparent drawback is the fact that both scientists and practitioners often ignore the interrelations between the physical environment, society and manmade environment when considering natural hazards. As a result, hazard protection is a socially isolated activity and disaster response an event-focused reaction – both highly professionalized but seldom viewed as an integral part of a larger development context (Weichselgartner, 2003).

A second group is social factors, such as cultural values, communication, understanding and mistrust. Scientists are often unable to tolerate the impreciseness of the 'big picture', whereas 'broad-brush' (but with specific recommendations) is often more useful to practitioners. As a result, scientists propose solutions that are often unworkable in practice, frequently due to a poor understanding of the institutional and other constraints to implementing such changes. It is important to determine how decision-makers can or will use the provided information to develop mitigation programmes and measures. This requires an understanding of the socio-economic and political/administrative context of the hazard planning process. Past studies have shown that scientific information can play no role in such planning processes. On the other hand, community boundaries are based on political/administrative areas for political decision-making processes and administrative convenience.

Traditionally, emergency managers consider a community to comprise a group of people who share the same geographically defined area, with the underlying assumption being that the group is relatively homogeneous and socially cohesive. In reality, the occupants of a spatially defined area are rarely homogeneous, but likely belong to a mosaic of communities who are inter-related and overlapping. Consequently, effective emergency management requires that decision-makers understand and respond to the diversity of communities.

Moreover, theoretical and conceptual research-based findings are rarely directly usable by policy-makers and practitioners and the language of science is often too complex and intimidating for many practitioners and policy-makers. As a result, authorities and scientists largely interact at the level of data transfer, but seldom at the level of equal partners who develop things together. Perhaps the scientific community will need to define the current and increasing risks more clearly, convey them in unison and, hence, more forcefully, without shying away from pointing out the inherent uncertainties. In particular, risk-related communication between science and its varied audiences is all too often structured on a 'deficit model' that assumes that the public simply does not know enough and that information flow should therefore be unidirectional, from knowledgeable experts to the ill-informed public. Mistrust plays an important role in disaster response, and the military style command-and-control response is especially problematic.

Third, structural factors, such as different institutional settings and standards, clearly restrict the co-production and transfer of knowledge across boundaries. Scientists and researchers have timeframes and deadlines that differ with decision-makers in policy and practice. Scientists involved in policy-making are constrained by political structure and agendas. Likewise, practitioners involved in scientific activities need to follow scientific standards. In contrast, scientists often do not take the time to explain fully how certain methods were conducted. Articles are written for those who already have a strong background and expertise in the topic, not for practitioners who have the desire to implement scientific ideas and findings. An important barrier is the reward system for scientists, which is based largely on products for the academic arena. Researchers are rewarded for scientific publications, with little acknowledgement for their work on brochures, reports and wider dissemination. One assessment producer theorized that they are rewarded for writing to their peers and not to the communities who need them. However, numerous development planners, disaster managers and emergency responders do not read academic journals. Decision-makers need specific conclusions and clear recommendations that they can act upon, rather than conceptual and theoretical arguments about frameworks and the terminology of disaster management.

Discussion

Coverage of knowledge systems solely in the domains of vulnerability and resilience is necessarily incomplete. According to findings from the Knowledge

Systems for Sustainable Development Project (2005), knowledge systems are more usefully conceptualized from an actor and institution focus than from the conventional information focus. In addition to the shortcomings of the acquired dataset, availability of time and workforce limited the sample size of case studies and the number of knowledge users included in the questionnaire survey. Hence, the study should be considered as non-exhaustive and explanatory.

Despite these limitations, focusing on generic assessment-related matters identifies a number of linkages between specific vulnerability and assessment determinants, as well as barriers and failures – functional, structural and social – that inhibit the co-production of used and applied knowledge. It is the quality of these relations that determines the grade of influence of research-based knowledge on action. Hence, the internal relationships of a knowledge/action system can be better understood as arenas of shared responsibility, embedded within larger systems of power and knowledge. Arguing in a similar way, Kerkhoff and Lebel (2006) provided valuable suggestions to gain some orientation to the issues of knowledge, action, power and engagement from within a shared research/action arena.

The findings of both the case study review and questionnaire survey indicate that divergent institutional settings and standards clearly restrict a cross-border co-production of knowledge. Not surprisingly, decision-makers do not always use the most appropriate available scientific information to influence policy decisions and, likewise, scientists do not primarily consider the needs of decision-makers when conducting research. Differences in approach and method in vulnerability and resilience research can often be attributed to the central questions of interest, the disciplinary composition and history of the knowledge production team, and the intended users. Difficulties exist in integrating disciplinary perspectives within the research team, as well as between the scientists and the various types of local decision-makers. According to this study, scientific assessments that have been designed in 'academic isolation' have significantly lower impact than assessments designed with input from internal and external people.

Influential scientific assessments cause changes in issue domains, defined as arenas in which interested actors seek to address an issue of common concern about which they have different beliefs and policy preferences (Clark et al, 2006). It is assumed that adjusting academic standards and settings to more resolution-exploring and problem-solving structures facilitates behavioural changes, and not just rational ones. Furthermore, faster and more effective transmission of existing and new knowledge to policy and decision-makers, as well as better communication of this knowledge to the public, can accelerate processes of change. Assuming that both formal and informal institutional settings constrain social learning and change, existing rules, laws, customs, norms and particularly reward systems for career advancement become focal points for triggering potential change. On an individual level, academic researchers need to strike a balance between pushing theoretical boundaries and generating information for practical use.

As a consequence of social barriers to understanding, language and trust, differences arise from ambiguities in framing problems and in the diverse ways in which the nature of problems are perceived. Divergent views exist regarding the perceptions of the problem character, the need for action, and the type and priority of actions that should be taken. Moreover, differences often result from uncertainties in the factual knowledge base of vulnerability and resilience. Since individual and collective cognitive processes have a strong tendency to maintain internal coherence and resist change, boundary work may be one feasible way to encourage successful dialogue and resolve conflicts between scientists and practitioners. Based on communication, translation and mediation, intermediary bodies can play an important role in the co-production of disaster-relevant knowledge, especially in developing countries where formal networks that provide bi-directional links across scales tend to be less dense and stable.

Assessments vary in the type of influence they have, ranging from no impact to attitudinal changes ('it will influence my beliefs') and behavioural changes ('I will consider the findings in the future'). Moreover, the influence of the same assessment varies across potential users. However, there seems to be a relationship between credibility and legitimacy. Users who considered assessments as untrue, for the most part also considered them as unfair. While publication date and gender are not influential factors with regard to knowledge production, both factors are more influential concerning the potential use of knowledge. Differences exist between male and female decision-makers regarding the use of information sources, the information considered to be most important and the credibility of assessments. Traditional knowledge transfer through the pipeline model requires time, resulting in delayed and, perhaps, not the most appropriate use of research-based knowledge. Furthermore, it seems that knowledge produced in the domains of vulnerability and resilience addresses more the needs of decision-makers working in the disaster management arena than it does for users dealing with other policy fields. Clearly, the former consider the assessment's findings as more relevant. In short, research-based knowledge in the domains of vulnerability and resilience is considered as highly credible and legitimate by multiple users. It is, moreover, considered as relevant by decision-makers – particularly by those working with disaster mitigation and response; but it addresses only to a low degree the needs of these users.

There is evidence that assessments of vulnerability and resilience increase their influence – or at least saliency – when they are co-produced for defined social–ecological systems in specific areas to meet local decision-making needs. A clear identification of the assessment's purpose, its unit and scale, and the intended audience followed by specific recommendations is obviously more important for decision-makers than addressing theoretical aspects of state-of-the-art methodology. In particular, scale should be a concern of both the unit of scientific analysis and of the administrative decision-making to avoid scientific assessments being implemented at geographic scales that are incompatible with the local management units (Cash and Moser, 2000). The goal is to understand systematically the complex process of interactions within and among societal arenas, spatial and temporal scales, and coupled human/environmental systems, and to integrate them within a more comprehensive analysis.

The fact that 12 case studies in our sample refer to a single hazard and 8 to multiple hazards supports the tendency of increasing acceptance that 'single-stressor/single-outcome' approaches fail to capture the reality of vulnerability and resilience for most social–ecological systems. However, only 20 per cent of the case studies analysed attempt to integrate practical elements and consider socio-economic and geophysical aspects across spatial scales. A shift towards more co-production of knowledge might improve data constraints and modelling capabilities, which still limit the tools for addressing the problem of up scaling and downscaling, and thus increases the low number of assessments addressing particular systems in multi-stressor contexts.

While scientific literature and data (i.e. statistics, census and maps) are often critical for scientific assessments, there is a trend among knowledge producers not only to use scientific information sources, but also to use a broad variety of other sources, including information provided by governments and NGOs. The fact that authors of recently published documents often disseminate their research findings to a higher degree through reports demonstrates that knowledge arenas are not stable but dynamic. This might be related to a relative increase of non-academic funding of scientific research. External sources of research funding increasingly seek evidence of applicability as an important indicator of performance. They expect the delivery of intermediate and final reports, and not of scientific papers. In contrast, the number of producers who particularly produce knowledge for and disseminate it to the general public, non-governmental organizations and the private sector is still very low.

By virtue of their multifaceted and dynamic character, the knowledge domains of vulnerability and resilience inevitably pose difficulties for discipline-based non-collective knowledge production. In both knowledge domains, many facts are uncertain, values are often in dispute and decisions are two edged. Such problems cannot be addressed by incompletely designed tools and programmes, which inevitably generalize, de-contextualize and reduce much of what is important about the place-specific character of vulnerability and resilience. Forms of knowledge production that entail making connections not only across discipline boundaries, but also between scholarly enquiry and policy and practice, are necessary. Hence, knowledge systems are needed that overcome technocratic reductionisms, integrate an extended range of sources and types of information, and engage in the co-production of knowledge through the interaction of producers and users, thus enhancing the quality of associated decision-making.

Propositions relating to core questions

On the basis of the empirical findings emerging from the case studies' analysis and questionnaire survey, the following propositions can be made with regard to the core questions of the VARIP project:

1 *How adequate is the knowledge base to support efforts for vulnerability reduction and building resilience? How is that knowledge distributed among actors?* While the production and quality of knowledge differ considerably, sufficient knowledge exists to reduce existing vulnerabilities to natural disasters. The transfer of existing knowledge into practical applications is inadequate. On the one hand, many scientists set the research agenda and do the research without input from other disciplines and practitioners, and then transfer the findings to potential users. On the other hand, many users make insufficient use of the existing possibilities to obtain research-based information. Both parties, however, are increasingly aware of the problem resulting from the use of limited information sources and distribution channels. The availability of consistent hazard and exposure data and technology to achieve integrated vulnerability assessments is often limited or beyond the resources of local governments.

2 *To what extent do the actors make use of the knowledge available to them? How relevant and pertinent is the knowledge to the needs of decision-makers?* Actors in policy and practice insufficiently use the research-based knowledge available to them; actors in science and research produce insufficient knowledge that is usable. The latter still consider scientific literature and data (i.e. statistics, census and maps) as the most critical information source for scientific assessments; decision-makers, by contrast, use primary governmental and internal institutional information sources. It seems that the trickle-down approach (i.e. eventually research will be taken up by users without additional effort by the producers) is still the default relationship between disaster-related scientists and practitioners. The practice of bringing research findings into the policy and practice arenas by publishing in peer reviewed journals is deeply embedded in the science system, manifested not only in attitudes, but also in incentive structures that reward publications in media with scientific impact and participation in forums with academic relevance. As a result, academic knowledge production is mostly 'career driven' ('publish or perish'). Although often relevant for practitioners, findings are rarely presented in a way that they can easily be used and applied by decision-makers.

3 *What barriers and failures limit the transfer of knowledge and feedbacks in the science/practice interface? Do the barriers and failures occur in the transfer of knowledge principally from science to practice, or from practice to science? How important are the intermediaries between science and practice and who are they?* The main barriers that limit the knowledge transfer are divergent objectives, needs, scopes, priorities, institutional

settings and standards. Hence, failures occur in both domains. Perhaps more important, however, are differences in the 'language' used, a lack of understanding of the counterpart's 'modes of operation', and mistrust, all hindering the co-production of knowledge. As a result, intermediary bodies and boundary organizations play an important role in 'translating' and 'coordinating' knowledge, particularly in developing countries with sparse networks and existing mistrust among actors. The case study analysis could not shed much light on the question of who the intermediaries are. Generally it can be said that boundary organizations, such as the International Red Cross and Red Crescent Movement, to mention just the prominent ones among many others, are more noticeable regarding disaster-response activities than in mitigation-related boundary work.

4 *How does the nature of institutions shape the science/practice interface? To what extent is institutional fragmentation a problem? How permeable are the boundaries of institutions to new information?* Researchers in the knowledge domains of vulnerability and resilience are constantly confronted with barriers, whether they rise up between the sciences and the humanities, between scientific disciplines, between the functional silos in a faculty, or between the scientist's world of ideas and the practitioner's world of action. Moreover, research grants increasingly encourage large interdisciplinary research teams, often located in institutions that are highly recognized and splendidly equipped. In contrast, institutions with lower recognition and fewer resources are limited in the scope of their research. Additionally, most funding schemes and research programmes do not allow comprehensive long-term research and are not equipped to deal with disaster studies that will require a generation to take effect. Fragmentation of responsibilities is also a problem for decision-makers in policy and practice. Issues related to vulnerability reduction are handled in different arenas within the political/administrative system: separate ministries, departments, programmes, budgets and time horizons, often with minimal relation to each other. Authorities and responsibilities are highly specialized by function and territory, and fragmentation of jurisdictions among federal, state and local governments hinders inter-agency communication and, thus, development of comprehensive disaster mitigation plans. Although there is an increasing number of cross-citations and papers classified in multiple knowledge domains, indicating an increasing integration among them (Janssen et al, 2006), the case study review suggests that research, particularly on ecological and social vulnerability and resilience, needs to be much better integrated.

5 *What major conflicts exist among actors and institutions in the interactions between science and practice? To what extent are the conflicts primarily about values or facts? Does social justice enter into the decision-making process?* Major conflicts result from functional, structural and social barriers that divide knowledge systems. Depending upon the quantity and quality of interactions among actors, conflicts range from a

'knowing better mentality' to mistrust and problems in understanding each others' language, needs and standards. Typically, vulnerability reduction and disaster mitigation have to compete with other societal needs, resulting in exclusion from governments' priority lists. The multidimensional and often invisible and infrequent nature of vulnerabilities contrasts with the clear responsibilities and time and financial budgets of authorities. The conflicts are not primarily about values and facts; but they are often hidden in the contextual surroundings, such as how much attention is paid to vulnerability, how politically contested it is, and how it is linked with other issues (e.g. security and terrorism). Given the analysis of environmental risks regarding issue attention, framing and actors, it can be assumed that a variety of cognitive and contextual factors influence progress in the area of vulnerability and resilience (Schreurs et al, 2001). Not surprisingly, different actors compete on responsibility and funds, seeking more to realize their needs and interests and less to meet the values and needs of the people at risk. In most case studies, no knowledge governance procedures are adopted to involve the less powerful and often more vulnerable actors in determining research objectives and disseminating research findings.

6   *What is the role of consensus in the science/practice interface? How is consensus built? How are conflicts resolved? Who are the consensus builders and mediators and what are the major processes that they use?* On the basis of this study, no propositions are possible. Achieving consensus might not be feasible, but also not required if differences in opinion are recognized in order to deal with them constructively (i.e. social learning takes place).

7   *What factors contribute most to adaptive capacity? How large is the gap between the capacity to adapt and the adaptation that actually occurs? What causes this gap and how can it be reduced? What new elements of enlarged capacity would contribute most to greater resilience in the face of environmental change over the short run and the longer term?* On the basis of this study, no propositions are possible.

8   *To what extent has social learning evolved among the principal actors? To what extent has social learning ameliorated or exacerbated vulnerability? What most limits or facilitates the ability to learn from one's own experience and the experience of others?* Little can be said from this study about social learning. Processes of social learning are usually codified in shared practices, tools and concepts. The fact that boundary organizations are both important and successful in knowledge transfer underlines the existing deficit in interacting and collaborating across the different barriers. One could even say that boundary organizations are the result or outcome of the lack of co-produced knowledge. Learning can be supported if scientists would more actively engage in activities that enhance self-reflection (e.g. assessing the assessment processes and products). The approach of adaptive management, which enables collaboration in developing and

communicating new ideas and practices to address uncertainty, may be a possible pathway for making policy and institutions more flexible in the face of global change (Weichselgartner and Sendzimir, 2004).
9   *Where is the science/practice interface vulnerable to failure? Where is the science/practice interface most vulnerable to failure to future risks?* Without a co-production of knowledge involving different actors, there is the danger of generalization and de-contextualization, resulting in undifferentiated 'communities at risk' or 'vulnerable countries'. Scientific approaches based on static thinking and linear causalities are contradictory to the multidimensional and dynamic structure of vulnerability and resilience. Likewise, single-discipline concepts are not capable of offering valuable solutions to the multilayered problems. On the practitioners' side, cooperation and collaboration across society are hindered when disaster schemes and programmes treat people as 'clients' in disaster management processes, ignoring the experience of those most at risk, and where 'paternalistic' science and technology do things *to* them and *for* them, rather than *with* them. Most hazard protection strategies lose effectiveness when top-down approaches ignore local social dynamics and are, in turn, ignored as socially isolated and irrelevant. This becomes even more important in times of rapid global change processes. Related is the fact that too little attention is devoted to the loss of local knowledge and traditional adaptation strategies.
10 *How can the science/practice interface be best improved?* Generally speaking, reducing the functional, structural and social barriers would certainly lead to improvements in the SPPI (specific recommendations are given in the following section). Consequently, there are many feasible ways to make improvements: broader joint programmes, flatter hierarchical networks, more cooperative actions, or a global vulnerability platform providing a library and database, thus facilitating information exchange between knowledge producers and users. Less frequent but promising could be the collaborative work on a common goal, such as decision support systems that are integrated with disaster planning, tying together spatial–temporal information with multiple perturbation scenarios that depict spatial–temporal reactions. Example communities in terms of vulnerability reduction could also be selected, observed and evaluated.

Recommendations

Despite a growing synergy among conceptual frameworks and increasing consensus on issues of importance, research on vulnerability and resilience is still facing theoretical and practical challenges – namely, to concurrently address and capture socio-economic and geophysical factors, multiple and interacting stressors, and cross-scalar influences and outcomes, as well as to confront aspects of governance, gender and social justice. Likewise, concepts to link research-based knowledge with action vary regarding their underlying assumptions, scope and influence, forming a fragmented and often contradictory set of approaches.

Given the rapid changes in coupled human/environment systems and the need to mitigate and adapt to changing conditions, the knowledge domains of social–ecological systems and knowledge systems are of crucial importance. It is time to interrelate and tie the two domains closer together, in both theoretical and practical terms. Not only disaster management, but also knowledge management, play an important role in reducing disaster vulnerability and enhancing resilience. Single-discipline concepts based on single-stressor focus, static thinking and linear causalities are contradictory to the socially divergent, multidimensional, dynamic, interactive and scale-dependent character of vulnerability and resilience. Single-discipline approaches and single-technical fixes are not capable of solving the multilayered natural disaster problem. Ultimately, the reduction of natural disaster impacts can be successful only if the structures and practices in disaster *and* knowledge management are properly adjusted, taking into account societal and natural conditions.

Needless to say that there is no magic bullet. One feasible way for knowledge producers and users to generate a deeper mutual understanding of each others' needs and constraints is to increase the amount and intensity of face-to-face interaction by creating institutional contexts where both are encouraged to interact. Obviously, designing such contexts necessitates precious resources – temporal, spatial and financial. Different research activities of the Science, Environment and Development Group at Harvard University's Center for International Development resulted in findings that indicate how effective systems have to be designed to harness research-based knowledge and know-how with action.

Major findings

The empirical evidence presented in this case study supports some of the foregoing findings. Hence, knowledge systems in the domains of vulnerability and resilience can increase effectiveness, we argue, by:

- creating dense social networks that provide bi-directional links across scales;
- strengthening integration of ecological and social research approaches and tools;
- creating mechanisms for early problem identification and framing in concert with practitioners in order to better meet the needs of various users;
- recognizing mutual dependencies and interactions and building trust for self-reflection to take account of individual mental frames and images;
- providing flexibility for experimentation to accumulate experiences from such trials in order to improve the capacity to cope with surprises, uncertainty and change;

- integrating understanding from multiple sources in order to discover what can be adapted to diverse local contexts and capacities without ignoring impacts across dimensions or peoples;
- engaging not in pipeline 'transfers' or linear 'outreach', but in the 'co-production' of knowledge through the close interaction of producers and users – hence, building a 'knowledge-action system';
- undertaking efforts that are salient, credible and legitimate as seen by multiple users in order that decisions and findings are more likely to be accepted and treated as socially just, even in the presence of differences of interest and conflict;
- including mechanisms to build social memory and learning by providing a reservoir of experience from which solutions to new problems can be drawn, as well as an open window for new practices which may be needed under changed contexts;
- providing mechanisms for linking solutions proposed by research with articulated needs and problems of practitioners.

Likewise, programmes aiming at reducing vulnerability and enhancing resilience should:

- involve a variety of actors in setting up the research agenda;
- establish a shared problem perception within this group of actors;
- be equally 'producer driven' and 'user driven';
- avoid, to the extent possible, the use of generalizing, de-contextualizing and reductionist approaches;
- include processes of participation, integration, learning and negotiation;
- engage both ends of the producer/user spectrum in a dialogue from which a negotiated view of what is both feasible and desirable emerges;
- take into account governance, equity, gender and social justice issues;
- engage with boundary organizations that have dual accountability to users and producers;
- develop possible scenarios with regard to future states of vulnerability and resilience practices and new management approaches.

References

Birkmann, J. (2006) *Measuring Vulnerability to Natural Hazards: Towards Disaster Resilient Societies*, United Nations University Press, Tokyo

Burton, I., Kates, R. W. and White, G. F. (1978) *The Environment as Hazard*, Oxford University Press, New York, NY

Cash, D. and Moser, S. C. (2000) 'Linking global and local scales: Designing dynamic assessment and management processes', *Global Environmental Change*, vol 10, no 2, pp109–120

Clark, W. C., Mitchell, R. B. and Cash, D. W. (2006) 'Evaluating the influence of global environmental assessments', in R. B. Mitchell, W. C. Clark, D. W. Cash, and N. M. Dickson (eds) *Global Environmental Assessments: Information and Influence*, MIT Press, Cambridge, pp1–28

Dikau, R. and Weichselgartner, J. (2005) *Der unruhige Planet: Der Mensch und die Naturgewalten*, Wissenschaftliche Buchgesellschaft, Darmstadt

Gallopín, G. C. (2006) 'Linkages between vulnerability, resilience, and adaptive capacity', *Global Environmental Change*, vol 16, no 3, pp293–303

Gunderson, L. H. and Holling, C. S. (2002) *Panarchy: Understanding Transformations in Human and Natural Systems*, Island Press, Washington, DC

Janssen, M. A., Schoon, M. L., Ke, W. and Börner, K. (2006) 'Scholarly networks on resilience, vulnerability and adaptation within the human dimensions of global environmental change', *Global Environmental Change*, vol 16, no 3, pp240–252

Jasanoff, S. (1993) 'Bridging the two cultures of risk analysis', *Risk Analysis*, vol 13, no 2, pp123–129

Kasperson, J. X., Kasperson, R. E., and Turner, B. L. II (1995) *Regions at Risk*, United Nations University Press, Tokyo

Kates, R. W., Colten, C. E., Laska, S. and Leatherman, S. P. (2006) 'Reconstruction of New Orleans after Hurricane Katrina: A research perspective', *Proceedings of the National Academy of Sciences Special*, vol 103, no 40, pp14653–14660

Kerkhoff, L. van and Lebel, L. (2006) 'Linking knowledge and action for sustainable development', *Annual Review of Environment and Resources*, vol 31, pp445–477

Knowledge Systems for Sustainable Development Project (2005) *Interim Progress Report*, Prepared by Nancy Dickson, www.ksg.harvard.edu/kssd/

Mileti, D. S. (1999) *Disasters by Design: A Reassessment of Natural Hazards in the United States*, Joseph Henry Press, Washington, DC

Mitchell, R. B., Clark, W. C., Cash, D. W. and Dickson, N. M. (2006) *Global Environmental Assessments: Information and Influence*, MIT Press, Cambridge, MA

Munich Re Group (2006) *Topics Geo Annual Review: Natural Catastrophes 2005*, Munich Re Group, Munich

Schreurs, M. A., Clark, W. C., Dickson, N. M. and Jäger, J. (2001) 'Issue attention, framing, and actors: An analysis of patterns across arenas', in The Social Learning Group (eds) *Learning to Manage Global Environmental Risks, Volume 1: A Comparative History of Social Responses to Climate Change, Ozone Depletion, and Acid Rain*, MIT Press, Cambridge, pp349–364

Turner, B. L. II, Kasperson, R. E., Matson, P. A., McCarthy, J. J., Corell, R. W., Christensen, L., Eckley, N., Kasperson, J. X., Luers, A., Martello, M. L., Polsky, C., Pulsipher, A. and Schiller, A. (2003) 'A framework for vulnerability analysis in sustainability science', *Proceedings of the National Academy of Sciences*, vol 100, no 14, pp8074–8079

Weichselgartner, J. (2003) 'Toward a policy-relevant hazard geography: Critical comments on geographic natural hazard research', *Die Erde*, vol 134, no 2, pp121–138

Weichselgartner, J. (2006) 'Soziale Verwundbarkeit und Wissen', *Geographische Zeitschrift*, vol 94, no 1, pp15–26

Weichselgartner, J. and Obersteiner, M. (2002) 'Knowing sufficient and applying more: Challenges in hazards management', *Global Environmental Change Part B: Environmental Hazards*, vol 4, nos 2–3, pp73–77

Weichselgartner, J. and Sendzimir, J. (2004) 'Resolving the paradox: Food for thought on the wider dimensions of natural disasters', *Mountain Research and Development*, vol 24, no 1, pp4–9

White, G. F., Kates, R. W. and Burton, I. (2001) 'Knowing better and losing even more: The use of knowledge in hazard management', *Global Environmental Change Part B: Environmental Hazards*, vol 3, nos 3–4, pp81–92

Wilbanks, T. J. and Kates, R. W. (1999) 'Global change in local places: How scale matters', *Climatic Change,* vol 43, no 3, pp601–628

Appendix 1: List of selected case studies

Adger, N. W. (1999) 'Social vulnerability to climate change and extremes in coastal Vietnam', *World Development*, vol 27, no 2, pp249–269

Anderson-Berry, L. and King, D. (2005) 'Mitigation of the impact of tropical cyclones in Northern Australia through community capacity enhancement', *Mitigation and Adaptation Strategies for Global Change*, vol 10, no 3, pp367–392

Buckland, J. and Rahman, M. (1999) 'Community-based disaster management during the 1997 Red River flood in Canada', *Disasters*, vol 23, no 2, pp174–191

Chang, S. E. and Falit-Baiamonte, A. (2002) 'Disaster vulnerability of businesses in the 2001 Nisqually earthquake', *Environmental Hazards*, vol 4, nos 2–3, pp59–71

Clark, G. E., Moser, S. C., Ratick, S. J., Dow, K., Meyer, W., Emani, S., Jin, W., Kasperson, R. E., Kasperson, J. X. and Schwarz, H. (1998) 'Assessing the vulnerability of coastal communities to extreme storms: The case of Revere, MA, USA', *Mitigation and Adaptation Strategies for Global Change*, vol 3, no 1, pp59–82

Collins, T. W. (2005) 'Households, forests, and fire hazard vulnerability in the American West: A case study of a California community', *Environmental Hazards*, vol 6, pp23–37

Cutter, S. L., Mitchell, J. T. and Scott, M. S. (2000) 'Revealing the vulnerability of people and places: A case study of Georgetown County, South Carolina', *Annals of the Association of American Geographers*, vol 90, no 4, pp713–737

DKKV (2003) 'Hochwasservorsorge in Deutschland: Lernen aus der Katastrophe 2002 im Elbegebiet', *Schriftenreihe*, no 29, Deutsches Komitee für Katastrophenvorsorge, Bonn

El-Raey, M., Frihy, O., Nasr, S. and Dewidar, K. (1999) 'Vulnerability assessment of sea level rise over Port Said Governorate, Egypt', *Environmental Monitoring and Assessment*, vol 56, no 2, pp113–128

Gillen, M. (2004) 'Urban governance and vulnerability: Exploring the tensions and contradictions in Sydney's response to bushfire threat', *Cities*, vol 22, no 1, pp55–64

Jacob, K., Gornitz, V. and Rosenzweig, C. (2007) 'Vulnerability of the New York City Metropolitan Area to coastal hazards, including sea level rise: Inferences for urban coastal risk management and adaptation policies', in L. McFadden, R. J. Nicholls and E. Penning-Rowsell (eds) *Managing Coastal Vulnerability: Global, Regional, Local*, Elsevier, New York, NY, pp139–156

Kuroiwa, J. (2002) 'Sustainable cities, a regional seismic scenario, and the 6-23-2001 Arequipa Peru earthquake', *Natural Hazards Review*, vol 3, no 4, pp158–162

Lewsey, C., Cid, G. and Kruse, E. (2004) 'Assessing climate change impacts on coastal infrastructure in the Eastern Caribbean', *Marine Policy*, vol 28, no 5, pp393–409

Luers, A. L., Lobell, D. B., Sklar, L. S., Addams, L. C. and Matson, P. A. (2003) 'A method for quantifying vulnerability, applied to the agricultural system of the Yaqui Valley, Mexico', *Global Environmental Change*, vol 13, no 4, pp255–267

Mileti, D. S. (1999) *Disasters by Design: A Reassessment of Natural Hazards in the United States*, Joseph Henry Press, Washington, DC

Mustafa, D. (2003) 'Reinforcing vulnerability? Disaster relief, recovery, and response to the 2001 flood in Rawalpindi, Pakistan', *Environmental Hazards*, vol 5, nos 3–4, pp71–82

O'Brien, K., Leichenko, R., Kelkar, U., Venema, H., Aandahl, G., Tompkins, H., Javed, A., Bhadwal, S., Barg, S., Nygaard, L. and West, J. (2004) 'Mapping vulnerability to multiple stressors: Climate change and globalization in India', *Global Environmental Change*, vol 14, no 4, pp303–313

Odeh, D. J. (2002) 'Natural hazards vulnerability assessment for statewide mitigation planning in Rhode Island', *Natural Hazards Review,* vol 3, no 4, pp177–187

Wood, N. J. and Good, J. W. (2004) 'Vulnerability of port and harbor communities to earthquake and tsunami hazards: The use of GIS in community hazard planning', *Coastal Management*, vol 32, no 3, pp243–269

Zoleta-Nantes, D. B. (2002) 'Differential impacts of flood hazards among the street children, the urban poor and residents of wealthy neighborhoods in Metro Manila, Philippines', *Mitigation and Adaptation Strategies for Global Change*, vol 7, no 3, pp239–266

Appendix 2: List of survey participants

No.	Producer	Organization
1	Guro Aandahl	Centre for Development and the Environment, University of Oslo
2	Neil Adger	Tyndall Centre, University of East Anglia
3	Linda Anderson-Berry	Disaster Mitigation Policy Planning Services, Australian Bureau of Meteorology
4	Suruchi Bhadwal	The Energy and Resources Institute, New Delhi
5	Jerry Buckland	Menno Simons College, University of Winnipeg
6	Stephanie Chang	Centre for Human Settlements, University of British Columbia
7	George Clark	Harvard College Library, Harvard University
8	Timothy Collins	Department of Geography, Arizona State University
9	Susan Cutter	Department of Geography, University of South Carolina
10	Wolf Dombrowsky	University of Kiel, Disaster Research Centre
11	Kirsten Dow	Department of Geography, University of South Carolina
12	Mohamed El-Raey	Institute of Graduate Studies and Research, University of Alexandria
13	Omran Frihy	Institute of Coastal Research, Alexandria
14	Mike Gillen	School of Geography Planning and Architecture, University of Queensland
15	Vivien Gornitz	National Aeronautics and Space Administration, Goddard Institute for Space Studies
16	Uwe Grünewald	University of Technology Cottbus, Hydrology and Water Resources Management
17	Klaus Jacob	Lamont-Doherty Earth Observatory, Columbia University
18	David King	Centre for Disaster Studies, James Cook University
19	Julio Kuroiwa	National University of Engineering, Lima
20	Robin Leichenko	Department of Geography, Rutgers University
21	Clement Lewsey	National Oceanic and Atmospheric Administration, National Ocean Service
22	David Lobell	Department of Geological and Environmental Sciences, Stanford University
23	Amy Luers	Department of Geological and Environmental Sciences, Stanford University
24	Bruno Merz	GeoForschungsZentrum Potsdam, Section Engineering Hydrology
25	Dennis Mileti	Institute of Behavioral Science, University of Colorado
26	Susanne Moser	Institute for the Study of Society and Environment, National Center for Atmospheric Research
27	Daanish Mustafa	College of Arts and Sciences, University of South Florida
28	Lynn Nygaard	Centre for International Climate and Environmental Research, Oslo
29	Karen O'Brien	Centre for International Climate and Environmental Research, Oslo
30	David Odeh	Odeh Engineers, Inc., North Providence, RI
31	Theresia Petrow	GeoForschungsZentrum Potsdam, Section Engineering Hydrology
32	Matiur Rahman	Department of Environment and Geography, University of Manitoba
33	Samuel Ratick	Department of Geography, Clark University
34	Michael Scott	Department of Geography, Salisbury University
35	Willi Streitz	University of Kiel, Disaster Research Centre
36	Annegret Thieken	GeoForschungsZentrum Potsdam, Section Engineering Hydrology
37	Henry Venema	International Institute for Sustainable Development, Winnipeg
38	Jennifer West	Centre for International Climate and Environmental Research, Oslo
39	Nathan Wood	US Geological Survey, Vancouver
40	Doracie Zoleta-Nantes	Department of Geography, University of the Philippines

No.	User	Organization
1	Jonathan Allan	Oregon State Department of Geology, Coastal Section (team leader)
2	Thomas Ambrosino	City of Revere (mayor)
3	William Bain	City of Newport (mayor)
4	Don Ballantyne	ABS Consulting, Operational Risk Consulting Division (director)
5	Gaby Breton	Canadian Centre for International Studies and Cooperation, Vietnam, Natural Disaster Mitigation Programme (leader)
6	Marc Bruyère	Manitoba Emergency Measures Organization, Eastern Region (regional emergency officer)
7	Alfredo Siu Delgado	National Institute for Civil Defense (INDECI), Regional Office Arequipa (regional director)
8	Gerry Delorme	Manitoba Health, Office of Disaster Management (director)
9	Lewis Dugan	Georgetown County, Emergency Management (manager)
10	El-Shinnawy	Coastal Research Institute (director)
11	Will Ewing	City of Toledo, Fire Department (fire chief)
12	David Farmer	Cairns City Council (chief executive officer)
13	John Fontaine	Ministry of Community Development, Dominica (local government commissioner)
14	Essam Hassan	Egyptian Environmental Affairs Agency (technical officer)
15	Blanche Higgins	Rhode Island Statewide Planning Program, Land Use Section (supervising planner)
16	Tony Hucks	Georgetown County, Fire Department (assistant fire/EMS chief)
17	Ricardo Hurtado	Commission for Ecology and Sustainable Development of Sonora (CEDES), Conservation Unit (director)
18	Peter Kammel	City of Pirna, Fire Department (chief)
19	James Kendra	Emergency Administration and Planning Program, University of North Texas (programme coordinator)
20	Vimal Khawas	Council for Social Development India (associate fellow)
21	Darryl Leggett	Fire Protection Association Australia (NSW state secretary)
22	MaryAnn Marrocolo	NYC Office of Emergency Management (assistant commissioner for planning and preparedness)
23	David McEntire	Emergency Administration and Planning Program, University of North Texas (associate professor)
24	Ross McKim	Cairns City Council, Cairns Local Disaster Management Group (executive officer)
25	Jacqueline Meszaros	National Science Foundation (programme officer)
26	Mruthyunjaya	Indian Council of Agricultural Research, National Agriculture Innovation Project (national director)
27	Michael Mulhare	Rhode Island, Department of Environmental Management, Office of Emergency Response (chief)
28	Shelley Napier	Manitoba Emergency Measures Organization, Interlake Region (regional emergency officer)
29	Cheryl-Lee Norris	Cairns City Council, Disaster Management Unit (coordinator)
30	Ines Pearce	City of Seattle Office of Emergency Management (Seattle Project impact director)
31	David Persaud	Ministry of Public Utilities and the Environment, Trinidad and Tobago (environmental manager)
32	Doug Peterson	Manitoba Floodway Authority (manager environmental services)
33	Joe Picciano	Federal Emergency Management Agency, Region II (acting director)
34	Jürgen Plaggenborg	State of Saxony, Ministry of Interior (officer of disaster protection)
35	Abdul Qayum	Government of Pakistan, Planning and Development Division (deputy chief)
36	Claudine Roberts	Ministry of Community Development, Dominica (assistant local government commissioner)

No.	User	Organization
37	Amenda Rutherford	Waccamaw Regional Council of Governments (land-use and transportation planner)
38	Muhammad Sabir	Ministry of Finance and Economic Affairs (senior economist)
39	Kim Seidler	City of Chico, Planning Division (director)
40	Jens Seifert	City of Dresden, Environmental Department (officer)
41	Phyllis Shulman	City of Seattle, Legislative Department (legislative assistant)
42	Robert Stallings	School of Policy, Planning, and Development, University of Southern California (professor emeritus)
43	Frank Stringi	City of Revere, Department of Planning and Community (city planner)
44	Russell Taylor	NSW Rural Fire Service (manager community education)
45	Kathleen Tierney	Natural Hazards Research and Applications Information Center, University of Colorado (director)
46	Damodar Tripathy	Research and Consultancy Pvt. Ltd, Orissa (managing director)
47	Scott Vail	State of California, Office of Emergency Services, Fire and Rescue Branch (deputy chief administrator)
48	Luis Rene Vallenas	National Institute for Civil Defense (INDECI), National Education and Training Management Unit (director)
49	Bruce Vild	Rhode Island Statewide Planning Program, Economic Development Section (supervising planer)
50	Sean Waters	Federal Emergency Management Agency, Region II (officer)
51	Jay Wilson	Oregon Emergency Management (coordinator of Earthquake, Tsunami, and Volcano Program)
52	Sarah Zaidi	Research and Information System for Earthquakes Pakistan (RISEPAK) (coordinator)

Appendix 3: Questionnaire for assessment producers

Name:
Institution:
Assessment Reference:

1. Who was involved in formulating the research questions?

Only the authors ☐ Colleagues of our institution ☐ External colleagues ☐
Others ☐ Please specify:

2. How regularly do you use the following information sources for your work and how important do you consider them? Please check all and rank them by importance.

	regularly	occasionally	rarely	not at all	rank
Internal institutional sources	☐	☐	☐	☐	☐
Governmental sources	☐	☐	☐	☐	☐
Non-governmental sources	☐	☐	☐	☐	☐
Scientific sources	☐	☐	☐	☐	☐
Media sources	☐	☐	☐	☐	☐
Personal communication	☐	☐	☐	☐	☐
Others, please specify:	☐	☐	☐	☐	☐

3. What sources and information were most influential and critical to the assessment?

1. Please name and specify why:
2. Please name and specify why:

4. Who was the intended audience for this assessment and what was its intended use?

Please specify:

5. What changes/additional information would you make/add to this assessment if its purpose was to be an immediately actionable plan for practitioner use?

Please specify:

6. Beside the above mentioned publication, were there any other ways of disseminating the findings of the assessment? Please check all that apply.

No ☐
Yes, scientific papers ☐
Yes, reports ☐ Please specify for whom:
Others ☐ Please specify:

7. **Do you think that the assessment addresses the need of users?**

   No ☐
   Please specify why:
   Yes ☐
   Please specify why and which:
      For whom might it be relevant? Please specify (if possible, include contact details):

8. **Are you aware of any impacts of your assessment, especially on policy and practice?**

   No ☐
   Yes ☐   Please specify:

9. **How do you evaluate the influence of the assessment? Please check all that apply.**

   It had only limited impact on science and practice ☐
   It had scientific impact, but not on practice ☐
   It influenced the beliefs of readers and thus will have impacts in the future ☐
   Others ☐   Please specify:

10. **Where do you see existing conflicts among actors and institutions in the interactions between science and practice?**

    Please specify:

11. **How can knowledge transfer at the science/practice interface be improved?**

    Please specify:

Appendix 4: Questionnaire for assessment users

Name:

Institution:
Responsibility:
Assessment Reference:

1. What information sources do you consult to obtain new information relevant to your work? How regularly do you use them and how important do you consider them? Please check all and rank them by importance.

	regularly	occasionally	rarely	not at all	rank
Internal institutional sources	☐	☐	☐	☐	☐
Governmental sources	☐	☐	☐	☐	☐
Non-governmental sources	☐	☐	☐	☐	☐
Scientific sources	☐	☐	☐	☐	☐
Media sources	☐	☐	☐	☐	☐
Personal communication	☐	☐	☐	☐	☐
Others, please specify:	☐	☐	☐	☐	☐

2. Were you aware of the assessment before we contacted you? Please check all that apply.

No, not at all ☐
No, but I heard that something was done ☐
  Please specify:
Yes, I knew about the assessment, but not this report ☐
  Please specify:
Yes, I knew about the report, but I did not read it ☐
  Please specify:
Yes, I knew the assessment and the report ☐
  Please specify:
Yes, I was involved ☐
  Please specify:

3. What sources and information have been most influential and critical to your work?

1. Please name and specify why:
2. Please name and specify why:

4. What kind of influence did/will the assessment have? Please check all that apply.

I did/will talk about it within my organization ☐
It did/will influence my beliefs ☐
I did/will use the work or refer to it to convince other people ☐
It did/will not influence my actions related to my work ☐

I did/will consider the findings in the future ☐
Please, specify reasons why this assessment did/will influence your work or not:

5. Does the assessment report address your needs? Please check all that apply.

No, not at all ☐
Please specify why:
No, but I see the strengths of the assessment ☐
Please specify:
For whom might it be relevant? Please specify (if possible, include contact details):
Yes, it addresses some of my needs ☐
Please specify which:
Yes, but the results and findings are not relevant ☐
Please specify why:
Yes, and the results and findings are relevant ☐
Please specify why:
What should the assessment have addressed to be more relevant and useful?
Please specify:

6. Do you believe that the assessment is accurate and technically sound?

No ☐ Please specify:
Yes ☐ Please specify:
What would have improved the accuracy and technical soundness of the assessment?
Please specify:

7. Do you find the assessment respectful of stakeholders and unbiased?

No ☐ Please specify:
Yes ☐ Please specify:
What could have improved the fairness of the assessment process?
Please specify:

8. What else would you like to tell us or might be of interest for us?

Please specify:

4

Linking Vulnerability, Adaptation, and Resilience Science to Practice: Pathways, Players and Partnerships[1]

Coleen Vogel, Susanne C. Moser, Roger E. Kasperson and Geoffrey D. Dabelko

Introduction

Climate-related catastrophes, such as the 2003 floods and heatwaves in Europe, the 2005 hurricanes in the US, Mexico and Cuba, and the persistent droughts and floods in Africa, Australia and Asia, as well as non-climatic high-impact events such as the 2004 Asian tsunami and the 2005 earthquake in Pakistan, hold a mirror up to the world showing its continued exposure to destructive natural forces. Perhaps more importantly, they also focus attention on the deep-seated patterns of underlying social vulnerability and limited coping capacity that make these natural forces so devastating. All of these examples, usually starkly portrayed via the media, bring to light the daily, real and complex interactions of vulnerability, adaptation and resilience, terms that scientists are grappling with in the global environmental change (GEC) and related scientific communities. They typically produce not just calls for better warning systems and improved scientific forecasting capabilities (although they are also needed), but increase the demand from the public and policy-makers for useful scientific information that could help to ameliorate the situations of those most at risk.

As the real world need and demand for actionable information on vulnerability, adaptation and resilience grows, many scientists in this expanding community have been drawn into this area of research through an interest in disasters, hunger and famine – others through an interest in global change. Many of them claim to produce policy-relevant or 'useful' information. Meanwhile, extensive research – including the social study of science and of science policy – has emerged over the past two decades or more to examine the social contract between science and society, to explore and improve the science/policy or (more generally) science/practice interface, and to make specific recommendations on how to improve communication and interaction between these two worlds. Frequently, the attempt to produce 'useful' science occurs separately from this study of the science/practice interface. Perhaps not surprisingly, then, policy-makers and managers often indicate that they do not receive the information they need, scientists are frustrated when their information is not being used, and, ultimately, communities remain vulnerable in the face of extreme events and environmental changes.

In this chapter, we examine the literature on science/practice interactions for useful insights that could inform a more effective exchange between researchers and potential information users of this field of science. We first examine the nature of the challenge of scientists and practitioners working together, especially highlighting the challenges that the knowledge/action interaction produces for both the scientific enterprise and for practice. We point to the need for improved communication and engagement, and highlight the specific challenges that arise for such interaction in the field due to the multiple disciplinary origins from which our knowledge base has emerged. Then we illustrate the challenges and instances of success of scientists working together with practitioners on such issues in a case study of a series of vulnerability assessments conducted in Southern Africa. Finally, we offer some conclusions and suggestions for further exploration (through research and trial on the ground) in the hope that many in the global environmental change and hazards communities become better prepared for practical engagement in real world problem-solving.

Challenges and opportunities of working at the science/practice interface of vulnerability, adaptation and resilience

Pathways to sustainability: Does it matter what we mean by vulnerability, adaptation and resilience?

Modern-day vulnerability, adaptation and resilience science is rooted in several decades of multidisciplinary research under a range of paradigms, theories and methodologies. Vulnerability, adaptation and resilience first became widely used in several assessments of environmental change (e.g. Timmerman, 1981; and later reflections in Kates et al, 2001). More recently such concepts are re-emerging and receiving renewed attention in discussions linked to global environmental change (see, Berkes et al, 2003; Gallopín, 2006; Thomalla et al, 2006). Some experts in this field consider vulnerability, adaptation and resilience through the lens of climate change, where the vulnerability approach is a modification of earlier work that focused on climate change impacts; this approach is best exemplified by the Intergovernmental Panel on Climate Change (IPCC) and the United Nations Framework Convention on Climate Change (UNFCCC) (see Adger, 1999, 2003; Downing and Patwardhan, 2003; Huq and Reid, 2004; Brooks et al, 2005). Others use a political ecology and sustainability perspective to examine vulnerability and resilience, in particular, in the context of a variety of global changes (e.g. Watts and Bohle, 1993; Bohle et al, 1994; Kasperson and Kasperson, 2001; Berkes et al, 2003; Ogunseitan, 2003; Turner et al, 2003a, 2003b; Folke et al, 2005). Still others emphasize a social justice perspective (e.g. Adger, 1999, 2003; Adger et al, 2003; Dow et al, 2006), while yet others approach vulnerability and enhanced resilience to change from a disaster risk-reduction orientation (e.g. Wisner, 1993; Wisner et al, 2003; Bankoff et al, 2004; Thomalla et al, 2006; see also the RADIX website, www.radixonline.org).

At the same time that the focus on vulnerability, adaptation and resilience has become central to the scientific debate in the global change community (including climate change scientists) and in the disaster risk-reduction community, there has also been a growing interest from the practitioner community (including, for example, a focus on various vulnerability assessment methodologies), that has been seeking ways to understand such approaches and concepts to better inform humanitarian interventions, such as the Regional Hunger and Vulnerability Programme (RHVP, 2006). The range of approaches to understanding these concepts has enriched our knowledge of the complex dynamics that produce vulnerability and adaptive capacity; but it also brings with it a variety of challenges, particularly in the application and use of these concepts in practice. Dilley and Boudreau (2001), for example, argue that vulnerability has become a term of art for assessment methods in several contexts, not just disaster or risk management. A case in point is food security, where the term 'vulnerability' has assumed a variety of connotations (Longhurst, 1994; Alwang et al, 2001; Füssel, 2004), a situation that Dilley and Boudreau (2001) argue has often been an impediment to food security interventions. Here, vulnerability is usually defined in relation to an outcome such as hunger. This definition may preclude employing the concept for the more specific task of evaluating the susceptibility of a population to explicitly identified exogenous events or shocks that could lead to these outcomes. Differences in understanding of 'vulnerability' also exist between climate impact assessors and hazard researchers. Subtle differences in the understanding of concepts such as 'resilience', 'coping capacity' and 'adaptation' are frequently lost in the course of a growing multidisciplinary discourse (Thomalla et al, 2006).

Clearly, the multidisciplinary nature of research in this growing field is not unique in having to face such linguistic, paradigmatic, theoretical and methodological tensions, and they are not necessarily negative (Miller, 2001). Rather, the range of insights that has enriched this field and the slow and mutual transformation that disciplinary approaches are experiencing as a result of learning from each other are deepening our scientific understanding. We argue here, however, that depending upon the approach used to examine vulnerability, adaptation and resilience, some opportunities for intervention will open and others will close. As we will show in our case study of vulnerability assessments in Southern Africa, involving practitioners from the outset in the research design, problem definition or framing, choice of approaches, negotiation of credible and legitimate knowledge systems (including 'expert knowledge' and 'local knowledge'), and communication and involvement with relevant stakeholders helps to make this choice consciously rather than by default. Despite the complexities and transaction costs incurred, such early and ongoing engagement will shape the science and application opportunities (Steinberg, 2005).

Conceptualizing the science/practice interface: Metaphors and implications

So what usually happens when scientific experts and policy-makers or practitioners interact on issues related to vulnerability, adaptation and resilience? What happens when practitioners ask for guidance from the scientific community? Answers to these questions typically uncover metaphors that are used to describe the interactions at the science/practice interface. They inadvertently reveal assumptions and biases about that interaction. They also hopefully challenge – at a deeper level – a rethinking of the very nature of the practice-oriented scientific enterprise.

For the inexperienced scientist, or a practitioner accustomed to 'outreach' thinking, for example, there is frequently an expectation that 'bridging' the science/practice 'gap' or 'gulf' is a fairly straightforward, unidirectional and simple process. Traditionally, the link between science and practice has been viewed as a linear process in which a set of scientifically vetted and legitimized findings are moved from the 'research sphere' to the 'policy sphere' (for some excellent critiques, see Crewe and Young, 2002; Court and Young, 2003; Devereux, 2003; Ellis, 2003; Jasanoff, 2003; Nowotny, 2003; Moll and Zander, 2006; Karl et al, 2007).

Those studying these interactions challenge our traditional notions of a simple, unidirectional message delivery or transfer. They question the claim that there is a clear divide between researchers and practitioners or users, and they contest that the exchange is merely a matter of transferring specialty knowledge to various target groups. They call into question the notion that science is transferred directly to policy with little or no interaction between user and producer groups (Ellis, 2003). What is emerging from this growing literature is a more complex and dynamic view of the activity and actions that are undertaken by those engaged in the science/practice interface. This view emphasizes a 'two-way' process that is shaped by multiple relations and reservoirs of knowledge, and a host of intermediaries and policy brokers. Rather than being a simple linear process, there is, instead, a very complex set of engagements and relationships that develop over time. So instead of needing 'bridges', 'pipelines' or 'highways of connectivity', it may be more appropriate to envision 'complex labyrinths of communication and engagement'. The interactions may more adequately be described as 'spider webs' of connectivity and exchange in which there are nodes and complex linkages, with old actors disappearing and new ones entering (see Chapter 1 in this volume).

Thus, a strong agreement is emerging that replaces the traditional paradigm of linear knowledge transfer to practitioners. It is built upon the aforementioned efforts to assist in negotiating between science and the practice communities. This newer paradigm describes that interface between science and practice as a *complex terrain that is best described as a multilevel system of governance and knowledge production* among a range of actors engaged in understanding and managing environment–society interactions (e.g. Cash and Moser, 2000; Cash et al, 2002a, 2002b; Cash et al, 2006).

Interestingly, when scientists and practitioners begin working together – through whatever type of networks, with or without intermediary boundary-spanning institutions – both the science and the practice change, sometimes in unexpected or unintended ways. For example, practitioners and policy-makers become more than mere recipients of scientific knowledge; they begin to help configure research agendas focusing on vulnerability, adaptation and resilience. Such outcomes can, however, blur the 'traditional' roles of scientist and practitioners, as the producer, user and brokering roles become more fluid and less compartmentalized. Knowledge thus flows in many directions and the distinction between 'pure' and 'applied' or modes I and II science can no longer be clearly made (Gibbons, 1999; Gibbons et al, 1994; Nowotny et al, 2001; Jasanoff, 2003; Nowotny, 2003; Owens, 2005; Moll and Zander, 2006). Consequently, the more traditional modes of scientific accountability (e.g. peer review) also require scrutiny:

> *Unlike the 'pipeline model', in which science [in this case, GEC science focusing on vulnerability, resilience and adaptation] generated by independent research institutions eventually reaches industry and government, Nowotny et al (2001) propose the concept of 'socially robust knowledge' as the solution to problems of conflicts and uncertainty. Contextualization, in their view, is the key to producing science for public ends. Science that draws strength from its socially detached position is too frail to meet the pressures placed upon it by contemporary societies. Instead, they imagine forms of knowledge that would gain robustness from their very embeddedness in society. The problem, of course, is how to institutionalize polycentric, interactive, and multipartite processes of knowledge-making within*

institutions that have worked for decades at keeping expert knowledge away from the vagaries of populism and politics.
(Jasanoff, 2003, p235)

The 'types of architecture' that may be required for effective science/practice engagement may, therefore, involve different types of networks and institutional arrangements that demand detailed understanding (Douglas, 1986; Wegerich, 2001a, 2001b; Karl et al, 2007). As Ellis (2003) neatly summarizes, these networks comprise various types of links that are embedded in wider contexts, including policy communities (Pross, 1986), policy streams (Kingdon, 1984), advocacy coalitions (Sabatier and Jenkins-Smith, 1999) and epistemic communities (Haas, 1991; Haas, 1992).

Depending upon the institutional context, these architectures can be quite stable or in flux over the duration of an interaction or policy process. For example, loss of institutional memory, clarity of roles, influx of new ideas, changes in the political saliency of the topics under investigation, and so on will all influence the quality of interactions, the stability of relationships and the degree of success achieved among those involved. Policy and practitioner brokers can play critical roles as intermediaries in framing policy choices and interpreting assessments for the decision-maker in such complex terrains (Cash and Moser, 2000; Cash, 2001; Farrell and Jäger, 2006; Mitchell et al, 2006). As a result, substantial scientific and practical interest has grown in 'boundary organizations' that can form a communication link and provide information-brokering services between the science and practice worlds (Guston, 2001; Fujimura, 1992; Gieryn, 1999; Miller, 2001; Jacobs et al, 2005; Lemos and Morehouse, 2005; Niederberger, 2005; van Kerkhoff, 2005; van Kerkhoff and Lebel, 2006).

The political context within which the interactions and communication is embedded can also strongly shape the science/practice interface (French and Geldermann, 2005). Researchers are sometimes concerned that this political context challenges the integrity of the scientific process (e.g. through a politically motivated selection of participants, or areas for research included or excluded). Of course, the research exercise itself is a political and institutionalized process shaped by the support for, and production of, research, questions over the initial 'agenda setting' and framing of the problem, and the final negotiation and implementation. In the negotiation of the nature of the problem and in the implementation process, the power relations among actors are often sharply brought into focus (Clay and Schaffer, 1984; van Kerkhoff and Lebel, 2006). Vulnerability to food security provides a good example. The use of interventions designed to reduce food insecurity are made more complex by decisions on the impact of food imports upon local food economies, the types of food used for aid (such as genetically modified foods), and governments' political motivations at the time – for example, the food crisis in the Southern African region (Ellis, 2003; Marsland, 2004). When scientists neglect – even if unintentionally – the political and strategic nature of scientific knowledge and the political context in which it is produced, they can be faced with uncomfortable and challenging situations for whose navigation many are ill equipped. This suggests that education, training and capacity-building in relevant skills for scientists working at the practice interface could prove very useful.

Where the science/practice interaction is not taken seriously or carefully designed, a number of disconnections can emerge that frustrate otherwise well-meaning measures to reduce vulnerability and enhance resilience: the scientific output is more likely to be mismatched to user requirements (i.e. not what practitioners need); it may not be delivered in time or in appropriate formats; those interacting do not communicate well; scientists feel their credibility is negatively affected by collaborating with practitioners; stakeholders do not feel their legitimate concerns are addressed; and so on. Thus, although there is a growing body of knowledge on vulnerability, adaptation and resilience and a variety of pressing application opportunities for that knowledge, all too often silos of knowledge are produced that fail to help make systems and communities more robust to extremes and to change.

Communication is the means and, indeed, the very foundation for engagement between the worlds of science and practice. It is also, sadly, often relegated to an afterthought of scientific practice. At other times scientists dismiss this communication and dialogue role as not their job. It is thus entirely possible that entire volumes of potentially valuable knowledge – such as the Intergovernmental Panel on Climate Change's assessment reports that focus on vulnerability and adaptation to climate change – can remain largely untapped by the practitioner community. The next section describes some of the challenges and opportunities involved in effective communication between scientists and practitioners in more detail.

Communication links at the science/practice interface: Challenges and opportunities

Communication plays a central role in any effort to improve the science/practice interface. Virtually everything in the interaction between these two worlds comes down to *what is communicated, how and when, through what channels, in what forums, for what purpose, by whom and for whom*. More fundamentally, communication touches on *the relationship between those engaged in the dialogue and their level of mutual understanding of (institutional) cultures, (professional) codes of conduct, modes of operation, information needs, decision contexts, including pressures, constraints, capacities* and so on.[2] While communication typically does not easily or directly translate information into policy or action, we argue that science has little chance to enter into decision-making or inform action at all when communication is poor or non-existent.

Our discussion here focuses on communication between science and two different practice communities, both relevant to, and interdependent in, the reduction of vulnerability and enhancement of adaptive capacity: the policy- and decision-making communities, on the one hand, and the broader public, on the other. Both are rather amorphous and not necessarily clearly separated from the world of science. To illustrate our arguments, we use examples of actions and policy-related communication that affect the three dimensions of adaptations commonly distinguished: those that reduce a person's or system's sensitivity to risk, alter exposure, and/or increase the resilience or coping capacity (e.g. Adger et al, 2005).

Science/practitioner communication in the context of policy-making and management

In the first context of policy- and decision-making practitioners,[3] science can play a number of roles (see Figure 4.1), ranging from assisting in problem identification and definition, to aiding in the search for and framing the response options and solutions, to the implementation and finally the evaluation of policy or management options (Lemmons and Brown, 1995; Moser, 2004, Moser, forthcoming). As Figure 4.1 suggests, we conceive of the policy- or decision-making process as cyclical, iterative and ongoing; thus, scientific input can occur at any or all stages, and to be most effective should equally be ongoing, even if the type of impact differs from stage to stage.[4]

Importantly, the temporal and spatial scales of the research and policy processes are not the same. Thus, effective input of science in policy could be

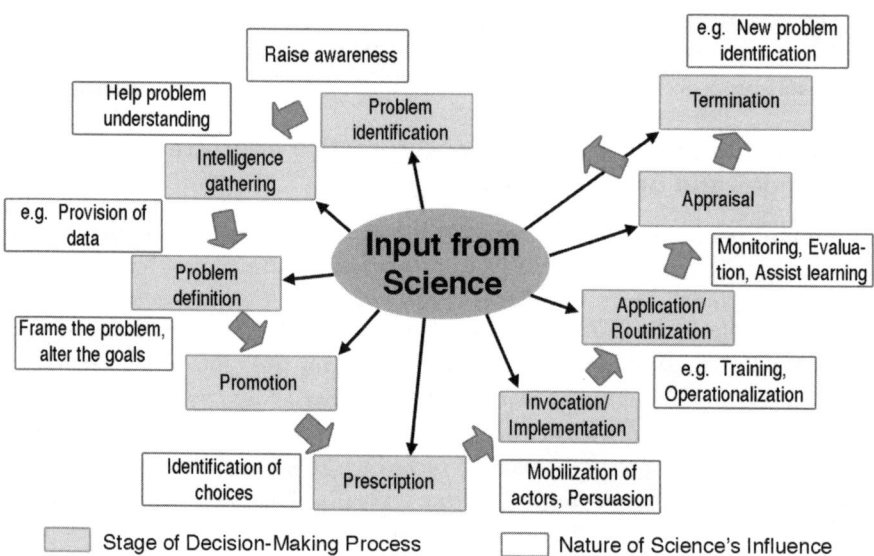

Figure 4.1 *Scientific input at various stages of the decision-making process and the nature of science's influence*

Source: adapted from Clark (2002) using insights from Mitchell et al (2006)

enhanced by taking into account the differences in temporal and spatial scales; different moments in time require different types of knowledge and different modes of communication (IAI, 2005). Each stage in the decision-making process depicted in Figure 4.1 has its own requirements in terms of how and when data or analyses are communicated, at what level of detail, and from whom to whom. Each stage also requires some degree of negotiation between the two sides. For example, when a water resources manager not yet cognizant of climate change first becomes aware of it, he or she may require some basic education. Communication at that point would need to illustrate at a relatively general level how climate change relates to water resources and why it is important to take climate change into account in water management (see Miller and Yates, 2005). Beyond the initial stage of getting someone's attention, the communication requirements become more specific. As that same water resources manager decides about supply and/or flood management at very specific times in the water year calendar, scientific information two weeks after the decision point – even if very interesting, regionally specific and relevant – is no longer helpful (Jones et al, 1999; Pulwarty and Melis, 2001; Pulwarty, 2003; Miles et al, 2006). If valuable information is offered in ways that do not easily integrate within decision-making procedures or are difficult to understand and interpret, managers may decide to ignore it (Hall and Paradice, 2005; Morss et al, 2005). Similarly, if relevant management information or policy choices are framed in ways that simply cannot garner political and public support (e.g. because the frame does not mobilize concern, does not suggest a sense of urgency or goes counter to deeply held values of the decision-makers), they may well fail to enter the policy debate or decision-making process (Gerhard, 1994; Pielke, 1997; Ogunseitan, 2000, 2003; Schreurs et al, 2001).

These problems of appropriately matching scientific information with decision situations and needs may seem trivial; yet they are all too often ignored (McNie and Elizabeth, 2007; Sarewitz and Pielke, 2007). Among the best strategies to match better scientific information with practitioners' actual information needs is to find out what a practitioner *does* exactly and what decisions are pending (rather than asking what kind of information he or she may *want* or *need*) (Altalo, 2005). In the case of a detailed analysis of vulnerability or adaptive capacity, for example, it helps little to offer the results to a local decision-maker to inform her actions if the results suggest intervention at higher levels of governance (Demeritt and Langdon, 2004).[5] Of course, many such problems can be avoided or minimized if scientists or information providers (e.g. extension agents) and decision-makers can build trustful relationships and mutual understanding of capacities and needs over time (Earle and Cvetkovich, 1995; Cash, 2001; Cash et al, 2002a). Typically, such trust will grow over time – with those involved characteristically having benevolent attitudes and plenty of social skills and experience – in a jargon-free, non-condescending communication environment, as both sides increase their mutual understanding of each others' language, institutional culture, the standards and expectations of professional conduct, and the larger decision context. This element of trust-building has been key in the Southern African example discussed below, as well as in many of the cases described in this volume, where various members of consortia working on vulnerability assessments have grown to trust each other and have begun shaping problem formulation and assessment methods in a collaborative spirit. Clearly, there and in other contexts, trust is not an expendable ingredient, nor one that can be handled lightly. As numerous studies in the risk literature have shown, trust can very easily be lost and, when that happens, is difficult to rebuild (Slovic, 1993; Cvetkovich et al, 2002; Poortinga and Pidgeon, 2004). Thus, great care needs to be given to this aspect of the interaction.

Challenges in communication across the science/policy/practice interface are also common. First, building effective communications is time consuming, and much of the necessary relationship-building that makes such communication effective occurs in the background, is not glamorous, is certainly not directly rewarded, and at times is penalized in some scientific circles (Chappell and Hartz, 1998; Jacobson et al, 2004). This lengthy process is exacerbated by the fact that scientists tend to lack specific training in non-scientific communication, while practitioners or media representatives frequently lack certain technical expertise (Gregory and Miller, 1998; Weigold, 2001; Kyvik, 2005). Even when 'language' barriers can be breached, scientists and practitioners have precious few opportunities for regular face-to-face interaction that can facilitate trust-building (Dabelko, 2005).

Second, while many in the scientific community hope that their research is relevant and some conduct research that is use inspired (Stokes, 1997), traditional 'applied science' tends to have less prestige and rewards than 'pure' curiosity-driven basic science (Moll and Zander, 2006). Third, achieving the balance between credibility and salience is not a trivial undertaking, as decision-makers frequently have high expectations as to how soon decision-specific information can be made available. Meanwhile, many scientists want to err on the side of caution and uncertainty by first vetting their findings in the peer review process (Van der Vink, 1997; Blockstein, 2002; Cash et al, 2002a). Fourth, scientists and practitioners can have very different notions of what constitutes 'legitimate' knowledge. Scientists frequently assume that knowledge that has emerged from a rigorous process of data-gathering, hypothesis-testing, empirical or model verification, and peer review is the 'truth' (or at least a superior truth) because of the 'expert' nature of scientific knowledge and, therefore, ready for transfer to and use by end users. To practitioners, legitimacy may be derived from considering and addressing key stakeholders' values and concerns and inclusion of non-scientific knowledges.[6] These differences need to be addressed carefully so as not to undermine successful science/practice interactions. And, finally, many scientists strongly object to, or at least resist, closeness to the world of politics for professional and ethical reasons (McNeely, 1999; Brooks, 2001; Freyfogle and Newton, 2002).

Taking these reasons together, it is not surprising that only a small percentage of scientists do the yeoman's share of communicating to non-scientific practitioners in the policy/management world and the public directly (Wellcome Trust, 2000; Jensen, 2005). Institutional and attitudinal changes are required to improve this situation, and in some cases (especially in large projects where funds are sufficient), it may be advisable to have a communications specialist involved (requiring dedicated funding for this responsibility). In addition, more social science research is needed to measure the effectiveness and outcomes of direct science/practitioner and boundary organization-mediated communication.

Communication with the lay public

Policy- and decision-makers in the public and private sectors are clearly important players in any effort to reduce vulnerability or increase a community's adaptive capacity and resilience to the synergistic impacts of environmental and socio-economic global changes (O'Brien and Leichenko, 2000; O'Brien et al, 2004). But such policy actions will not occur, nor will the implementation take place, without public support or consent, and in many cases active participation by individuals and communities. This requires at least some understanding of the issues at hand, if, perhaps, not deep knowledge. Moreover, individual action is fostered and/or constrained by the policies implemented at higher levels and also influenced by communication coming from governmental and scientific sources (or other mediating channels), as we highlight in the cases from Southern Africa below. In fact, communication with non-scientists can be viewed as a form of adaptive capacity-building that enhances resiliency to the impacts of climate change (Adger et al, 2005; Moser and Luganda, 2006). It is thus critical to also examine the role of communication in the context of this second cohort of the practice community with the lay public (Moser, 2004).

A vast body of literature, developed in numerous disciplines and interdisciplinary fields, has contributed to our understanding of how the lay public does and does not pick up, process, understand and act on information about environmental risks (Slovic, 2000; Dietz and Stern, 2002; Cox, 2006), how their perceptions and mental models are shaped and change, and what effect certain forms of communication have on behaviour.[7] Much of this communication between science and the public is mediated by the news media, and as such is subject to the political–economic imperatives of the media enterprise and the media's filter on 'newsworthiness' (e.g. stories that have drama, emotional appeal, novelty and human interest). Many scientists are not accustomed or trained to present their findings in ways that appeal to reporters, and, in fact, resist having their work 'sensationalized' to fit the media's (the public's) taste.

On the other hand, insights from communication studies should be of significant interest to those who grapple with how to alert, inform, motivate, mobilize and support effectively public action on global change risks. Many scientists are increasingly frustrated that these risks (which are difficult to perceive, understand and – for now – directly experience) have not generated a sufficient sense of urgency among lay publics (Biodiversity Project, 1998; Moser and Dilling, 2004). Classic examples include climate change and the loss of biodiversity (and related ecosystem services), both of which are highly likely – for the vast majority of the world's population – to increase vulnerability and reduce adaptive capacity (Millennium Ecosystem Assessment, 2005; IPCC, 2007).

While the public tends to perceive such global change problems as real, under way and sufficiently established by science, many people are still unclear about causes and solutions, and – generally – do not view the issues as immediate, urgent or amenable to personal or collective action (Immerwahr, 1999). In the case of natural disasters, lay people frequently view such events as 'acts of God' and thus also not amenable to personal or policy intervention. Once an immediate threat is averted or past and 'normal life' re-established, both the periodic extremes and the gradual long-term changes get placed far below the more pressing daily concerns of food, jobs, safety, healthcare and education.

The implications of these findings are tremendous: if individuals are to be involved in mitigating and adapting to climate and other global changes, the problems need to be meaningful and relevant; people need help to understand both causes and solutions; communicators must – despite uncertainty – create a sense of appropriate urgency (but not irrational fear); and they must enable and empower people to act in sustainable ways and support relevant public policy (Uzzell, 2004; Vlek and Steg, 2004). As Stone (1989, p281) put it: 'difficult conditions become problems only when people come to see them as amenable to human action'. Importantly, scientists and other communicators must avoid (or abandon) the assumption that better information and understanding alone will lead to such environment-friendly behaviour or policy support (McKenzie-Mohr and Smith, 1999; Gardner and Stern, 2002; National Research Council, 2002). A clear understanding of habits, barriers, identities and pragmatic support needs, as well as institutions, laws and social norms that constrain or facilitate certain behaviours, must accompany any effort that seeks to use communication in support of individual and collective behaviour change (Moser and Dilling, 2007).

Rather than having experts convey specialized knowledge to a non-expert audience, and focusing on how to best 'get the message across', the alternative communication model is considerably more 'democratic', mutual and interactive (Communication and Cognition, 1996; Servaes et al, 1996; Mcleod et al, 1999). Borrowing from the notion and practice of 'participatory communication' (Rockefeller Foundation, 1997; Gray-Felder, 1999; Figueroa et al, 2002), this model suggests that experts and lay individuals are equally involved in a dialogue over challenging issues, together defining problems and solutions, not aiming at persuasion but empowerment, and not at individual behaviour change but shifts in social norms, policies and culture (Figueroa et al, 2002).

Very similar arguments are detailed in the US National Conference on Risk Communication of 1989 (see National Research Council, 1989) and by other risk scholars (Fischhoff, 1989, 1995).

It is difficult to see how vulnerability can be reduced and adaptive capacity increased without such active involvement of those most directly concerned. As the example from Southern Africa below shows, until dialogue and a greater understanding of the context in which policy, science, practitioners and other stakeholders become part of vulnerability assessment, useful interventions can be delayed and the pace of the much needed intervention retarded.

Various tools or methodologies are currently being used as 'communication entry points' into a wider dialogue between science and practice on vulnerabilities to changes in the Southern African region. The case study on Southern Africa is illustrative as it shows the evolution of complex interactions between science and the world of policy and practice and demonstrates how knowledge production develops through co-production. Initially configured around rather simple modes of communication (e.g. roundtables and workshops), a number of rather complex interactions, consequences, responses and outcomes have resulted that, in turn, have begun to drive, shape and reconfigure the engagement process. Institutional alliances have become challenged, modes of knowledge production have moved from simple institutional interactions to more polycentric modes of interaction (Jasanoff, 2003; Cash et al, 2006), and a range of interesting cleavages in institutional arrangements has emerged as new questions and challenges to traditional and acceptable methods are being revealed.

Multiple players, multiple knowledge systems: An example from Southern Africa

Several vulnerability assessments have recently been undertaken in the Southern African region to inform current policy and humanitarian interventions in the region.[8] Achievement of the Millennium Development Goals, a major set of development targets agreed on by all member states of the United Nations in 2000 at the Millennium Summit, is core to livelihood and ecosystem 'health' in Southern Africa. Southern Africa is one of the only regions in the world, however, facing chronic, recurrent food insecurity and persistent threats of famine (Devereux, 2000; Devereux and Maxwell, 2001), and one where the number of extreme poor has arguably risen during the last few years compared with other regions of the world (e.g. Asia), where it has decreased (Sachs, 2005). Southern Africa is also a region that clearly illustrates how multiple users and producers of knowledge have been brought together to address a persistent vulnerability crisis in the region characterized by repeated humanitarian appeals for food and other resources. Different 'spaces' for knowledge interaction on vulnerability, adaptation and resilience have been created (including expert and lay knowledge interaction, and universal, contextual, technical and cultural interactions) through the formation of vulnerability assessment committees (VACs) and a regional vulnerability assessment committee (Maunder, 2005; Owens, 2005; Maunder and Wiggins, 2007). The 'players', in this case, were drawn together by the 'food and humanitarian crisis', heightened in 2001, but still persisting in the region. Such interactions, while successful in some instances, as highlighted below, have also raised some frustrations: inability to fully comprehend the dimensions of the 'problem', including the multiple causes of the crisis; inability to fully capture this multiple causation in shaping the crisis; and failure to come to an understanding of the political context in which the evidence to inform interventions that would reduce vulnerability was, and still is, currently embedded. The quality of the derived evidence to inform interventions in the region and the limited engagement of civil society, including the non-governmental and civil society organizations and sectors, are factors as well (Tschirley et al, 2004; Drimie and Misselhorn, 2005; Wiggins, 2005).

In the case at hand, we can begin to analyse the role of interactions around applied and conceptual knowledge production on vulnerability, adaptation and resilience, and assess how that knowledge was used in practical applications and also, in turn, fed back into knowledge production. The analysis of this case study begins to answer the call made by Owens (2005, p290) for 'more research into and at the "boundary", where a number of issues ... are under-theorized, and [where] there is considerable scope for careful empirical research in a variety of policy contexts'. By positioning such knowledge into the context of the humanitarian crisis in Southern Africa, some degree of practical robustness in our understanding of vulnerability and adaptation is gained; but as we show below, the emerging problems of how to institutionalize and sustain such interactions, which Jasanoff (2003, p235) calls 'polycentric, interactive, and multiple processes of knowledge making' away from the vagaries of populism and politics, we suggest remain central challenges in the science/practice interaction in the region. Such challenges in this case can, and often do, frustrate interventions that are ultimately designed to increase robustness and resilience in highly vulnerable communities.

Some background to the Southern African case

Chronic and persistent vulnerability prevails in Southern Africa. During 2002 to 2003, for example, the severe drought in Southern Africa contributed to food shortages for an estimated 14 million people (House of Commons, 2002–2003). This chronic situation is the product of a number of factors besides climate variability, including the growing spread of human immunodeficiency virus/acquired immune deficiency syndrome (HIV/AIDS), weakened and eroded social safety nets, weakened capacity and poor governance (Boudreau and Holleman, 2002; De Waal and Whiteside, 2003; Benson and Clay, 2004; Drimie, 2004; Marsland, 2004). A situation of below-normal rainfall for two to three agricultural seasons aggravated conditions in many parts of the region. As a result of this growing crisis, the United Nations issued an appeal for US$611 million to address the crisis in the Southern African Development Community region (Lesotho, Malawi, Swaziland, Zambia, Zimbabwe and Mozambique) in July 2002 (SARPN, 2004; House of Commons, 2002–2003). Many thus call the vulnerability crisis a chronic situation, triggered by droughts and food insecurity but aggravated by other 'livelihood' factors as mentioned above (e.g. Maunder, 2005; Maunder and Wiggins, 2007).

Prompted by this growing crisis, the Kariba High-Level Vulnerability Assessment Technical Consultation brought together over 100 technical early-warning and food-security professionals from the region in September 2000 (Mock, 2005a; Jackson et al, 2006) in order to find improved and, where possible, integrated ways of targeting interventions and reducing vulnerability in the region. The meeting highlighted a number of scientific approaches to livelihood and vulnerability assessments. The Food, Agriculture and Natural Resources Directorate (FANR) of the Southern African Development Community (SADC) established a number of multiagency VACs. Their mandate was to use vulnerability assessments for the purposes of food security planning (Jere, 2005b). The importance of these vulnerability assessments to inform better interventions for the region cannot be overstated and they are receiving heightened attention from the SADC, local governments and various institutions, donor agencies and humanitarian organizations worldwide – for example, the Food and Agricultural Organization (FAO), the Food Insecurity and Vulnerability Information and Mapping System (FIVIMS), the Famine and Early Warning System Network (FEWSNET), the World Food Programme (WFP) and the UK Department for International Development (DFID).

Successes, problems and 'food for further thought'

The detailed vulnerability assessments undertaken by the VACs and Regional Vulnerability Assessment Committee (RVAC) throughout the SADC region (in August 2002, December 2002, May/June 2003 and May 2004, and currently ongoing) highlight several successes in the science/practice interaction. Recent assessment of the VACs concluded that a number of positive attributes have been associated with their activities (Marsland, 2004; Tschirley et al, 2004; Maunder, 2005; Tango International Inc, 2005a, 2005b). They have, for example, enabled each of the countries in which they have been established to create a forum for all relevant stakeholders to come together and learn more about and to better understand vulnerability issues; to provide a key information source for the humanitarian assistance community to respond to complex emergencies; and to create the opportunity to influence policies related to emergency and poverty responses.

What is clear from such science/practice interactions has been the enhanced understanding of the actual crisis in the region. This understanding has been 'informed by the context' in which the science/policy/practice engagement has occurred (what Gibbons, Nowtony and others refer to as the 'embeddedness' and 'social contextualization' of the interactions). This enhanced understanding of the problem at hand has been the product of multiple knowledge interactions – for example, scientists and consultants discovering the inadequacy of some vulnerability methods and frameworks used to address the multiple complexities of the various vulnerability contexts in the region. Through these interactions, participants uncovered the multidimensional, complex nature of food insecurity and vulnerability in the region – not easily addressed by the simpler intervention models used by practitioners or common in the scientific literature.

Most of the VACs were initially focused on collecting information and data to provide a deeper understanding of food insecurity, prompted by the apparent food crisis and emergency of 2002. It was soon realized, however, that the humanitarian crisis was, and is, embedded in a socio-economic context that includes the role of macro-economic failures in the region dating back to the 1970s, the liberalization of domestic markets, and the role that HIV/AIDS plays as it intersects and interacts with food crises (SADC–FANR–RVAC, 2003; Marsland, 2004; Maunder, 2005; SADC-RVAC, 2005; Wiggins, 2005). Today the focus, in most instances, of vulnerability assessments is on a more nuanced view of vulnerability and the requirements for efforts to build sustainable resilience.

One of the institutional advantages of the consultative process encouraged by the VACs was the ability to include the lessons and insights emerging from the field into the planning and thinking of the design of vulnerability assessments to better capture the unfolding situation in the field. In the December 2002 round of assessments, various countries began to adapt the VACs' original vulnerability assessment approach to better fit their own local realities. Various VACs (e.g. in Mozambique) moved towards a 'multi-sectoral' questionnaire using the rapid rural appraisal technique. The focus was still on food security but included additional analyses on issues related to water and sanitation, HIV/AIDS, and agricultural prospects. Developed at the local level, the assessments departed from earlier notions of the vulnerability assessment by using information for longer-term planning and decision-making that was seen as a priority and relevant by local stakeholders (for more details, see Marsland, 2004). The role of stakeholder engagement and consultation thus became key to 'informing' the types of vulnerability information that was collected.

Likewise, in the Zimbabwean vulnerability assessment case (Tango International Inc, 2005a), it was realized early on that a focus on food balance sheets and food production was not really capturing the vulnerability that many in the country were and are experiencing (Tango International Inc, 2005a). In the early round of several assessments, the goal was to identify 'food gaps' at national and household levels to better inform food interventions, including a series of 'snapshots' of data on food shortages from field assessments. As with many other cases and country examples, the stresses and risks were driven by multiple causes, and as information continued to be exchanged and methodologies debated, the focus of the vulnerability assessments began to shift (Marsland, 2004). In this case the VAC process enabled the switch in focus based on the consensus of key stakeholders and motivated by the stakeholders' interest in improving the breadth of VAC data (Tango International Inc, 2005a). The active engagement and interaction of knowledge brokers and stakeholders turned into a particular strength in the vulnerability assessment process: 'the greatest accomplishment of the VAC was to bring the stakeholder community together to begin the process of "harmonizing" statistical estimates of food aid and discovery of the nature and causes of food insecurity' (Mock, 2005a, p12). In these rounds of interactions, not only was knowledge about vulnerability being applied, but participants brought about and shared fundamental breakthroughs in the conceptualization of vulnerability and resilience to the causes of change in the region.

Certain weaknesses have also become apparent, however, in the science/practice engagements around vulnerability in the region. A variety of stakeholders have been involved in the VAC processes. The urgency of the persistent 'food' crisis has meant that those involved are usually drawn from a range of groups, including government staff, non-governmental representatives and scientists. Several scientists in the region are working with various non-governmental organizations (NGOs) and humanitarian agencies as fieldworkers or consultants to the various organizations. In Southern Africa the mode of operation involves building new 'institutional designs' or 'architectures', such as consortia where scientists, practitioners and other stakeholders all bring together their respective knowledge systems to assist in assessments, to better understand and manage food insecurity – for example, the Food Insecurity and Vulnerability Information Management System for South Africa (FIVIMS-ZA), the Regional Hunger and Vulnerability Programme (RHVP) and the VACs. Such consortia have enabled information to feed into the vulnerability assessment process, improved the use of methods and data, and helped to effectively target the outcomes to the most sensible points of intervention. The legitimacy and future sustainability of such institutional arrangements have also been brought under the spotlight.

The production and uses of knowledge via these modes of interaction thus revealed a number of problems. Recent critiques (e.g. Jackson et al, 2006, p9) of the vulnerability assessments list the following challenges, many echoing the issues raised by Jasanoff (2003) and Nowotny (2003) and those discussed above on communication flows:

- Relevant policies are poorly understood by stakeholders, which then limits their use in very applied contexts of heightened vulnerability, reduced resilience, and poor adaptation;
- Definitions and perceptions of vulnerability vary within the region.
- Assessments are critically dependent upon the flow of data between information providers and users.
- While VACs are considered to be among the few forums that coordinate information on vulnerability that channel recommendations to govern-

ments and collaborating partners, they have not been fully institutionalized. VACs, therefore, exist as informal committees with unclear institutional roles and responsibilities and suffer from *ad hoc* financing arrangements (adapted from Jackson et al, 2006). For more discussion, see Jackson et al (2006, p9) and other relevant publications on the RHVP website, www.wahenga.net.

While stakeholder engagement at the science/practice interface was useful in some aspects of the process, the downside of such an approach, argued by some, is the 'creation' of a rather loose network as opposed to a formal 'institutionalized process'. While many of the VACs were chaired by government (e.g. the Ministry of Agriculture), the operational side of the VAC was characterized more by a 'loose alliance of interested organizations' (Jere, 2005b, p15). The somewhat loose relation between stakeholders and various agencies has meant that, in some cases, tensions have arisen between 'information sharing' and a devolution of tasks between partners. The lack of 'leadership' may also have led to the generation of 'false consensus' and a spirit that discouraged dissent (Darcy and Hofmann, 2003). In some assessments of the process, a common frustration with such a loose arrangement was revealed in most countries: 'The Zimbabwe Vulnerability Assessment Committee (ZimVAC) needs to find an institutional home within the government structure' (Tango International Inc, 2005b, p19). The absence of an institutional frame was seen by many as impeding the progress of the VACs. At the local level, as is the case in Mozambique, a more harmonized assessment process with activities of FEWSNET and FIVIMS activities, housed in the facilities of the Technical Secretariat for Food Security and Nutrition (SETSAN), illustrated, for example, how a coordinated vulnerability assessment process may be more sustainable. In the wider regional context, however, the science/practice interaction around the VAC experience demonstrated the fractures that exist, and remain, in region-wide coordination (Drimie and Misselhorn, 2005).

How does one, then, design a forum or create 'spaces' for such interaction and coordination? One of the findings from much of the interaction in the region is that 'there is currently no formal forum for civil society to engage with SADC at the regional level' on issues of food security (see, for example, results from a workshop presented by Drimie and Misselhorn, 2005, p6). One possible arrangement to overcome these tensions and difficulties is through 'boundary organizations' (Cash, 2001; Guston, 2001; Miller, 2001). In Southern Africa, for example, the UN established a form of a 'boundary organization' in the Regional Inter-Agency Coordination Support Office (RIACSO), a light model of coordination both within and between country coordination (i.e. between those focusing on food security and those focusing more on health issues and the implementation of the emergency response) (Darcy and Hofmann, 2003). Such an organization may also help to overcome problems of a lack of 'trust' by groups working on behalf of certain donors, and may ensure credibility of evidence where various sources of information are gathered. Assessments of several VAC activities also point to continued attempts at integration that all involve a dialogue and engagement across user and producer groups of knowledge:

> *More attention needs to be given to district-level audiences and ideally results should be tied to administrative and programming units. Whenever possible, there should be an* interactive analysis that brings together district councils, local government authorities and agencies *such as district and area executive committees.* (Jackson et al, 2006, p53, emphasis added)

In summary, the vulnerability assessments in Southern Africa provide an example of the science/practice interaction 'in action' that is designed to reduce vulnerability to change and climate-driven extreme events in the region. They show both some of the benefits and challenges involved in the attempt to bring together multiple sources of knowledge rooted in the science and action domains. Despite some successes, a number of issues persist that require further examination, particularly if such efforts are to be made more sustainable and not seen as 'one-off' engagements to support relief efforts. One is the tension between developing consensus on the methodologies used by a range of stakeholders across a wide region (posing particular challenges for comparability and regional integration) when their applicability in local contexts demands idiosyncratic adjustments. Such debate is, however, ultimately healthy and may lead to better methodologies and framings of the problems in the region. Another issue is the slowness of the delivery of products that reflects not only a lack of data and need for scientific credibility, but also the time-consuming process of coming to negotiated understanding in science/practice interactions and the need to clarify the role of 'external' agencies, stakeholders and scientists at the outset of the process. Finally, the absence of some form of 'organizational base' or institutional 'frame', as well as clear 'rules of engagement' at the outset of such information exchanges, research activities and interactions continue to frustrate many operating the VACs, and can inhibit the building of longstanding trusted relationships. Such problems have prompted some (e.g. Maunder, 2005, p12) to argue that as the next round of vulnerability assessments proceeds through the RHVP:

> *Major challenges [remain] to the future of the VACs. The current analysis is failing to deepen the understanding of the causes of food security. The overlapping crises remain largely confounded. Consequently, responses continue to address symptoms rather than causes.*

Conclusions

Linking science to practice, as we try to show in this chapter, is not a simple task. It involves a variety of possible pathways and players, but always depends upon a spirit of partnership and, perhaps, a convergence of interests. The responsibility for making this linkage work is by necessity mutual; it rests on the shoulders of those in both the scientific and practice communities. As Ralph Cicerone, president of the US National Academy of Sciences, has stated so aptly: 'Science must be useful and science must be used' (Cicerone, 2005). While we focus in this chapter on the perspectives with which we are most familiar – those from the science side – there are misunderstandings and misconceptions about science, practice and the connection between the two on both sides. Many practical and logistical reasons exist for the observed failures and challenges and resistance to making that connection, as the case of Southern Africa (as well as other cases in this volume) demonstrates.

Our examination of 'resistance' in this chapter leads us to identify a number of areas for further exploration in future deliberations and research. They are as follows. Our current metaphors for characterizing science/practice interactions are sometimes faulty, sometimes simply insufficient to reflect the complexity of interactions. Discourse in this arena typically evokes the image of 'bridges', 'highways' or 'pipelines' of connectivity between science and practice. In reality, there is almost never a clearly defined route or connection. Rather, something like 'spider webs', composed of nodes and a multitude of ephemeral linkages (as elaborated upon in Chapter 1 of this volume), lie between science and practice. Frequently, policy brokers and intermediaries traverse these nets, shaping notions of the science that is needed for policy decisions, or of the societal interventions required to reduce vulnerability and build resilience, and helping to form coalitions of participants required to achieve the reservoirs of support to make things happen. Our language needs to better reflect the dynamics that actually operate.

In these nets, boundary organizations may play a particularly important role, though they may not represent the only institutionalized or less formal processes of working at the science/policy interface. Boundary organizations can provide communications and brokerage services and 'signal' systems that alert and shape the perceptions of scientists, practitioners and interested publics. Despite the interest in such organizations, we still know relatively little about where and how they may be important, and who the relevant players are in the spider webs of connections.

Communication breakdowns, it is clear, are a central barrier to better coordination and integration between science and practice. Addressing these communication breakdowns touches on a host of issues: how do we best match scientific information with decision needs? How can the time-consuming and high resource needs required for an informed public be met? What is 'legitimate' knowledge for needed decisions and where should that knowledge come from? How may it be validated? How can the incentives be refocused and the communication and outreach capacities of scientists be increased to meet the growing information and knowledge resource demands from practitioners? Over the past several decades, as societies have struggled with these issues, an alternative communication model has emerged from a range of fields to replace the traditional, linear one-way notions that have previously underlain the science/practice interaction. It is a more democratic model of communication in which different experts, risk-bearers and local communities all have something to bring to the table. But we have much to learn as to how to do this effectively.

Science – as a social institution – has always been in a position to play a potentially significant role in detecting and defining global environmental problems, framing and shaping the public and policy debates around them, helping to identify socially and ecologically appropriate solutions, and informing the social learning process. It is for this reason that we see an important emerging possibility for the scientific and practitioner communities to engage, when required, on a number of themes. As this chapter has shown, we see important emerging opportunities for practical engagement with the wider community, particularly in the field of global environmental change, to work more effectively with those who would enhance the science on such themes, and develop and implement policies that could reduce vulnerability, increase adaptive capacity, and build the resilience of people and the environment in the face of global change.

Acknowledgements

Some of the contents of this chapter were first discussed at a workshop organized by the International Human Dimensions Programme (IHDP) at a workshop in Arizona in February 2005 on the state of the science of vulnerability, adaptation and resilience in global environmental change. Several ideas were captured and we acknowledge the useful inputs of the members of the group, including Emilio Moran, Aromar Ravi, Daniel Sarewitz, Michael Schoon, Barry Smit and Maureen Woodrow. We also appreciate comments and inputs from Scott Drimie and Alison Misselhorn on earlier drafts of this chapter, and the careful editorial review by Mimi Berberian. The opinions expressed and any remaining oversights are our own.

Notes

1   This chapter is adapted from Vogel, C., Moser, S. C., Kasperson, R. E. and Dabelko, G. D. (2007) 'Linking vulnerability, adaptation and resilience science to decisions: Pathways, players and partnerships', *Global Environmental Change*, vol 17, no 304, pp349–364.
2   This broad conceptualization differs from the fairly common notion that communication simply means the conveyance of technical information.

3 For the purposes of this very truncated discussion, we make no explicit distinction between policy-makers in the public realm and decision-makers and (resource) managers in the private sectors. At the same time, we recognize that the types of decisions they make differ, and scientists may feel differently about interacting with one group versus the other.
4 Science is depicted here as if it were the central and only input into the decision-making process, which, of course, is not the case. We simply focus here on the relevant interaction between science and decision-making, fully aware that there are other, and frequently competing, forms of information and input into that process.
5 Cross-scale challenges in the science/practice interaction are treated in-depth in Cash and Moser (2000), Wilbanks (2002), Adger et al (2005) and Cash et al (2006).
6 Differently 'constructed' knowledges, contestations between 'scientific/expert' and local or indigenous knowledge (itself highly contested) have been a subject of interest for many years and are growing in importance as many different knowledge holders seek to engage in the science/practice interface.
7 The literature is too vast to cite even just a representative sample here. Readers are directed to studies in environmental sociology; cognitive, environmental, eco- and political psychology; behavioural geography; behavioural economics and economic psychology; (mass) communication and media studies; (social) marketing; and various interdisciplinary fields, such as studies of risk perception and communication, or that of the public understanding of science.
8 Much of the discussion that follows is based on a number of reviews (e.g. Darcy and Hofmann, 2003; Jere, 2005a, 2005b; Maunder, 2005; Mock, 2005a, 2005b; SADC, 2005; Tango International Inc, 2005a, 2005b; Tschirley et al, 2004) of the vulnerability assessment process that was commissioned by the Southern African Development Community–Food, Agriculture and Natural Resources–Regional Vulnerability Assessment Committee (SADC–FANR–RVAC).

References

Adger, W. N. (1999) 'Social vulnerability to climate change and extremes in coastal Vietnam', *World Development*, vol 27, no 2, pp249–269

Adger, W. N. (2003) 'Social capital, collective action, and adaptation to climate change', *Economic Geography*, vol 79, no 4, pp387–404

Adger, W. N., Huq, S., Brown, K., Conway, D. and Hulme, M. (2003) 'Adapting to climate change in the developing world', *Progress in Development Studies*, vol 3, no 3, pp179–195

Adger, W. N., Arnell, N. W. and Tompkins, E. L. (2005) 'Successful adaptation to climate change across scales', *Global Environmental Change*, vol 15, pp77–86

Altalo, M. (2005) 'Don't ask me what I want, ask me what I do', Paper presented at the US Climate Change Science Program Workshop, Climate Science in Support of Decision Making, 14–16 November, Arlington, VA

Alwang, J., Siegel, P. and Jorgenson, S. (2001) *Vulnerability: A View From Different Disciplines*, Social Protection Discussion Paper Series No 115, World Bank Social Protection Unit, Washington, DC

Bankoff, G., Frerks, G. and Hilhorst, D. (2004) *Mapping Vulnerability: Disasters, Development and People*, Earthscan, London

Benson, C. and Clay, E. (2004) *Understanding the Economic and Financial Impacts of Natural Disasters*, Disaster Risk Management Series No 4, World Bank, Washington, DC

Berkes, F., Colding, J. and Folke, C. (eds) (2003) *Navigating Social Ecological Systems: Building Resilience for Complexity and Change*, Cambridge University Press, New York, NY

Biodiversity Project (1998) *Engaging the Public on Biodiversity: A Road Map for Education and Communication Strategies*, The Biodiversity Project, Madison, WI

Blockstein, D. E. (2002) 'How to lose your political virginity while keeping your scientific credibility', *BioScience*, vol 52, no 1, pp91–96

Bohle, H., Downing, T. E. and Watts, M. (1994) 'Climate change and social vulnerability: The sociology and geography of food security', *Global Environmental Change*, vol 4, no, pp37–48

Boudreau, T. and Holleman, C. (2002) 'Household food security and HIV/AIDS: Exploring the linkages', The Food Economy Group, www.fews.net/resources/gcontent/pdf/1000087.pdf

Brooks, H. (2001) 'Autonomous science and socially responsive science: A search for resolution', *Annual Review and the Energy Environment*, vol 26, pp29–48

Brooks, N., Adger, W. N. and Kelly, M. K. (2005) 'The determinants of vulnerability and adaptive capacity at the national level and the implications for adaptation', *Global Environmental Change*, vol 15, pp151–163

Cash, D. W. (2001) 'In order to aid in diffusing useful and practical information: Agricultural extension and boundary organizations', *Science, Technology & Human Values*, vol 26, pp431–453

Cash, D. W. and Moser, S. C. (2000) 'Linking local and global scales: Designing dynamic assessment and management processes', *Global Environmental Change*, vol 10, pp109–120

Cash, D. W., Clark, W. E., Alcock, F., Dickson, N., Eckley, N., Guston, D. H., Jäger, J. and Mitchell, R. B. (2002a) 'Knowledge systems for sustainable development', *Proceedings of the National Academy of Sciences*, vol 100, pp8086–8091

Cash, D. W., Clark, W. C., Alcock, F., Dickson, N., Eckley, N. and Jäger, J. (2002b) *Salience, Credibility, Legitimacy and Boundaries: Linking Research, Assessment and Decision Making*, RWP02–046, J. F. Kennedy School of Government, Harvard University, Cambridge, MA

Cash, D. W., Adger, W. N., Berkes, F., Garden, P., Lebel, L., Olsson, P., Pritchard, L. and Young, O. (2006) 'Scale and cross-scale dynamics: Governance and information in a multilevel world', *Ecology and Society*, vol 11, no 2, p8, www.ecologyandsociety.org/ vol 11/iss2/art8

Chappell, C. R. and Hartz, J. (1998) 'The challenge of communicating science to the public', *Chronicle of Higher Education*, www.physics.ohio-state.edu/wilkins/writing/Resources/essays/sci_comm.html

Cicerone, R. (2005) 'Enhancing the scope and utility of climate science', Keynote address at the US Climate Change Science Program Workshop, Climate Science in Support of Decision Making, 14–16 November, Arlington, VA

Clark, T. W. (2002) *The Policy Process: A Practical Guide for Natural Resource Professionals*, Yale University Press, New Haven, CT

Clay, E. J. and Schaffer, B. B. (1984) *Room for Manoeuvre: An Exploration of Public Policy in Agricultural and Rural Development*, Heinemann Educational Books, London

Communication and Cognition (1996) 'Editorial: Is popularization possible? Special issue on popularization of science', *Communication and Cognition*, vol 29, no 2, pp149–152

Court, J. and Young, J. (2003) *Bridging Research and Policy: Insights from 50 Case Studies*, ODI Working Paper 213, Overseas Development Institute, London

Cox, R. (2006) *Environmental Communication and the Public Sphere*, Sage Publications, Thousand Oaks, CA

Crewe, E. and Young, J. (2002) *Bridging Research and Policy, Context, Evidence and Links*, ODI Working Paper 173, Overseas Development Institute, London

Cvetkovich, G., Siegrist, M., Murray, R. and Tragesser, S. (2002) 'New information and social trust: Asymmetry and perseverance of attributions about hazard managers', *Risk Analysis*, vol 22, no 2, pp359–367

Dabelko, G. (2005) 'Speaking their language: How to communicate better with policy-makers and opinion shapers – and why academics should bother in the first place', *International Environmental Agreements: Politics, Law and Economics*, vol 15, no 4, pp381–386

Darcy, J. and Hofmann, C.-A. (2003) *According to Need? Needs Assessment and Decision-Making in the Humanitarian Sector*, Humanitarian Policy Group (HPG), Report No 15, Overseas Development Institute, London

De Waal, A. and Whiteside, A. (2003) 'New variant famine: AIDS and food crisis in southern Africa', *The Lancet*, vol 362, pp1234–1237

Demeritt, D. and Langdon, D. (2004) 'The UK Climate Change Programme and communication with local authorities', *Global Environmental Change*, vol 14, pp325–336

Devereux, S. (2000) 'Famine in Africa', in S. Devereux and S. Maxwell (eds) *Food Security in Sub-Saharan Africa*, ITDG, London, pp117–148

Devereux, S. (2003) *Policy Options for Increasing the Contribution of Social Protection to Food Security*, Forum for Food Security in Southern Africa

Devereux, S. and Maxwell, S. (eds) (2001) *Food Security in Sub-Saharan Africa*, ITDG Publishing, London

Dietz, T. and Stern, C. (eds) (2002) *New Tools for Environmental Protection: Education, Information and Voluntary Measures*, National Academy Press, Washington, DC

Dilley, M. and Boudreau, T. (2001) 'Coming to terms with vulnerability: A critique of the food security definition', *Food Policy*, vol 26, pp229–247

Douglas, M. (1986) *How Institutions Think*, Routledge and Kegan Paul, London

Dow, K., Kasperson, R. E. and Bohn, M. (2006) 'Exploring the social justice implications of adaptation and vulnerability', in W. N. Adger et al (eds) *Fairness in Adaptation to Climate Change*, MIT Press, Cambridge, MA, pp79–96

Downing, T. E. and Patwardhan, A. (2003) 'Vulnerability assessment for climate adaptation', in B. Lim and E. Spanger-Siegfried (eds) *Adaptation Policy Frameworks for Climate Change: Developing Strategies, Policies and Measures*, APF Technical Paper 3, United Nations Development Programme, New York, pp67–89, www.undp.org/gef/undp-gef_publications/ publications/apf%20technical%20paper03.pdf

Drimie, S. (2004) *The Underlying Causes of the Food Crisis in Southern Africa – Malawi, Mozambique, Zambia and Zimbabwe*, Oxfam-GB Policy Research Paper, HSRC, www.sarpn.prg.za

Drimie, S. and Misselhorn, A. (2005) *Enhancing Civil Society Participation in SADC Food Security Processes*, Conference report, Look, Listen and Learn Project, Regional Conference, 14–15 November 2005, SARPN, ODI, FANRPAN

Earle, T. C. and Cvetkovich, G. T. (1995) *Social Trust: Toward a Cosmopolitan Society*, Praeger, Westport, CT

Ellis, F. (2003) 'Human vulnerability and food security: Policy implications', Forum for Food Security in Southern Africa, Overseas Development Institute, London, www.odi.org.uk/Food-SecurityForum/docs/vulnerability_theme3.pdf

Farrell, A. E. and Jäger, J. (eds) (2006) *Assessments of Regional and Global Environmental Risks: Designing Processes for the Effective Use of Science in Decisionmaking*, Resources for the Future, Washington, DC

Figueroa, M. E., Kincaid, D. L., Rani, M. and Lewis, G. (2002) *Communication for Social Change: An Integrated Model for Measuring the Process and Its Outcomes*, Communication for Social Change Working Paper, Johns Hopkins University, Center for Communication Programs, Baltimore, MD and The Rockefeller Foundation, NY

Fischhoff, B. (1989) 'Risk: A guide to controversy', Appendix to National Research Council (ed) *Improving Risk Communications*, National Academy Press, Washington, DC, pp211–319

Fischhoff, B. (1995) 'Risk perception and communication unplugged: Twenty years of process', *Risk Analysis*, vol 15, pp137–145

Folke, C., Hahn, T., Ollson, P. and Norberg, J. (2005) 'Adaptive governance of social-ecological systems', *Annual Review of Environmental Resources*, vol 30, pp441–473

French, S. and Geldermann, J. (2005) 'The varied contexts of environmental decision problems and their implications for decision support', *Environmental Science and Policy*, vol 8, pp378–391

Freyfogle, E. T. and Newton, J. L. (2002) 'Putting science in its place', *Conservation Biology*, vol 16, no 4, pp863–873

Fujimura, J. (1992) 'Crafting science: Standardized packages, boundary objects, and 'translation' ', in A. Pickering (ed) *Science as Culture and Practice*, University of Chicago Press, Chicago, IL, pp168–211

Füssel, H.-M. (2004) 'Coevolution of the political and conceptual frameworks for climate change vulnerability assessments', in F. Biermann, S. Camp and K. Jacob (eds) *Proceedings of the 2002 Berlin Conference on the Human Dimensions of Global Environmental Change: Knowledge for the Sustainability Transition. The Challenge for Social Science*, Global Governance Project, Amsterdam, The Netherlands, pp302–320

Gallopín, G. C. (2006) 'Linkages between vulnerability, resilience, and adaptive capacity', *Global Environmental Change*, vol 16, pp293–303

Gardner, G. T. and Stern, P. C. (2002) *Environmental Problems and Human Behavior*, second edition, Pearson Custom Publishing, Boston, MA

Gerhard, L. C. (1994) 'Framing policies on resources and the environment', *Geotimes*, vol 39, pp20–22

Gibbons, M. (1999) 'Science's new social contract with society', *Nature*, vol 402, ppC81–C84

Gibbons, M., Limoges, C., Nowotny, H., Schwartzman, S., Scott, P. and Trow, M. (1994) *The New Production of Knowledge: The Dynamics of Science and Research in Contemporary Societies*, Sage, London

Gieryn, T. F. (1999) *Cultural Boundaries of Science: Credibility on the Line*, University of Chicago Press, Chicago, IL

Gray-Felder, D. (1999) *Communication for Social Change: A Position Paper and Conference Report*, Johns Hopkins University, Center for Communication Programs, Baltimore, MD, and The Rockefeller Foundation, New York, NY

Gregory, J. and Miller, S. (1998) *Science in Public: Communication, Culture and Credibility*, Plenum Press, New York, NY

Guston, D. H. (2001) 'Boundary organizations in environmental policy and science: An introduction', *Science, Technology, and Human Values*, vol 26, pp87–112

Haas, E. B. (1991) *When Knowledge is Power: Three Models of Change in International Organizations*, University of California Press, Berkeley, CA

Haas, P. M. (1992) 'Introduction: Epistemic communities and international policy coordination', *International Organization*, vol 46, no 1, pp1–35

Hall, D. J. and Paradice, D. (2005) 'Philosophical foundations for a learning-oriented knowledge management system for decision support', *Decision Support Systems*, vol 39, no 3, pp445–461

House of Commons (2002–2003) *The Humanitarian Crisis in Southern Africa, Third Report of Session 2002–03*, International Development Committee, The Stationery Office Limited, London

Huq, S. and Reid, H. (2004) 'Mainstreaming adaptation in development', *IDS Bulletin*, vol 35, no 3, pp15–21

IAI (Inter-American Institute for Global Change Research) (2005) *Linking the Science of Environmental Change to Society and Policy: Lessons from Ten Years of Research in the Americas* (Group D on Communicating Science), Workshop report from a meeting in Ubatuba, Brazil, 27 November–2 December, www.icsu-scope.org

Immerwahr, J. (1999) *Waiting for a Signal: Public Attitudes Toward Global Warming, the Environment and Geophysical Research*, American Geophysical Union, www.ago.org/sci_soc/attitude_study.pdf

IPCC (Intergovernmental Panel on Climate Change) (2007) *Climate Change 2007: The Physical Science Basis*, Contribution of Working Group I to the Fourth Assessment Report of the Intergovernmental Panel on Climate Change, Cambridge University Press, New York, NY

Jackson, J., Mkwende, G. and Mathule, L. (2006) *A Rapid Appraisal of the Predictive Performance of the 2005 Annual VAC Assessments in Lesotho and Malawi*, Regional Hunger and Vulnerability Programme, Johannesburg, South Africa, www.wahenga.net

Jacobs, K., Garfin, G. and Lemart, M. (2005) 'More than just talk: Connecting science and decisionmaking', *Environment*, vol 47, no 9, pp6–21

Jacobson, N., Butterill, D. and Goering, P. (2004) 'Organizational factors that influence university-based researchers' engagement in knowledge transfer activities', *Science Communication*, vol 25, pp246–259

Jasanoff, S. (2003) 'Technologies of humility: Citizen participation in governing science', *Minerva*, vol 41, pp223–244

Jensen, P. (2005) 'Who's helping to bring science to the people?', *Nature*, vol 434, p956

Jere, P. (2005a) *Vulnerability Assessment Methodology Review*, Namibia Country Report, Commissioned by SADC–FANR–RVAC, PJ Development Consultancy, Lilongwe, Malawi

Jere, P. (2005b) *Vulnerability Assessment Methodology Review*, Swaziland Country Report, Commissioned by SADC–FANR–RVAC, PJ Development Consultancy,

Lilongwe, Malawi
Jones, S. A., Fischhoff, B. and Lach, D. (1999) 'Evaluating the science–policy interface for climate change research', *Climatic Change*, vol 43, pp581–599
Karl, H. A., Susskind, L. E. and Wallace, K. H. (2007) 'A dialogue, not a diatribe: Effective integration of science and policy through joint fact finding', *Environment*, vol 49, pp20–34
Kasperson, J. X. and Kasperson, R. E. (eds) (2001) *Global Environmental Risk*, United Nations University Press, Tokyo
Kasperson, R. E. (2005) 'Bridging vulnerability science and practice', Paper presented at the Open Meetings of the International Human Dimensions Programme on Global Environmental Change, October 2005, Bonn, Germany
Kates, R. W., Corell, R., Hall, J. M., Jaeger, C. C., Lowe, I., McCarthy, J. J., Schellnhuber, H. J., Bolin, B., Dickson, N. M., Faucheux, S., Gallopín, G. C., Grubler, G. C., Huntley, B., Jäger, J., Jodha, N. S., Kasperson, R. E., Magobunje, A., Matson, P., Mooney, H., Moore III, B., O'Riordan, T. and Svedin, U. (2001) 'Sustainability science', *Science*, vol 292, pp641–642
Kingdon, J. W. (1984) *Agendas, Alternatives, and Public Policies*, Harper Collins, New York, NY
Kyvik, S. (2005) 'Popular science publishing and contributions to public discourse among university faculty', *Science Communication*, vol 26, pp288–311
Lemmons, J. and Brown, D. A. (1995) 'The role of science in sustainable development and environmental protection decisionmaking', in J. Lemmons and D. A. Brown (eds) *Sustainable Development: Science, Ethics, and Public Policy*, Kluwer Academic Publishers, Dordrecht, The Netherlands, pp11–38
Lemos, M. C. and Morehouse, B. J. (2005) 'The co-production of science and policy in integrated climate assessments', *Global Environmental Change*, vol 15, pp57–68
Longhurst, R. (1994) 'Conceptual frameworks for linking relief and development', *IDS Bulletin*, vol 25, no 4, pp17–23
Marsland, N. (2004) *Development of Food Security and Vulnerability Information Systems in Southern Africa: The Experience of Save the Children UK*, www.sarpn.org.za
Maunder, N. (2005) 'Regional food security policy issues: Challenges and opportunities for civil society engagement', Background paper for the Conference on Enhancing Civil Society Participation in SADC Food Security Processes, SARPN, ODI and FANRPAN, 14–15 November 2005, Johannesburg, South Africa
Maunder, N. and Wiggins, S. (2007) 'Food security in southern Africa: Changing the trend? Review of lessons learnt on recent responses to chronic and transitory hunger and vulnerability', *Natural Resources Perspectives*, no 106, www.odi.org.uk/Publications/natural_resource_perspectives.asp
McKenzie-Mohr, D. and Smith, W. (1999) *Fostering Sustainable Behavior: An Introduction to Community-Based Social Marketing*, New Society Publishers, Gabriola Island, BC
Mcleod, J. M., Scheufele, D. A. and Moy, P. (1999) 'Community, communication, and participation: The role of mass media and interpersonal discussion in local political participation', *Political Communication*, vol 16, no 3, pp315–336
McNeely, J. A. (1999) 'Strange bedfellows: Why science and policy don't mesh and what can be done about it', in J. Cracraft and F. T. Grifo (eds) *The Living Planet in Crisis: Biodiversity, Science and Policy*, Columbia University Press, New York, NY, pp275–285

McNie, L. and Elizabeth, C. (2007) 'Reconciling the supply of scientific information with user demands: An analysis of the problem and review of the literature', *Environmental Science and Policy*, vol 10, pp17–38

Miles, E. L., Snover, A. K., Whitley Binder, L. C., Sarachik, E. S., Mote, P. W. and Matua, N., (2006) 'An approach to designing a national climate service', *Proceedings of the National Academy of Sciences*, vol 103, no 52, pp19616–19623

Millennium Ecosystem Assessment (2005) *Millennium Ecosystem Assessment Synthesis Report*, Island Press, Washington, DC, www.millenniumassessment.org/en/index.aspx

Miller, C. (2001) 'Hybrid management: Boundary organizations, science policy, and environmental governance in the climate regime', *Science, Technology, and Human Values*, vol 26, no 4, pp478–500

Miller, K. and Yates, D. (2005) *Climate Change and Water Resources: A Primer for Municipal Water Providers*, AWWA Research Foundation, Denver, CO

Mitchell, R., Clark, W. C., Cash, D. W. and Dickson, N. M. (eds) (2006) *Global Environmental Assessments: Information and Influence*, MIT Press, Cambridge, MA

Mock, N. (2005a) *Vulnerability Assessment Methodology Review*, Lesotho Country Report, Commissioned by SADC–FANR–RVAC, Southern African Development Community, Gaborone, Botswana

Mock, N. (2005b) *Vulnerability Assessment Methodology Review*, Mozambique Country Report, Commissioned by SADC–FANR–RVAC, Southern African Development Community, Gaborone, Botswana

Moll, P. and Zander, U. (2006) *Managing the Interface: From Knowledge to Action in Global Change and Sustainability Science*, Oekom Verlag, München

Morss, R. E., Wilhelmi, O. V., Downton, M. W. and Gruntfest, E. (2005) 'Flood risk, uncertainty, and scientific information for decision making: Lessons from an interdisciplinary project', *Bulletin of the American Meteorological Society*, vol 86, no 11, pp1593–1601

Moser, S. C. (2004) 'Climate change and the sustainability transition: The role of communication and social change', *IHDP UPDATE*, vol 4, pp18–19

Moser, S. C. (forthcoming) 'The contextual importance of uncertainty in climate-sensitive decision-making: Toward an integrative decision centered screening tool', in T. Dietz and D. Bidwell (eds) *Climate Change and the Great Lakes: Decision-Making Under Uncertainty*, Michigan State University Press, East Lansing

Moser, S. C. and Dilling, L. (2004) 'Making climate hot: Communicating the urgency and challenge of global climate change', *Environment*, vol 46, no 10, pp32–46

Moser, S. C. and Dilling, L. (2007) *Creating a Climate for Change: Communicating Climate Change and Facilitating Social Change*, Cambridge University Press, Cambridge, UK

Moser, S. C. and Luganda, P. (2006) 'Talk for a change: Communication in support of societal response to climate change', *IHDP UPDATE*, vol 1, pp17–20

National Research Council (1989) *Improving Risk Communication*, Committee on Risk Perception and Communication, Commission on Behavioral and Social Sciences and Education and Commission on Physical Sciences, Mathematics, and Applications, National Academy Press, Washington, DC

National Research Council (2002) *New Tools for Environmental Protection: Education, Information, and Voluntary Measures*, National Academy Press, Washington, DC

Niederberger, A. A. (2005) 'Science for climate policy change policymaking: Applying theory to practice to enhance effectiveness', *Climate Change Science and Public Policy*, vol 32, no 1, pp2–16

Nowotny, H. (2003) *The Potential of Transdisciplinarity, Rethinking Interdisciplinarity*, www.interdisciplines.org/interdisciplinarity/papers/5

Nowotny, H., Scott, P. and Gibbons, M. (2001) *Re-Thinking Science: Knowledge and the Public in an Age of Uncertainty*, Polity Press, Cambridge, UK

O'Brien, K. L. and Leichenko, R. (2000) 'Double exposure: Assessing the impacts of climate change within the context of economic globalization', *Global Environmental Change*, vol 10, pp221–232

O'Brien, K., Leichenko, R., Kelkar, U., Venema, H., Aandahl, G., Tompkins, H., Javed, A., Bhadwal, S., Barg, S., Nygaard, L. and West, J. (2004) 'Mapping vulnerability to multiple stressors: Climate change and globalization in India', *Global Environmental Change*, vol 14, pp303–313

Ogunseitan, O. A. (2000) *Framing Vulnerability: Global Environmental Assessments and the African Burden of Disease*, Belfer Center for Science and International Affairs, John F. Kennedy School of Government, Harvard University, Cambridge, MA

Ogunseitan, O. A. (2003) 'Framing environmental change in Africa: Cross-scale institutional constraints on progressing from rhetoric to action against vulnerability', *Global Environmental Change*, vol 13, pp101–111

Owens, S. (2005) 'Making a difference? Some perspectives on environmental research and policy', *Transactions of the Institute of British Geographers*, vol 30, pp287–292

Pielke Jr., R. A. (1997) 'Reframing the US hurricane problem', *Society and Natural Resources*, vol 10, pp485–499

Poortinga, W. and Pidgeon, N. F. (2004) 'Trust, the asymmetry principle, and the role of prior beliefs', *Risk Analysis*, vol 24, no 6, pp1475–1486

Pross, P. (1986) *Group Politics and Public Policy*, Oxford University Press, Toronto

Pulwarty, R. S. (2003) 'Climate and water in the West: Science, information and decision-making', *Water Resources Update*, vol 124, pp4–12

Pulwarty, R. S. and Melis, T. S. (2001) 'Climate extremes and adaptive management on the Colorado River: Lessons from the 1997–1998 ENSO event', *Journal of Environmental Management*, vol 63, pp307–324

RHVP (Regional Hunger and Vulnerability Programme) (2006) *Review Paper on Community Coping, Vulnerability and Social Protection*, Final version, February 2006, http://www.wahenga.net

Rockefeller Foundation (1997) *Communication for Social Change: Forging Strategies for the 21st Century*, Johns Hopkins University, Center for Communication Programs, Baltimore, MD, and The Rockefeller Foundation, New York, NY

Sabatier, P. and Jenkins-Smith, H. C. (1999) 'The advocacy coalition framework: An assessment', in P. Sabbatier (ed) *Theories of the Policy Process*, Westview Press, Boulder, CO

Sachs, J. D. (2005) *The End of Poverty: Economic Possibilities for Our Time*, Penguin Press, New York, NY

SADC–FANR–RVAC (Southern African Development Community–Food, Agriculture and Natural Resources–Regional Vulnerability Assessment Committee) (2003) *Towards Identifying Impacts of HIV/AIDS on Food Insecurity in Southern Africa and Implications for Response: Findings from Malawi, Zambia and Zimbabwe*, SADC–FANR, Zimbabwe

SADC–RVAC (Southern African Development Community Directorate) (2005) *Strengthening Vulnerability Assessments and Analysis in the SADC Region: A Five-Year Programme 2005–2009*, SADC Regional Vulnerability Assessment Committee, Gaborone, Botswana

Sarewitz, D. and Pielke Jr., R. A. (2007) 'The neglected heart of science policy: Reconciling supply of and demand for science', *Environmental Science and Policy*, vol 10, pp5–16

SARPN (Southern African Regional Poverty Network) (2004) *Scoping Study Towards DFIDSA's Regional Hunger and Vulnerability Programme*, UK Department for International Development (DFID), Prepared by SARPN, Pretoria, South Africa

Schreurs, M., Clark, W. C., Dickson, N. and Jäger, J. (2001) 'Issue attention, framing, and actors: An analysis of patterns across arenas', in W. C. Clark, J. Jäger, J. van Eijndhoven and N. Dickson (eds) *Learning to Manage Global Environmental Risks: A Comparative History of Social Responses to Climate Change, Ozone Depletion and Acid Rain*, MIT Press, Cambridge, MA, pp349–364

Servaes, J. T., Jacobson, L. and White, S. A. (1996) *Participatory Communication for Social Change*, Sage Publications, Thousand Oaks, CA

Slovic, P. (1993) 'Perceived risk, trust and democracy', *Risk Analysis*, vol 13, no 6, pp675–682

Slovic, P. (ed) (2000) *The Perception of Risk*, Risk Society and Policy Series, Earthscan, London

Steinberg, P. (2005) 'Is anyone listening? The impact of research on global environmental practice', *International Environmental Agreements: Politics, Law and Economics*, vol 15, no 4, pp377–379

Stokes, D. E. (1997) *Pasteur's Quadrant*, Brookings Institute, Washington, DC

Stone, D. A. (1989) 'Causal stories and the formation of policy agendas', *Policy Science Quarterly*, vol 104, no 2, pp281–300

Tango International Inc (2005a) *Vulnerability Assessment Methodology Review*, Zimbabwe Country Report, Tango, www.wahenga.net

Tango International Inc (2005b) *Vulnerability Assessment Methodology Review Synthesis*, Paper commissioned by the Southern African Development Community–Food, Agriculture and Natural Resources–Regional Vulnerability Assessment Committee (SADC–FANR–RVAC), Tango, www.wahnega.net

Thomalla, F., Downing, T., Spanger-Siegfried, E., Han, G. and Rockstrom, J. (2006) 'Reducing hazard vulnerability: Towards a common approach between disaster risk reduction and climate adaptation', *Disasters*, vol 30, pp39–48

Timmerman, P. (1981) *Vulnerability, Resilience and the Collapse of Society*, Environmental Monographs no 1, Institute for Environmental Studies, University of Toronto, Toronto

Tschirley, D., Nijhoff, J. J., Arlindo, P., Mwinga, B., Weber, M. T. and Jayne, T. S. (2004) *Anticipating and Responding to Drought Emergencies in Southern Africa: Lessons from the 2002–2003 Experience*, Report prepared for the NEPAD Regional Conference on Successes in African Agriculture, 22–25 November 2004, Nairobi, Kenya

Turner II, B. L., Kasperson, R. E., Matson, P. A., McCarthy, J. J., Corell, R. W., Christensen, L., Eckley, N., Kasperson, J. X., Luers, A., Martello, M. L., Polsky, C., Pulsipher, A. and Schiller, A. (2003a) 'A framework for vulnerability analysis in sustainability science', *Proceedings of the National Academy of Sciences*, vol 100, pp8074–8079

Turner Jr., B. L., Matson, P. A., McMcarthy, J. J., Corell, R. W., Christenson, L., Eckely, N., Hovelsrud-Broda, G., Kasperson, J. X., Kasperson, R. E., Luers, A., Martello, M. L., Mathieson, R., Naylor, C., Polsky, A., Pulsipher, A., Schiller, A., Selin, H. and Tyler, N. (2003b) 'Illustrating the coupled-human environment systems for vulnerability analysis: Three case studies', *Proceedings of the National Academy of Sciences*, vol 100, no 14, pp8080–8085

Uzzell, D. (2004) 'From local to global: A case of environmental hyperopia', *IHDP UPDATE*, vol 4, pp6–7

Van der Vink, G. E. (1997) 'Scientifically illiterate vs. politically clueless (editorial)', *Science*, vol 276, p1175

van Kerkhoff, L. (2005) 'Integrated research: Concepts of connection in environmental science and policy', *Environmental Science and Policy*, vol 8, pp452–463

van Kerkhoff, L. and Lebel, L. (2006) 'Linking knowledge and action for sustainable development', *Annual Review of Environmental Resources*, vol 31, pp12.1–12.33

Vlek, P. and Steg, L. (2004) 'Solid environmental policy needs valid environmental psychology', *IHDP UPDATE*, vol 4, pp11–13

Watts, M. J. and Bohle, H. G. (1993) 'The space of vulnerability: The causal structure of hunger and famine', *Progress in Human Geography*, vol 17, no 1, pp43–67

Wegerich, K. (2001a) *Determining Factors of Local Institutional Change in Countries in Transition*, Occasional Paper no 38, Water Issues Study Group, School of Oriental and African Studies, University of London, London

Wegerich, K. (2001b) *Institutional Change: A Theoretical Approach*, Occasional Paper no 30, Water Issues Study Group, School of Oriental and African Studies, University of London, London

Weigold, M. F. (2001) 'Communicating science: A review of the literature', *Science Communication*, vol 23, pp164–193

Wellcome Trust (2000) *The Role of Scientists in Public Debate*, Research study conducted by MORI for The Wellcome Trust, www.wellcome.ac.uk/doc%5Fwtd003429.html

Wiggins, S. (2005) 'Success stories from African agriculture: What are the key elements for success, new directions for African agriculture', *IDS Bulletin*, vol 36, no 2, pp17–22

Wilbanks, T. J. (2002) 'Geographic scaling issues in integrated assessments of climate change', *Integrated Assessment*, vol 3, pp100–114

Wisner, B. (1993) 'Disaster vulnerability: Geographical scale and existential reality', in H.-G. Bohle (ed) *Worlds of Pain and Hunger*, Universitat Freiburg, Saarbrlicken, Breitenbach, pp13–52

Wisner, B., Blaikie, P., Cannon, T. and Davis, I. (2003) *At Risk: Natural Hazards, People's Vulnerability and Disasters*, Routledge, London

Part III

Growing Political Urgency: Climate Change

5

The US Climate Change Science Program

Zachary Christman

Introduction and brief history of the programme lineages

With an increased public awareness and scientific uncertainty associated with a changing earth system, the US has crafted policy and programmatic structures to increase understanding and preparedness for variations that may occur as a result of changes to our biophysical surroundings. The US Global Change Research Act of 1990 created mandates for protocols and products through which the scientific and practitioner communities were to make recommendations to the US government and determine how these recommendations should be implemented (see Appendix 2, p149). During the nearly three decades since the act was designed, the original US Global Change Research Program and the preceding Climate Change Science Program underwent many structural and functional changes. By 2010, with the new Obama administration, the Climate Change Science Program was defunct, reverted back to the Global Change Research Program and awaited new initiatives by the Obama administration and the US Congress, following the international meeting in Copenhagen in December 2009.

Through these transitions, the mission of the overarching strategy to understand global change has shifted from iterative recommendations to comprehensive synthesis, with implications for the effectiveness of the interface between the understood science and the implemented actions. This case study seeks to characterize the science/policy interface, using the Climate Change

Science Program (CCSP) to evaluate the changes over its history and to understand the implications associated with these changes.

The stated mission of the Climate Change Science Program of 2003 was to facilitate the creation and application of knowledge of the Earth's global environment through research, observations, decision support and communication. While serving primarily as a liaison among scientific researchers, government organizations and decision-makers, the CCSP crafted informational products to translate the findings, applications and legislation among these groups in the spirit of, but with many notable differences from, the original Climate Change Research Act of 1990. Through a series of infrastructure reorganizations and reinterpretations of its mission, the CCSP served a different role in the interface between scientists and practitioners than its predecessor (and now subsidiary), the US Global Change Research Program, shifting the focus of the organization from iterative recommendations and assessment initiatives to informational synthesis (see Appendix 1, p147).

Case context: Initial mandate of the US Global Change Research Program and its development

Throughout the administration of President George H. W. Bush, a number of policies were established to investigate and better understand the changing Earth system and the role of human influence upon its current state. In 1989, a formal administrative entity, the US Global Change Research Program (USGCRP), was established, coordinating the research efforts of 15 governmental departments and agencies to recommend policy implications for managing the US response to global change. Specifically, the governmental entities were to produce information readily usable by policy-makers attempting to formulate effective strategies for preventing, mitigating and adapting to the effects of global change. Overseeing the USGCRP was the Subcommittee on Global Change Research, a division of the Committee on Environment and Natural Resources within the National Science and Technology Council. While it did not directly lead research investigations, the USGCRP sought to coordinate the pertinent activities of the 15 government agencies, which included the US Agency for International Development (USAID), the National Oceanic and Atmospheric Administration (NOAA), the Environmental Protection Agency (EPA), the National Aeronautics and Space Administration (NASA), the National Institutes of Health (NIH), the National Science Foundation (NSF), the US Department of Agriculture (USDA), the Department of Defense (DOD), the Department of Energy (DOE), the Department of Health and Human Services (HHS), the Department of the Interior/US Geological Survey (USGS), the Department of Transportation (DOT), the Department of Commerce, the Department of State, and the Smithsonian Institution.

Shortly after the presidential directive establishing the USGCRP, Congress passed the US Global Change Research Act of 1990, with the expressed purpose of providing for the development and coordination of a comprehensive and integrated US research programme which would assist the nation and the world to understand, assess, predict and respond to human-induced and natural processes of global change. Inherent to this act were mandates for two recurring reports assessing the state of knowledge on global change. The first involved a scientific assessment of the knowledge and uncertainties pertaining to global change and offered specific insights on the impacts of global change upon a variety of sectors, including agriculture, the natural environment, biological diversity, land and water resources, energy production and use, transportation, human health and welfare, and human social systems. Proactively, this report also investigated the current trends associated with each of these factors and offered a projection into the next 25 to 100 years. Although the act mandated that this scientific assessment be issued no less than every four years, the first report stemming from this mandate, the *U.S. National Assessment of the Potential Consequences of Climate Variability and Change* (USGCRP, 2001), was not published until 2001. The second recurring report was an annual report to be submitted to Congress, including a summary of the achievements of the programme over the past year, with an evaluation of the progress made toward accomplishing each goal and an analysis of the budgetary assessment. Importantly, this report was also required to make annual recommendations concerning any changes in roles or responsibilities, as well as any new legislation that might aid in the accomplishment of these goals. These reports are published annually as the volume *Our Changing Planet*, the first of which was issued in 1995. The CCSP, meanwhile, decided not to undertake another national assessment of impacts, but instead to commission and issue 21 so-called *Synthesis and Assessment Product* (SAP) reports, which together were viewed as constituting the second national assessment (a third national assessment in 2010 is being contemplated to meet the congressional requirement of a national assessment over four years).

During the 1990s, a number of volumes pertaining to the effects and implications of climate change were produced by USGCRP member organizations. In accordance with the United Nations Framework Convention on Climate Change (UNFCCC), *Climate Action* reports were submitted in 1994, 1997 and again in 2002, with the 21 SAP reports appearing over 2007 to 2009. The Office of Technology Assessment of the US Congress in 1993 prepared a two-volume report entitled *Preparing for an Uncertain Climate* to assess the following questions: what is at risk and over what timeframes? How can we best plan for an uncertain climate? Will we have the answers when we need them, and does the USGCRP reflect the short- and long-term needs of the decision-makers? Although this report highlighted the need to weigh the long-term vulnerability and preparedness of the national social and ecological systems, it also contained some ambivalence regarding the urgency of the issue, asking: 'Why adopt a policy today to adapt to a climate change that may not occur ... effort put into adopting the measure could well be wasted ... future generations may have more sophisticated technologies and greater wealth that can be used for adaptation' (USOTA, 1993, p4), reflecting the Bush administration's stance on climate change.

On 11 June 2001, shortly after ascending to the presidency, George W. Bush issued a public statement that the US would no longer adhere to the terms set forth by the Kyoto Treaty on Climate Change, citing the high degree of uncertainty associated with the climate change issue. In doing so, he also established the Climate Change Research Initiative to investigate the uncertainties associated with climate change and to identify priority areas where investments could make a difference. Additionally, he created the National Climate Change Technology Initiative to improve methods of measuring the effects of climate change and to innovate upon new technologies for ameliorating these effects. On 14 February 2002 in an address before NOAA, the president announced the Clear Skies initiative, which aimed to reduce power plant emissions, but also set forth a number of tangible changes to the US policy on global change

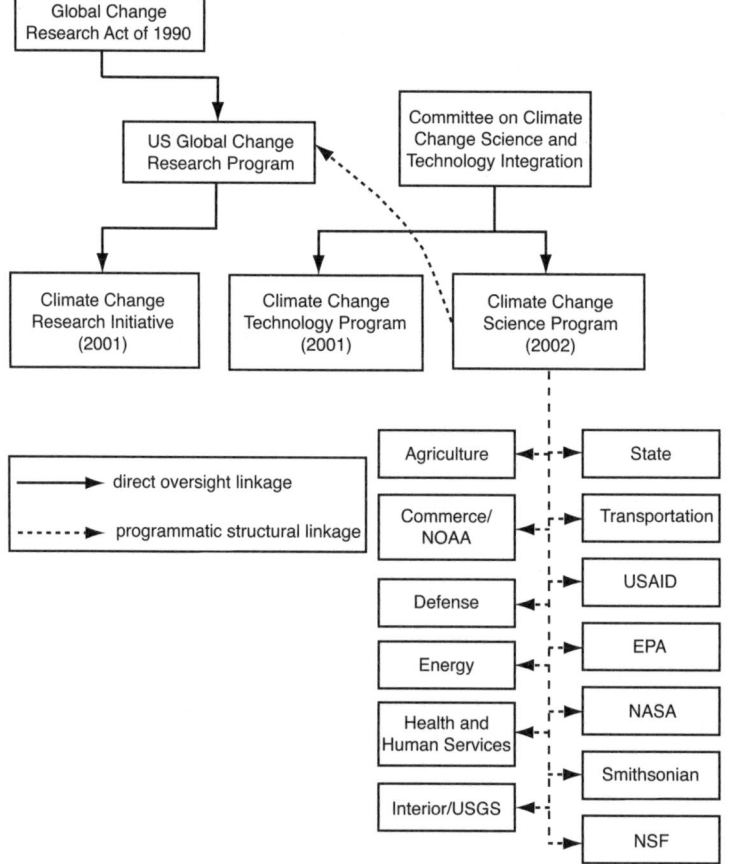

Figure 5.1 *Structure of the US organizations related to climate change activities*

Source: chapter author

research, effectively establishing the Climate Change Science Program under the direction of the Committee on Climate Change Science and Technology Integration (see Figure 5.1). This new programme, founded on the vision of a nation and the global community empowered with science-based knowledge to manage the risks and opportunities of change in the climate and related environmental systems, effectively enveloped the existing US Global Change Research Program in order to 'facilitate the creation and application of knowledge of the Earth's global environment through research, observations, decision support, and communication' (USCCSP, 2003, p4).

Vulnerability and resilience within the Climate Change Science Program

The basic question proposed by the *Strategic Plan* by the Climate Change Science Program is twofold. How will variability and potential change in our climate and related systems affect natural environments and our way of life? How can we use and improve this knowledge to protect the global environment and to provide a better living standard for all (USCCSP, 2003, p4)? The organization listed its five core goals as follows:

1 Improve knowledge of the Earth's past and present climate and environment, including its natural variability, and improve understanding of the causes of observed variability and change.
2 Improve quantification of the forces bringing about changes in the Earth's climate and related systems.
3 Reduce uncertainty in projections of how the Earth's climate and related systems may change in the future.
4 Understand the sensitivity and adaptability of different natural and managed ecosystems and human systems to climate and related global changes.
5 Explore the uses and identify the limits of evolving knowledge to manage the risks and opportunities related to climate variability and change.

Additionally, the organization outlined its four core approaches of scientific research, observation, decision support and communications with which to achieve these core goals (USCCSP, 2003, p6).

Of the core goals, the first three can broadly be classified as assessing the current climate situation, understanding the forces that are acting upon it, and evaluating the uncertainty associated with these observations and conclusions. Goal five serves to synthesize the available information in order to prepare for the future. It is primarily within the fourth goal that implications of global change upon human systems are evaluated, with three key research foci:

1 Improve knowledge of the sensitivity of ecosystems and economic sectors to global climate variability and change.

2   Identify and provide scientific outputs for evaluating adaptation options in cooperation with mission-oriented agencies and other resource managers.
3   Improve understanding of how changes in ecosystems (including managed ecosystems such as croplands) and human infrastructure interact over long periods of time (USCCSP, 2003, pp23–25).

The objectives have been widely criticized for relegating human dimensions issues largely to goals 4 and 5, although it is clear that human interactions are present in all five goals. Although the concepts of 'vulnerability' and 'resilience' are not expressly used within the *Strategic Plan*, the final two goals approximate these concepts through a focus on the sensitivity of humans to global changes. While they share the language of the field of vulnerability, the references to 'sensitivity and adaptability' in goal 4 seem to invoke the ecological context of these words, referring to the resistance to an alteration of the system from a stable state. The fourth goal claims relevance to the limits of the human and natural systems; but many of the 'actors' in these systems are entirely biophysical. Notable exceptions include the stated priorities of societal vulnerability to sea-level rise, changes in natural resources management, and the safety and efficiency of the transportation infrastructure.

The fifth goal is described with language that seeks to identify methods of ameliorating or responding to the risks and opportunities related to climate variability and change; but its tangible effects outlined in the *Strategic Plan* are rooted in information management. The priorities of this goal explore the potential uses of observations and forecasts for purposes of decision support, as well as the best practices for managing data and their statistical uncertainty. These scenario-building exercises explore the range of possibilities; but the decisions derived are within a federal context, rather than under state or local purview. This gap has been significantly filled by the National Research Council report *Informing Decisions in a Changing Climate* that appeared in 2008 (USNRC, 2009a).

By advocating for postponed climate action, an option weighed in the volume *Preparing for an Uncertain Climate* (USOTA, 1993), the US at that time effectively engaged in risk-taking, rather than formulating a risk-averse strategy for the future, something that is being redressed by the Obama administration. The subsequent stated strategies of the Climate Change Science Program to 'improve knowledge,' 'improve quantification,' 'reduce uncertainty,' 'understand sensitivity and adaptability' and 'explore uses and ... limits' (USCCSP, 2003, p4) all reiterate a prolongation of the status quo at the time, feigning action through continued inquiry.

Budgetary and personnel allocations for climate change

With nearly US$2 billion annually distributed among its agencies, the Climate Change Science Program has commanded an impressive amount of research

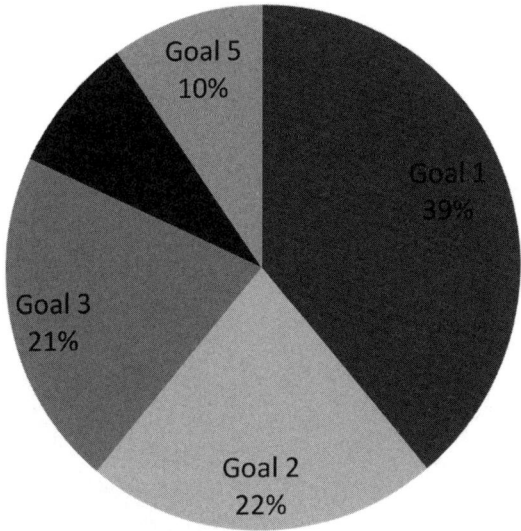

Figure 5.2 *US Climate Change Science Program (CCSP) budget allocated by goal (2009 estimated budget)*

Source: adapted from Our Changing Planet 2010, www.downloads.globalchange.gov/ocp/ocp2010/ocp2010.pdf, accessed 20 April 2010

monies over its brief history. Figures 5.2 and 5.3 detail the financial resources available per goal and agency, with Figure 5.4 demonstrating the change in total CCSP research budget through the duration of the programme.

Although the programme has operated with around US$1.5 to $1.8 billion per year, these are largely financial resources that have already been allocated to the member agency in each fiscal year's budget. Total funds available to the accomplishment of each goal range from 14 per cent of the total CCSP research budget for goal 4 to 27 per cent for goal 2. However, the allocated resources per goal, and the research matter of the goal itself, have often represented ongoing projects within each agency that happened to coincide with the spirit of the goals, rather than projects specifically aimed at achieving them. Despite its budget, the programme has been criticized for being under-supported and commanding products without allocating resources to create them. The National Academy of Sciences/ National Research Council report evaluating the CCSP, *Restructuring Federal Climate Research to Meet the Challenges of Climate Change* (USNRC, 2009b) concluded, for example, that the CCSP has suffered over its history from inadequate authority over budgets and programmes.

Since the foundation of the USGCRP in 1989, budgetary spending has increased from US$135 million (US$209 million in 2005 value) to a height of US$1760 million in 1995 (US$2234 million in 2005 value), with per-agency percentages remaining roughly equal through the six budgetary cycles of the Climate Change Science Program. But in its 2009 report, the US National

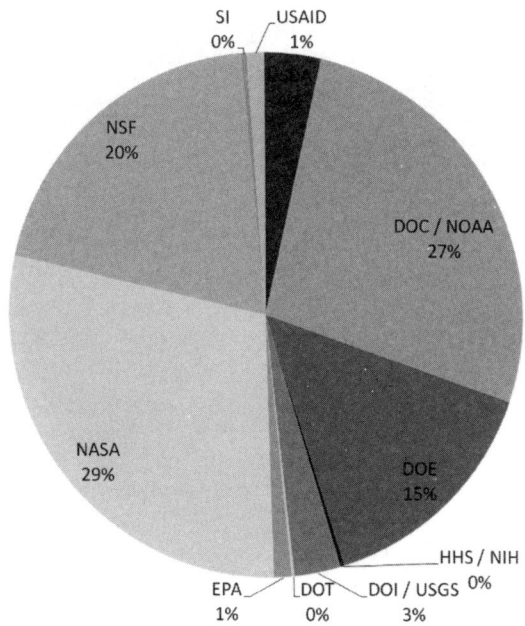

Figure 5.3 *CCSP budget allocated by agency (2009 estimated budget, excluding the NASA space budget)*

Source: adapted from Our Changing Planet 2010,
www.downloads.globalchange.gov/ocp/ocp2010/ocp2010.pdf, accessed 20 April 2010

Research Council concluded that CCSP funding had declined over time by about 20 per cent (USNRC, 2009b, p9; see Figure 5.4).

While the majority of the funds has been allocated for quantitative assessment of the climate as a natural system, approximately US$94.8 million (11 per cent of the total non-space budget) was allocated to the focus on human contributions and responses, with an additional US$23.9 million (approximately 3 per cent) allocated for investigation of the coupled human/environmental impacts of land-use and land-cover change (see Figure 5.5). Although these tasks should be noted for their attention to the interrelationship of humans with their environments, the listed activities offer few feedbacks that might contribute to understanding the vulnerabilities and adaptations of societies to these changes.

Although they were sought, the per-project allocations and the association of projects with each overarching goal were not available for analysis either to this analyst or to the NRC review committee, further testifying to the autonomous nature of the activities of the member agencies of the CCSP in practice.

THE US CLIMATE CHANGE SCIENCE PROGRAM | 139

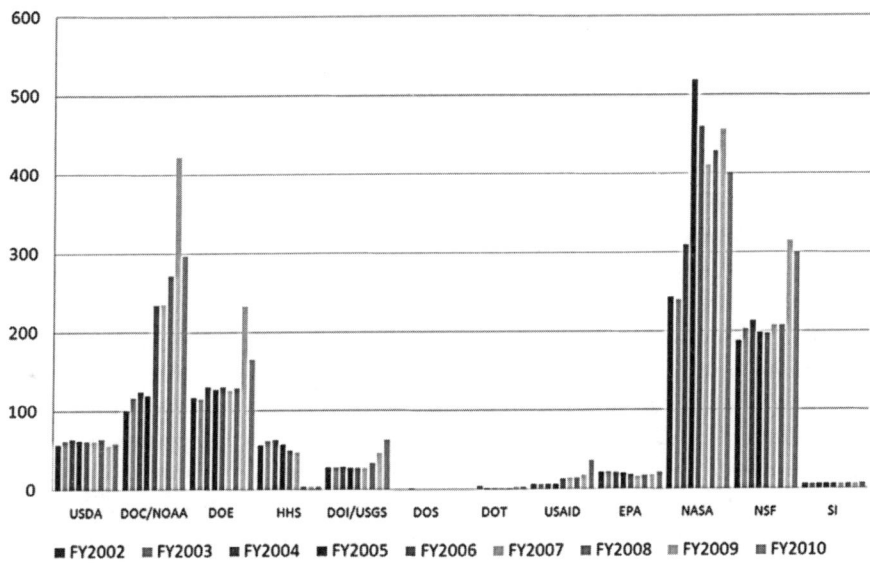

Figure 5.4 *Change in CCSP agency budgets, 2002–2010*

Source: funding for Global Change Research under the CCSP and USGCRP, fiscal years 1989–2009, Our Changing Planet 2010, www.climatescience.gov/infosheets/ccsp-8/, accessed 20 April 2010

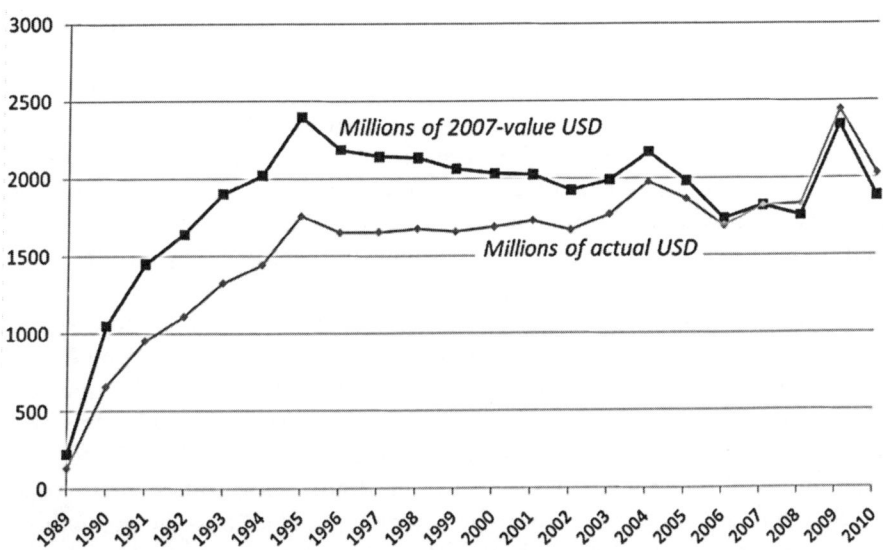

Figure 5.5 *Change in allocated resources for US Global Change Research Program (USGCRP)- and US Climate Change Science Program (CCSP)-related activities*

Source: funding for Global Change Research under the CCSP and USGCRP, fiscal years 1989–2009, Our Changing Planet 2010, www.climatescience.gov/infosheets/ccsp-8/, accessed 20 April 2010; data from 2008–2010 updated from Our Changing Planet 2010, www.downloads.globalchange.gov/ocp/ocp2010/ocp2010.pdf, accessed 20 April 2010

Products and metrics for accomplishment

The synthesis and assessment products

Envisioned as a continuation of the synthetic assessment mandated by the US Global Change Research Act of 1990, these 21 *Synthesis and Assessment Product* reports were designed to separate activities by research theme and sector in order to enlist experts in the field to manage each question. These research projects sought to address the seven climate and related systems of CCSP-related research: atmospheric composition, climate variability and change, global water cycle, land-use/land-cover change, global carbon cycle, ecosystems, and human contributions and responses. The core approaches recognized the need to improve upon the knowledge of metrics and uncertainties inherent to these studies, as well as distilling the information into a format usable for decision-makers. Importantly, the fourth core approach stressed openness and transparency in communicating these results to scientific and stakeholder communities. Despite the initial mandate for these products in the CCSP revised *Strategic Plan*, which set goals for the accomplishment of many of these projects within two years and the balance in two to four years, these deadlines have been amended, such that the first product (*Temperature Trends in the Lower Atmosphere*) was made public only in May 2006 and the final product was published in January 2009. Furthermore, continuing a problem of the 2001 *National Assessment*, no plan has yet (as of 2010) been developed to continue and update the various synthesis products.

SAP Product 1.1 *Temperature Trends in the Lower Atmosphere: Steps for Understanding and Reconciling Differences* indicates the contributions and limitations of these products (USCCSP, 2006). In the same format that is proposed for many of the first three categories of reports, this report describes patterns in temperature records, the uncertainty associated with these patterns, and possible explanations and interpretations of these phenomena. The 127-page document (excluding appendices) is written in the style of a popular-science magazine article, referencing case studies and professional interpretations in a generalized manner for a seemingly non-scientific audience. Explanatory sections such as 'Why do we need statistics?' and 'Definition of a linear trend' offer background information for a broad audience, but also restrict the utility of these documents for the more scientifically informed community. Other than a basic primer in atmospheric forces, there is no section that addresses the response by a human population to the possibilities explored; however, the final section of the report offers recommendations for the improved collection and use of temperature-related climate information. The other products reveal a more mixed record, with a number of useful assessments between coverage over the CCSP's five goals.

A major focus of the Climate Change Science Program has been to reduce uncertainty in metrics pertaining to the understanding of the Earth system. Metrics describing uncertainty were evaluated in the National Research

Council's report *Thinking Strategically: The Appropriate Use of Metrics for the Climate Change Science Program* (USNRC, 2005). This report emphasized the importance of evaluating metrics by the information gained through measurement, whether or not the specific end result was achieved. In doing so, the results of an enquiry could be used to guide future planning and subsequent evaluations. The National Research Council strongly urged the CCSP to incorporate applications and future planning within the metrics used to craft the 21 *Synthesis and Assessment Product* reports. It also cautioned against a CCSP goal for reducing uncertainty as a measure of progress, noting that new uncertainties often appear with increased scientific understanding.

Agents and feedbacks within the science/practice interface

As a result of the changes established in 2002, the hierarchical structure of the organizations addressing climate change in the US government has shifted (see Figure 5.1). Accordingly, the actors participating in this arrangement have experienced a change in organization, affecting the flow of communication and information products. Figure 5.6 suggests the relationships among members of this network.

With this organization, the central nodes of activities connecting science and practice were the product project teams established to address research

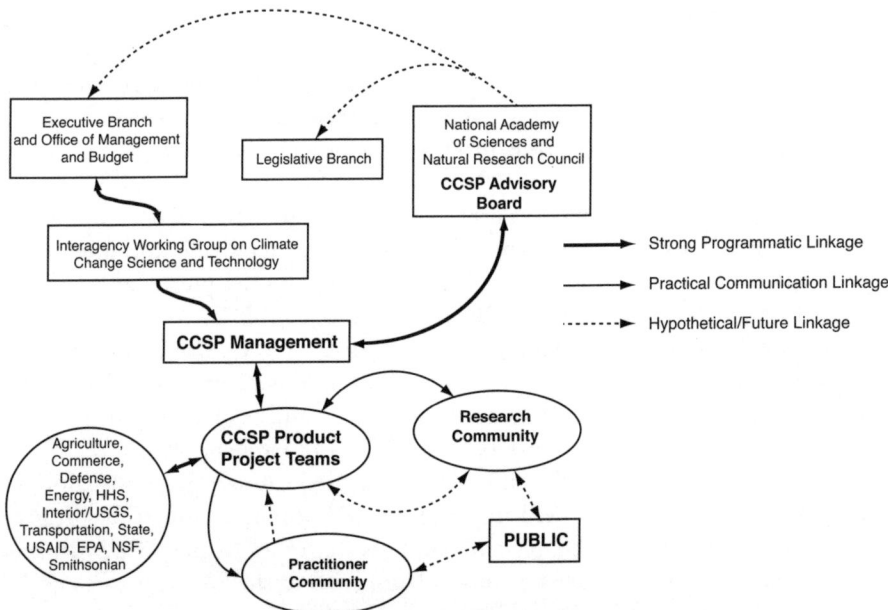

Figure 5.6 *Idealized connections for exchange of information and recommendations*

Source: chapter author

questions pertaining to each CCSP goal. These project teams, composed of researchers and administrators from CCSP member organizations and governmental divisions, directly reported progress on CCSP goals to a CCSP management office, which, in turn, reported to the Interagency Working Group on Climate Change Science and Technology and was ultimately responsible to the executive branch and the Office of Management and Budget. The CCSP was also overseen by an advisory committee established by the National Research Council, which evaluated the products and publications of the programme and made recommendations for its future activities (see USNRC, 2007, 2009a). The interface between the scientific research community and the community of practitioners was largely coordinated through each CCSP product project team, with the understanding that the researchers were responsible for representing the range of known information and opinions. The interface with the public seemed to occur exclusively through the delivery of these SAPs, with possible indirect interactions of the practitioner and research communities with the public. These figures represent an idealized version of the network in which information was to flow in both directions from agencies to researchers of the scientific and practitioner communities, at large, and, ultimately, on to the public. In reality, the programme infrastructure largely extends only as far as the participating members of each product project team and coordination among the 15 federal governmental agencies.

After the initial 2002 release of the *Strategic Plan* of the Climate Change Science Program, in 2004 the National Research Council conducted a formal review of the programme and its aims. It emphasized the need for the CCSP to establish tangible goals and clear financial allotments, as well as emphasizing interactions between the national and international communities of researchers and practitioners. Their suggestions were directly targeted towards the effective management of this large programme, which they saw as crucial to achieving a usable result.

Two major public events occurred early in the growth and development of the Climate Change Science Program, allowing for feedback and interaction with the communities of scientists, practitioners and the public at large. The first event, held in December 2002 in Washington, DC, was the Planning Workshop for Scientists and Stakeholders. Designed to receive public feedback regarding the draft version of the *Strategic Plan*, this meeting involved representatives from the 15 affiliated agencies and consisted of plenary presentations with breakout groups to evaluate the *Strategic Plan*.

Another workshop in November 2005 entitled Climate Science in Support of Decision Making was held in Arlington, Virginia, to illustrate progress towards accomplishing the decision support activities of the CCSP *Strategic Plan*. The workshop featured members from each of the SAP teams, as well as representatives from comparable national and international efforts. Again, researchers affiliated with the product project teams, as well as participants in comparable efforts, met to discuss the strategy and scope of the SAPs. Several researchers representing working groups of the Intergovernmental Panel on

Climate Change and previous participants in comparable efforts, such as the Millennium Ecosystem Assessment and Arctic Climate Impact Assessment, presented challenges and strategies of their projects. Within the course of the meeting, several presentations invoked the concept of adaptive management; but this was presented in the context of the iterative nature of the scientific method, rather than strategically adapting to vulnerability.

One implementation of the CCSP which effectively served to deliver streamlined information from scientists and SAP authors to the practitioner population and public at large was the series of *Infosheets* prepared to distil the lengthy reports into two- to four-page explanations. Although these reports necessarily simplified the details associated with the longer reports, they offered executive summaries, delivering information to individuals who might not otherwise have received the information. Within the public-oriented website (www.climatescience.gov), these documents were made available for free download and distribution.

While it remains premature to say with any certainty how the synthetic assessments will influence the communities of researchers and practitioners, a number of preliminary observations and contrasts can be made. A crucial mandate of the Global Change Research Act of 1990 was the iterative nature called for in comprehensive reports. These were intended to reflect the state-of-the-science understandings at each interval, but also to utilize this knowledge to generate short- and long-term projections intended to aid in crafting policies that might ameliorate these effects. Additionally, the comprehensive nature of these reports, of which the *U.S. National Assessment of the Potential Consequences of Climate Variability and Change* (USGCRP, 2001) was the only example, offered a holistic representation of all of the related factors. Rather than merely organizing them by research topic, these reports offered reports tailored to ten regional groups as well as the five major sectors of agriculture, water, human health, coastal/marine areas and forests. The annual editions of *Our Changing Planet* also reflected an updated set of key facts and foci related to global environmental change; but they have primarily served as anecdotal case studies, rather than projecting the understanding of trends into the future.

Relevance to the core questions established by the VARIP project

The case of the Climate Change Science Program offers a number of insights regarding the nature of science/practice interfaces. With respect to the ability to support efforts for vulnerability reduction and the building of resilience and adaptation, the CCSP demonstrates that there is a vast knowledge base available; but much of this can remain untapped by the structure of the enquiry. The goals and research questions included in these concepts represent a useful first step; but limited resources, budget and lack of leadership and presidential support undermined these efforts. Currently, the knowledge seems to be distributed in a unidirectional fashion, from knowledge producer to a group of largely federal knowledge users with few feedbacks, across a complex spider web. However, it has been suggested that practitioners need to be included in the continued development of these products, and the Obama administration plan to establish a new climate services programme seeks such a development.

Although it represented a subtle shift in the organizational title, the conversion from the far-reaching Global Change Research Program to a more targeted Climate Change Science Program did relate to an intentional avoidance of other types of global changes, notably those such as agricultural expansion, which are undeniably human induced and controlled. This change in lexicon, however, included the potential to introduce a dialogue of uncertainty into the programme structure, such that significant efforts had to be invested in statistical proof of the conclusions, rather than the record of possibilities.

In two reports, the Committee on Strategic Advice on the US Climate Change Science Program of the National Research Council issued important evaluations. In the first, *Evaluating Progress of the U.S. Climate Change Science Program* (USNRC, 2007), the committee reached six major conclusions:

1 Discovery science and understanding of the climate system are proceeding well; but use of the knowledge to support decision-making and to manage the risks and opportunities of climate change is proceeding slowly.
2 Progress in understanding and predicting climate change has improved more at global, continental and ocean-basin scales than at regional and local scales.
3 Our understanding of the impact of climate change upon human well-being and vulnerabilities is much less developed than our understanding of the natural climate system.
4 Science-quality observation systems have fuelled advances in climate change science and applications; but many existing and planned observing systems have been cancelled, delayed or degraded, which threatens future progress.
5 Progress in communicating CCSP results and engaging stakeholders is inadequate.
6 The separation of leadership and budget authority presents a serious obstacle in the CCSP.

In its second report, *Restructuring Federal Climate Research to Meet the Challenges of Climate Change* (USNRC, 2009b), the committee followed these evaluation findings with a number of key recommendations aimed at a substantial upgrading of the US Climate Change Research Program, with its top priorities to include the following:

- Reorganize the programme around integrated scientific/societal issues.
- Strengthen research on adaptation, mitigation and vulnerability.
- Initiate a national assessment process with broad stakeholder participation to determine the risks and costs of climate change impacts upon the US and to evaluate options for responding.
- Coordinate federal efforts to provide climate services (scientific information, tools and forecasts) routinely to decision-makers.

In its first five months in 2009, the Obama administration appointed a 'climate czar', a secretary of energy and a science adviser, all highly knowledgeable about climate change. Efficiency requirements for the US auto fleet were increased, and further initiatives to reduce greenhouse gas emissions were expected at the time of writing.

Recommendations and conclusions

For the Climate Change Science Program to become an effective agent in managing the US response to vulnerability to climate and other types of global environmental changes, the programme must be restructured. In the continued flux of scientific knowledge, recommendations often built upon previous knowledge, but sometimes replaced or substantially augmented these established understandings with new insights.

Although great effort was invested in quantitatively recording the ranges of climatic variability and the associated uncertainty, little attention was previously invested in the human feedbacks hinted at by goals 4 and 5. While the *Synthesis and Assessment Product* reports have made some contributions, their limited scope is a major concern.

The CCSP could have as a strong model the Intergovernmental Panel on Climate Change, whose *Fifth Assessment Report* is already under way. Along with extremely publicized press packets and public summaries, the *Fourth Assessment Reports* have targeted many of the same themes as the CCSP, on a global scale and with a much more public presence.

Despite its challenges, a recast Climate Change Science Program and the new Climate Services effort of the National Oceanic and Atmospheric Administration have the potential to comprehensively address many varied sectors of human interactions with the environment. The annual *Our Changing Planet* along with the SAPs offer a mechanism for the US government to compile the information and interpretations regarding our increased understandings of climate change. But whatever new creation evolves in the Obama administration, much greater integration between science and decision-making will need to be achieved, beginning with high-level support in the presidential and congressional offices.

References

USCCSP (US Climate Change Science Program) (2003) *Strategic Plan for the U.S. Climate Change Science Program*, Final Report, July 2003, USCCSP, Washington, DC

USCCSP (2006) *Temperature Trends in the Lower Atmosphere: Steps for Understanding and Reconciling Differences*, Report by the US Climate Change Science Program and the Subcommittee on Global Change Research (eds T. R. Karl, S. J. Hassol, C. D. Miller and W. L. Murray), USCCSP, Washington, DC

USCCSP (2008) *Revised Research Plan for the U.S. Climate Change Science Program*, USCCSP, Washington, DC

USCCSP and the Subcommittee on Global Change Research (1995–2010) *Our Changing Planet*, Annual Reports, USCCSP, Washington, DC

US Climate Change Technology Program (2006) *Strategic Plan*, DOE/PI0005, Washington, DC

USGCRP (US Global Change Research Program) (2001) *U.S. National Assessment of the Potential Consequences of Climate Variability and Change*, Foundation Report, www.usgcrp.gov/usgcrp/nacc

USNRC (US National Research Council) (2005) *Thinking Strategically: The Appropriate Use of Metrics for the Climate Change Science Program*, National Academies Press, Washington, DC

USNRC (2007) *Evaluating Progress of the U.S. Climate Change Science Program: Methods and Preliminary Results*, National Academies Press, Washington, DC

USNRC (2009a) *Informing Decisions in a Changing Climate*, National Academies Press, Washington, DC

USNRC (2009b) *Restructuring Federal Climate Research to Meet the Challenges of Climate Change*, National Academies Press, Washington, DC

USOTA (US Office of Technology Assessment) (1993) *Preparing for an Uncertain Climate*, Volume 1, OTA-0-567, US Government Printing Office, Washington, DC

Appendix 1: Initial US Climate Change Science Program (CCSP) Strategic Plan Summary of Synthesis and Assessment Products

Product number (completion date)	Title	Lead agency (and contributors)
CCSP Goal 1: Extend knowledge of the Earth's past and present climate and environment, including its natural variability, and improve understanding of the causes of observed changes		
Product 1.1 (May 2006)	Temperature trends in the lower atmosphere: steps for understanding and reconciling differences	NOAA (with NASA, DOE, NSF)
Product 1.2 (June 2008)	Past climate variability and change in the Arctic and at high latitudes	USGS (with NSF, NOAA, NASA, DOE)
Product 1.3 (June 2008)	Reanalysis of historical climate data for key atmospheric features: implications for attribution of causes of observed change	NOAA (with NASA, DOE)
CCSP Goal 2: Improve quantification of the forces bringing about changes in the Earth's climate and related systems		
Product 2.1 (May 2007)	Scenarios of greenhouse gas emissions and atmospheric concentrations and review of integrated scenario development and application	DOE (with EPA, NOAA, NASA)
Product 2.2 (May 2007)	North American carbon budget and implications for the global carbon cycle	NOAA (with DOE, NASA, USDA, USGS)
Product 2.3 (September 2007)	Aerosol properties and their impacts upon climate	NASA (with NOAA)
Product 2.4 (June 2008)	Trends in emissions of ozone-depleting substances, ozone layer recovery, and implications for ultraviolet radiation exposure and climate change	NOAA (with NASA)
CCSP Goal 3: Reduce uncertainty in projections of how the Earth's climate and related systems may change in the future		
Product 3.1 (June 2007)	Climate models: assessment of strengths and limitations for user applications	DOE (with NOAA, NASA, NSF)
Product 3.2 (December 2007)	Climate projections for research and assessment based on emissions scenarios developed through the Climate Change Technology Program	NOAA (with NSF, DOE)
Product 3.3 (June 2008)	Climate extremes: analysis of the observed changes and variations and prospects for the future	NOAA (with NASA, USGS, DOE)
Product 3.4 (June 2008)	Risks of abrupt changes in global climate	USGS (with NOAA, EPA, DOE, NSF)
CCSP Goal 4: Understand the sensitivity and adaptability of different natural and managed ecosystems and human systems to climate and related global changes		
Product 4.1 (September 2007)	Coastal elevation and sensitivity to sea-level rise	EPA (with USGS, NOAA, NASA, DOE)
Product 4.2 (December 2007)	State of knowledge of thresholds of change that could lead to discontinuities (sudden changes) in some ecosystems and climate-sensitive resources	USGS (with EPA, NOAA, USGS, DOE, NSF)
Product 4.3 (December 2007)	Analyses of the effects of global change on agriculture, biodiversity, land and water resources	USDA (with EPA, NOAA, NASA, NSF, USGS, DOE, USAID)
Product 4.4 (December 2007)	Preliminary review of adaptation options for climate-sensitive ecosystems and resources	EPA (with USDA, NOAA, NASA, USGS, DOE, USAID)

Product number (completion date)	Title	Lead agency (and contributors)
Product 4.5 (June 2007)	Effects of climate change on energy production and use in the US	DOE
Product 4.6 (December 2007)	Analyses of the effects of global change on human health and welfare and human systems	EPA (with NOAA, NASA, DOE)
Product 4.7 (December 2007)	Impacts of climate variability and change on transportation systems and infrastructure – Gulf Coast study	DOT (with USGS, DOE, NASA)

CCSP Goal 5: Explore the uses and identify the limits of evolving knowledge to manage risks and opportunities related to climate variability and change

Product 5.1 (July 2007)	Uses and limitations of observations, data, forecasts and other projections in decision support for selected sectors and regions	NASA (with EPA, NOAA, USGS, DOE)
Product 5.2 (July 2007)	Best-practice approaches for characterizing, communicating and incorporating scientific uncertainty in decision-making	NOAA (with EPA, UGSG, DOE, NSF, NASA)
Product 5.3 (December 2007)	Decision support experiments and evaluations using seasonal to inter-annual forecasts and observational data	NOAA (with NASA, EPA, USGS)

Source: adapted from http://www.climatescience.gov/Library/sap/sap-summary.php, accessed 26 April 2007

Appendix 2: CCSP goals and focus areas

Focus area	Description	Agencies involved
Goal 1	**Improve knowledge of the Earth's past and present climate and environment, including its natural variability, and improve understanding of the causes of observed variability and changes**	
Focus 1.1	Better understand natural long-term cycles in climate – e.g. Pacific Decadal Variability (PDV), North Atlantic Oscillation (NAO)	DOE, NASA, NOAA, NSF, USGS
Focus 1.2	Improve and harness the capability to forecast El Niño-La Niña and other seasonal to inter-annual cycles of variability	DOE, NASA, NSF, USGS
Focus 1.3	Sharpen understanding of climate extremes through improved observations, analysis and modelling, and determine whether any changes in their frequency or intensity lie outside the range of natural variability	DOE, NASA, NOAA, NSF, USGS
Focus 1.4	Increase confidence in the understanding of how and why climate has changed	DOE, NASA, NSF, USGS, SI
Focus 1.5	Expand observations and data/information system capabilities	DOE, NASA, NOAA, NSF, USGS, SI, EPA
Goal 2	**Improve quantification of the forces bringing about changes in the Earth's climate and related systems**	
Focus 2.1	Reduce uncertainties about the sources and sinks of greenhouse gases (GHGs), emissions of aerosols and their precursors, and their climate effects	DOE, NASA, NOAA, NSF, DOT
Focus 2.2	Monitor the recovery of the ozone layer and improve the understanding of the interactions of climate change, ozone depletion, tropospheric pollution, and other atmospheric issues	DOE, NASA, NSF, USDA, SI
Focus 2.3	Increase knowledge of the interactions among emissions, long-range atmospheric transport and transformations of atmospheric pollutants, and their response to air quality management strategies	NASA, NSF, USDA
Focus 2.4	Develop information on the carbon cycle, land cover and use, and biological/ecological processes by helping to quantify net emissions of carbon dioxide, methane and other greenhouse gases, thereby improving the evaluation of carbon sequestration strategies and alternative response options	DOE, NASA, NOAA, NSF, USDA, USGS, SI
Focus 2.5	Improve capabilities to develop and apply emissions and related scenarios for conducting 'If … then … ' analyses in cooperation with CCTP	DOE
Goal 3	**Reduce uncertainty in projections of how the Earth's climate and related systems may change in the future**	
Focus 3.1	Improve characterization of the circulation of the atmosphere and oceans and their interactions through fluxes of energy and materials	DOE, NASA, NOAA, NSF
Focus 3.2	Improve understanding of key 'feedbacks', including changes in the amount and distribution of water vapour, extent of ice and the Earth's reflectivity, cloud properties, and biological and ecological systems	DOE, NASA, NSF, USGS

Focus area	Description	Agencies involved
Focus 3.3	Increase understanding of the conditions that could give rise to events such as rapid changes in ocean circulation due to changes in temperature and salinity gradients	NASA, NSF, USGS
Focus 3.4	Accelerate incorporation of improved knowledge of climate processes and feedbacks within climate models to reduce uncertainty in projections of climate sensitivity, changes in climate, and related conditions such as sea-level rise	DOE, NASA, NOAA, NSF
Focus 3.5	Improve national capacity to develop and apply climate models	DOE, NASA, NOAA, NSF
Goal 4	**Understand the sensitivity and adaptability of different natural and managed ecosystems and human systems to climate and related global changes**	
Focus 4.1	Improve knowledge of the sensitivity of ecosystems and economic sectors to global climate variability and change	DOE, NASA, NSF, USDA, USGS, SI, EPA
Focus 4.2	Identify and provide scientific inputs for evaluating adaptation options, in cooperation with mission-oriented agencies and other resource managers	NSF, DOT, NIH, EPA, SI
Focus 4.3	Improve understanding of how changes in ecosystems (including managed ecosystems such as croplands) and human infrastructure interact over long periods of time	DOE, NASA, NSF, USDA, SI
Goal 5	**Explore the uses and identify the limits of evolving knowledge to manage risks and opportunities related to climate variability and change**	
Focus 5.1	Support informed public discussion of issues of particular importance to US decisions by conducting research and providing scientific synthesis and assessment reports	DOE, NASA, NSF, USDA, USGS, SI, STATE, EPA
Focus 5.2	Support adaptive management and planning for resources and physical infrastructure sensitive to climate variability and change; build new partnerships with public- and private-sector entities that can benefit both research and decision-making	NASA, NOAA, NSF, USDA, USGS, USAID, EPA
Focus 5.3	Support policy-making by conducting comparative analyses and evaluations of the socio-economic and environmental consequences of response options	NASA, USDA, SI

Source: adapted from Our Changing Planet 2006, http://www.usgcrp.gov/usgcrp/Library/ocp2006/, accessed 26 April 2007

6

Linking Climate Change Science with Policy in California[1]

*Guido Franco, Dan Cayan, Amy Luers,
Michael Hanemann and Bart Croes*

Introduction

The California Climate Change Scenarios assessment (the Scenarios Project) was initiated to explore possible climate change impacts in the state to help inform important state climate policy decisions such as:

- How much climate change is acceptable?
- What actions are needed to enhance our ability to cope with future change?

Californian policy-makers have long recognized the critical role that science can play in the policy process by characterizing the risks and alternatives for managing climate change. In this chapter we describe how the Scenarios Project (along with a series of other climate assessments) has helped to establish California as one of the most significant climate policy actors in the US.

Across the globe, science has been crucial for raising awareness of climate change and building support for climate policy. At the international level, the Intergovernmental Panel on Climate Change (IPCC) assessments have helped to move international policy with the signing of the United Nations Framework Convention on Climate Change (UNFCCC) and the Kyoto Protocol. At the national level, the US Climate Change Science Program was similarly developed to inform policy and management decisions around climate issues.

Over the last decade a number of scientific assessments have been undertaken to build awareness and inform regional climate policy decisions within the US. For example, the Climate Impacts Group (CIG) at the University of Washington, established as part of the National Oceanic and Atmospheric Administration (NOAA) Regional Integrated Sciences and Assessments (RISA) programme, has prepared a series of reports on the recent and possible future climate changes in the Pacific Northwest and their impacts upon sensitive systems, including water resources, fisheries and forests.[2] CIG, along with partners from King County, Washington and Local Governments for Sustainability, has also prepared a guide book providing information for local governments on how to adapt to climate change.[3] In Colorado, the Rocky Mountain Climate Organization has prepared reports on how climate disruption threatens[4] the west's snow and water.[5] Many of these regional assessments have been the outgrowth of, or seeded, formal or informal regional climate research networks.

In California, the state has established the California Climate Change Center[6] as a regional interdisciplinary research programme to help inform relevant climate science. In response to Governor Arnold Schwarzenegger's Executive Order of 2005, the Climate Change Assessment Office established the Scenarios Project. The research for this project was carried out by over 70 scientists from within and outside of the California Climate Change Center, many of whom were also connected to other formal or informal research networks focused on regional climate-related issues. The findings from this assessment were included in a high-profile state report to the governor and legislature outlining a strategy for climate action.[7]

As regional and national efforts to manage climate variability and change expand, climate research programmes such as the California Climate Change Center and the National Oceanic and Atmospheric Administration's RISA programmes are likely to become increasingly important. We believe these regional decision-support efforts could benefit by sharing experiences among programmes. As a result, we have written this chapter to document experiences from the Californian assessment activities.

This chapter is divided into three main sections. First, we present a brief history of how climate change research in California has interacted with the policy arena and, in particular, highlight those events that led up to the Scenarios Project. Next, we point to the characteristics of the Scenarios Project that made it a challenging assessment to carry out and describe facets that contributed to its successful completion. Finally, we conclude with some lessons learned from this climate assessment experience.

A brief history

Climate change has been on the minds of policy-makers in California for at least 18 years. In 1988 State Senator Byron Sher spearheaded the adoption of the state Assembly Bill 4420 (AB 4420) which called on the California Energy

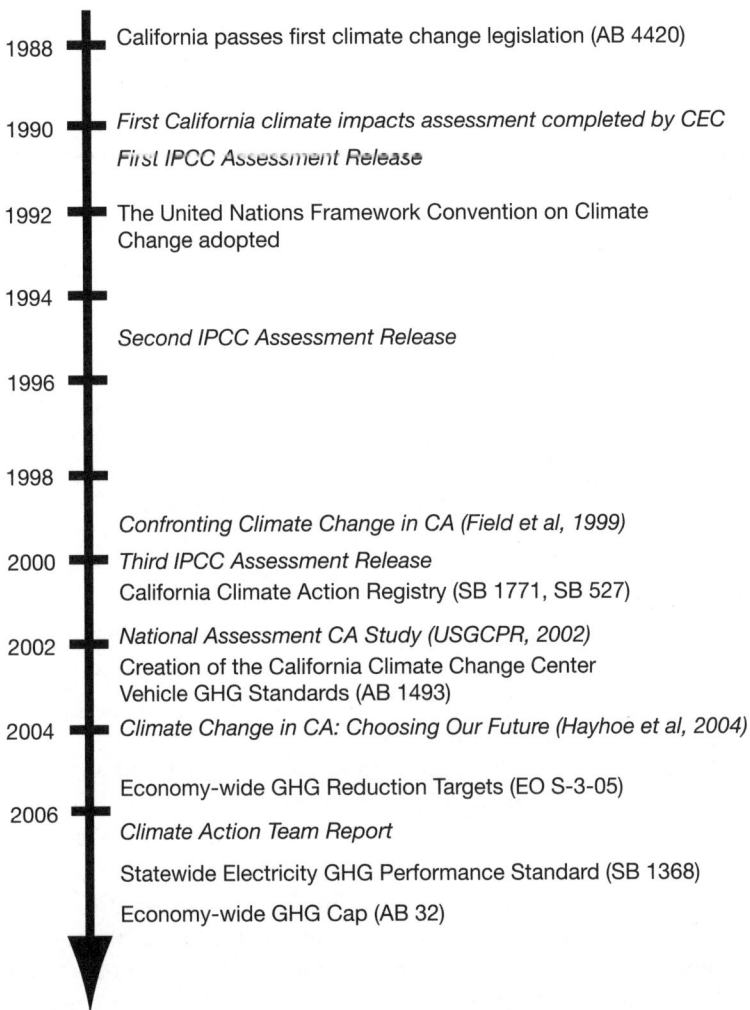

Figure 6.1 *Climate science and Californian climate policies: A timeline*

Note: Analytical reports are listed in italicized text.
Source: chapter authors

Commission (CEC) to lead in preparing the first assessment of the potential impacts of climate change upon California and of the options for reducing greenhouse gas (GHG) emissions in the state. The 1988 law led to two high-profile climate reports: *The Impacts of Global Warming on California* (CEC, 1989) and *Global Climate Change: Potential Impacts and Policy Recommendations* (CEC, 1991). These reports helped to generate public discussion around climate change in the state; however, it was over a decade before the state implemented its first policies taking real action to address climate change (see Figure 6.1). This lack of political action for over a decade can likely be attributed to at least two reasons:

1 the absence of a unified 'voice' from the scientific community in California conveying their concerns about potential impacts directly to decision-makers;[8] and
2 the absence of leadership in Sacramento in the governor's office on climate change issues.[9]

In this section, we highlight California's most significant climate policies and describe the central role of regional climate science research efforts in building support for climate action in the state.

California climate policies

Over the last six years California has passed some of the strongest climate policies in the US. These new policies have been motivated, in part, by increasing concern over the risks of climate-related impacts and facilitated by the state's existing framework of energy and air quality policies.

California took its first steps to regulate GHG emissions in 2000 by passing Senate Bill 1771, which specified the creation of the non-profit California Climate Action Registry (CA Registry). The CA Registry allows state organizations to register their emissions and to track voluntary emissions reduction measures. Senate Bill 1771 was a mild legislative response to a report issued in 1998 by the California Energy Commission updating the state-wide inventory of greenhouse gas emissions (CEC, 1998a), describing the strategies that could be implemented to reduce emissions (CEC, 1998b), and suggesting that companies reducing emissions should be protected against future potential mandatory requirements. This was followed in 2002 by the passage of Assembly Bill 1493 (Pavley), the precedent-setting law which required the Air Resources Board to develop regulations limiting the amount of greenhouse gases emitted from automobiles. Although automakers have filed suit to block the implementation of the vehicle GHG standards, arguing, in part, that these state pollution standards are pre-empted by federal fuel economy standards, many other states have moved forward to adopt California's standards. On 2 April 2007, the US Supreme Court ruled that carbon dioxide is a pollutant and that the US Environmental Protection Agency (EPA) has the authority to regulate this and other greenhouse gases. This rule will bolster the case that California is making regarding its authority to regulate greenhouse gas emissions from automobiles.

In 2005, California once again demonstrated its leadership on climate issues when Governor Schwarzenegger signed an Executive Order (S-3-05) on 1 June establishing greenhouse gas emission targets for California. These targets are to bring emission levels in 2010 and 2020 to what the state emitted in the years 2000 and 1990, respectively, and to reduce emissions by 80 per cent from the 1990 levels by 2050. The Executive Order also transferred the coordination of climate change activities in California from the CEC to the secretary of the California Environmental Protection Agency (CalEPA), requiring the secretary to prepare a report delineating how the state would be able to comply with the GHG emission targets. The CalEPA secretary formed the Climate Action Team (CAT) with representatives from the California Resources Agency, Air Resources Board, the Energy Commission and four other state agencies to oversee the production of this report. The Executive Order also mandated the preparation of a biennial science report describing the impacts that climate change would have on water supply, forestry, public health, agriculture and the coastline, and discussing coping and adaptation strategies that the state should consider. This resulted in the Scenarios Project described in the section 'California Climate Scenarios Project'.

In September 2006, the 2020 emissions reduction target set by the 2005 Executive Order was put into law with Assembly Bill 32. AB 32 established mandatory reporting of GHG emissions from major sources in California and requires the Air Resources Board to develop regulations to cap GHG emissions to 1990 levels by 2020.

Climate science

Although advances in climate policy in California were relatively modest during the late 1980s and 1990s, a number of significant coordinated research efforts and research programmes were initiated during this period that would lay the foundation for future climate action. In particular, a series of high-profile assessments highlighted for California policy-makers the severity of the risks posed by unabated climate change and helped to raise public support for climate action. The growing support for climate action in California has been documented by a series of polls led by the Public Policy Institute of California (PPIC). In 2006 the PPIC poll indicated that 70 per cent of likely voters favoured actions to address global warming – this was up from 35 per cent reported in a 2000 poll (PPIC, 2000, 2006).

Government and academic institutions

During the late 1990s the National Oceanic and Atmospheric Administration (NOAA), a federal agency whose mission includes predicting and understanding weather and climate, created the RISA research programme to better understand information needs and to provide for both short- and long-term operation and planning purposes for regional and local resource managers. In California, under the NOAA RISA programme, the California Applications Program (CAP) was established at the Scripps Institution of Oceanography, University of California San Diego. The CAP has focused on climate variability and climate change impacts upon water resources, wildfire and human health, and has been involved in climate impact studies and assessments produced in the state.[10] During this same period the US Department of Energy (DOE) funded the Accelerated Climate Prediction Initiative (ACPI) as a demonstration project, which on the US West Coast involved the Scripps Institution of Oceanography, the University of Washington at Seattle, Pacific Northwest

National Laboratory and other research institutions. The main goal of this programme was to increase capabilities to produce decade-to-century scale projections of climate at regional scales and to use these projections for some climate impact and adaptation analyses. The results of this regional ACPI study were disseminated in a special issue of *Climatic Change* (Barnett and Pennell, 2004). This study used one global climate model, the National Center for Atmospheric Research's (NCAR) Parallel Climate Model, and one global GHG emission scenario to investigate potential impacts of climate change upon water resources and other water-related phenomena.

Meanwhile at the national level, the US Global Change Research Program published the first national 2001 assessment report (USGCRP, 2001), which has been praised for its inclusiveness and high stakeholder involvement (Morgan et al, 2005). As part of this work, several regional assessments reports were produced, including one dealing exclusively with California (USGCRP, 2002). The California contribution to the national assessment was widely disseminated and Robert Wilkinson, director of the California Climate Assessment, provided several presentations to high-level officials in the state and to other audiences at multiple meetings and forums.

At the same time as the national assessment was under preparation, California initiated its own state-supported integrated climate research programme. The California Energy Commission's Public Interest Energy Research (PIER) programme initiated climate change research with an exploratory project with the Electric Power Research Institute (EPRI) involving scientists from California as well as outside California to investigate the potential impacts of climate change upon water resources, forestry, agriculture, coastal properties and ecosystems. This project culminated with a report in 2003 suggesting that human adaptation to climate change would be costly for California and that impacts upon ecosystems would be severe, with little room for adaptation considering other stressors such as increased unabated urbanization. This study served as the foundation for future studies and some features of the study (e.g. the use of estimated future urban growth patterns still remains a unique feature of this study) (Wilson et al, 2003).

Following this initial report, PIER commissioned roadmaps of research on different topics such as regional climate modelling, GHG inventory methods, water resources and carbon sequestration. The roadmaps were designed to identify research gaps not adequately covered by existing research programmes at the national or international levels, but of high importance for California. Technical staff from different state agencies and researchers associated with universities and research institutions in California participated in the review process for these roadmaps. This effort culminated with an integrated five-year research plan on climate change for California released at the end of 2003 and aimed at addressing the following policy-relevant questions:

- How is climate changing in California and what are plausible climate change scenarios for California?

- How would climate change (physical impacts) affect California's environment and economy?
- What are the merits of different mitigation and adaptation strategies?
- How would climate change affect energy supply and demand?
- How would climate change policies affect the economy?

The integrated plan extracted the critical areas of work that PIER could support with a funding level of about US$6 million a year and that could substantially assist PIER in addressing the above questions. To implement this plan, the PIER programme created a virtual research centre, the California Climate Change Center, with core research at the University of California at Berkeley and at the Scripps Institution of Oceanography, University of California San Diego, as well as an equally important and substantial complementary set of diverse research activities at other research institutions. This research centre is remarkable in being one of the first state-sponsored climate research programmes in the US. An important underpinning of the PIER programme is that it uses ongoing national and international research efforts as the foundation for its research programme. PIER mostly funds applied research projects with a goal of informing policy formulation in the state.

This chapter only reports on the activities of the California Climate Change Center that are related to the Scenarios Project. It is important to indicate, however, that the centre has and will continue to generate information on other climate change topics, such as regional climate detection and attribution, regional climate projections, improved methods to estimate greenhouse gas emissions, and mitigation, impacts and adaptation studies.[11] Research results generated by the centre have been used to prepare the official state-wide inventory of GHG gases in the state (CEC, 2002, 2005a) and to identify preliminary mitigation strategies in different policy forums (CEC, 2005b; CAT, 2006).

Among the many high-impact products that were produced from the various coordinated research programmes were those highlighting the potentially severe threats that climate change posed to California's water resources. This threat was first brought to the attention of the scientific community when the chief hydrologist of the California Department of Water Resources (DWR) released a study (Roos, 1991) showing that the contribution to annual water-year runoff occurring in the spring and summer months has been on a declining trend since records began during the early 1900s. Peter Gleick and others had hypothesized that warming would result in an early melt of snow in the Sierra Nevada, reducing the capacity of the snowpack to serve as a natural water reservoir in California (Gleick, 1986, 1987). Then, in 2002, CAP scientists produced an analysis that graphically showed how 1 April snow levels in the Sierra Nevada would be increasingly and substantially reduced as climate warms (Knowles and Cayan, 2002). When this report was released the executive branch of state government had different players, most of them unaware or only partially aware of the work done on climate change during prior years, including the early Energy Commission's report on the subject (CEC, 1989). The CAP study, along with others reported by researchers associated with the University of Washington at Seattle, produced as part of the ACPI project, raised the awareness of top water managers in California to the potential serious effects of a warming climate on water resources in the state. These managers decided that the 2005 version of the *State Water Plan*, which is a major water policy document in California prepared every five years, should include a discussion on climate change. Given the limitations of time and resources available from when this decision was made to the public release of the draft version of this plan, as much information as possible was included. In its final form, the 2005 plan contained a literature review on climate change and water resources in the state and some side notes with relevant information taken from the 2003 PIER study (DWR, 2006).

The emergence of information with increasing levels of detail about potential climate change impacts in the region helped to create momentum necessary for meaningful climate policy formulation in California. At the same time, the continuity and focus of the science effort to elucidate regional climate change phenomena were enabled by a somewhat informal, but tight-knit, collaboration among several climate scientists in the state and the region. Support from several respective academic and government institutions and at least a modicum of funding for longstanding climate problems was key in sustaining this involvement. The existence of forums such as the Pacific Climate (PACLIM) Workshop series, soon to hold its 23rd annual or semi-annual meeting, is one example of the cohesiveness of the science network.

Non-governmental organizations

An important part of the climate–science interface in California includes the non-governmental organization (NGO) community that has taken the lead in coordinating research reports and a couple of influential climate impact assessments. These assessments differed from many of the previous studies in that they focused, from their inception, on communicating the risks of climate change to the non-technical audience. For example, in 1999 the Union of Concerned Scientists (UCS) and the Ecological Society of America (ESA) initiated and coordinated an assessment that was lead by Professor Chris Field and other prominent Californian scientists on the potential impacts of climate change upon ecosystems in California (Field et al, 1999). The final report was distributed widely to state agencies and has been referred to as the 'green book' that helped to educate California policy-makers about climate change (Boyd, 2006).

This 1999 assessment drew much attention in the state government and helped to reinitiate discussions about climate change. The California Resources Agency coordinated these discussions; but the effort did not result in concrete implementation of new climate change initiatives, in part due to a lack of interest in the governor's office. However, the report by Field and colleagues was likely influential in building support in the legislature for Assembly Bill 1493 (Pavley), the California vehicle GHG standard.

In 2004, UCS initiated a second California assessment that was again focused from its inception on both outreach and scientific credibility. The resulting study was conducted by 18 scientists, including the principal scientist associated with the California Climate Change Center. The results of the analysis were first published in a paper in 2004 in the *Proceedings of the National Academy of Sciences* (PNAS) (Hayhoe et al, 2004). The major contribution of this paper was to show that future global greenhouse gas emission levels could play a substantial role in determining the severity of impacts in California. UCS worked with the science team to produce a summary of the PNAS paper for outreach to non-technical audiences. The public version of this report, entitled *Choosing Our Future: Climate Change in California*, was released in the weeks leading up to the public hearings around the rule makings for AB 1493 to help provide support for climate action in the state. Following the release, a series of presentations were conducted by report authors with high-level officials at the different state agencies, policy-makers and the governor's office.

This 2004 PNAS paper and the public version of this report (*Choosing our Future*) received wide media attention and have had a major impact upon climate policy in the state. Its findings were made part of the rule-making record with testimony provided by some of the researchers during the public hearings that the Air Resources Board conducted for consideration of the, at that time, proposed regulations for AB 1493 (Pavley). It was also included as an appendix to the Western Governor's Association Recommendations.[12] Even outside California the PNAS study and the *Choosing our Future* outreach brochure drew attention in the policy world. Authors were asked to report on the analysis in the US Congress, as well as international conferences on climate policy.

California Climate Scenarios Project

The California Climate Scenarios Project was prepared in response to a 2005 state Executive Order's call for biennial science reports. The secretary of CalEPA entrusted the leadership of the preparation of these science reports to the PIER programme and the California Climate Change Center. However, CalEPA was also aware of the importance of communicating the assessment to non-technical audiences and therefore also recruited the UCS to play a lead role in synthesizing the assessment findings.

The final science report, which became known as the Scenarios Project, was prepared as a multi-institutional collaboration among CEC, other state agencies (including the Air Resources Board, the Department of Water Resources and the California Department of Forestry and Fire Protection) and the UCS. The Climate Action Team adopted the findings of the scientific team and included the full set of reports prepared by the science team in the report that the CalEPA Secretary and the Climate Action Team submitted to the governor and legislature (CAT, 2006).

The Scenarios Project is unique among the climate assessments that have been conducted in California. Several distinguishing factors include the following:

- The Scenarios Project was built into a state-mandated comprehensive report that outlined a set of strategies for managing climate change through aggressive mitigation and adaptation approaches.
- The Scenarios Report was initiated and completed in less than eight months.
- The dissemination of its key findings to the scientific community, policy-makers and the public (California Climate Change Center, 2006) was planned from the beginning.
- The Scenarios Project assessment was broader than previous efforts, including the efforts of 70 physical science and social science experts from academic, government and other organizational units.
- Connections between the scientists and the technical staff at different state agencies were established or reinforced, which will improve the quality of future long-term planning in California prepared by these agencies.

All of these factors contributed to making this a high-profile assessment with the potential of influencing climate policy in the state. According to Eileen Tutt, assistant secretary for Climate Change Activities, California Environmental Protection Agency: 'The findings of the report contributed greatly to our understanding of the effects of climate change emissions in California. These findings were the basis of the scientific evidence reflected in the March 2006 Climate Action Team report and in AB 32, the California Global Warming Solutions Act of 2006' (CEC, 2007, p72). However, several aspects added to the challenges of completing a successful product. Below we highlight some of the challenges of the Scenarios Project and some of the key factors that we attribute to the success of the assessment.

Challenges in preparing the Scenarios Project

The conflicting pace of politics and science

The Executive Order called for the final science 'scenarios' report by the end of January 2006, providing only six months for the preparation of this report. In practice, additional time was made available to the research team because the final report was released in March 2006. A substantial amount of time was consumed in non-research activities, such as assembling the team and developing the scope of work. A logistic challenge was the process of putting in place many necessary contracts to fund different research groups. Given the tight timeline, work started immediately after the research team was assembled. CalEPA, with some of the agencies under its purview, and the PIER programme funded the study. The contracts were in place in September, well after the work was under way.

Due to the extremely tight deadline some of the Scenarios Project assessments relied on work that had already been started under funding by PIER or that which other researchers had produced or were close to completing.

Mixed expectations

What Californian policy-makers wanted was sometimes beyond the bounds of what the science team could deliver. CalEPA requested an assessment of physical impacts and, importantly, the translation of these impacts to economic outcomes and discussions about adaptation strategies. However, given the lack of a comprehensive body of work on this subject, the lack of necessary datasets and the extremely short timetable, the economic analyses were only able to partially achieve these goals.

The challenge of eliciting stakeholder involvement

Given the extremely short schedule for the Scenarios Project, only very minor stakeholder involvement was possible. As part of the CAT activities, three public workshops were organized first to discuss the scope of the analyses and then to report draft and final results. Relatively little public input was received, which might be attributed to the following factors:

1. The stakeholders may have been more interested in the strategies that the CAT was developing to control GHG emissions from state sources.
2. Very little time was available to digest the results presented.
3. The strong track record of the research team may have created a perception of high credibility and acceptance.
4. Results were perceived to be highly uncertain.
5. There was a perception that significant impacts are far in the future.
6. Few impacts were perceived to affect the stakeholders directly.

Ideally, as the awareness of decision-makers is raised concerning the potential severity of climate change impacts, the dialogue between them and the scientific community will increase.

Factors that contributed to the success of the Scenarios Project

Scientific credibility

The research was based upon 'mainstream' climate model simulations delivered to the IPCC climate assessment, and was carried out by active scientists from major research institutions within and outside the state. All of the papers were submitted to a peer-review process and their major findings were summarized in an overview paper (Cayan et al, 2006). The Office of the President of the University of California was in charge of the peer-review process with at least two, and usually three, reviewers per paper.

Presentation and communication

The authors of the paper upon which this chapter is based, with assistance from the different researchers involved in this study, also prepared a brochure summarizing the major findings of the study in simple language (California Climate Change Center, 2006). The target audience for this brochure comprised policy-makers in state government and the private sector, and the public in general.

Project management

The Scenarios Project was managed by a committee established by the Climate Action Team. The committee was composed of representatives from PIER, the Air Resources Board, the UCS, the DWR, the California Department of Forestry and Fire Protection, the California Department of Food and Agriculture, the California Department of Finance, and the California Department of Transportation. This committee met four or five times during the execution of the project to discuss general progress and links to other components of the Climate Action Team report. Additionally, the science team formed a 'gang of five' (the authors of this chapter) who coordinated efforts and kept the individual study efforts in contact and on schedule in delivering results.

Scientific network

A large, strongly-to-weakly collaborative network of climate science researchers within and outside of California provided the technical expertise, models and databases to complete an extensive analysis in an extremely short interval. The importance of local scientists, however, cannot be overemphasized. Because they were local, they were able to participate in multiple meetings with representatives from CalEPA and the CAT team during the design phase of the study and to communicate the scientific findings to decision-makers and the public in general at multiple forums. Their availability and accessibility to local news networks also resulted in 'tailored' reporting in local news outlets about the implications of climate change to their communities.

Lessons learned

There are several lessons learned from the effort that culminated with the release of the Scenarios Project report. These lessons become more illuminating, however, if they are internalized in the context of the larger climate change science–policy efforts that have been taking place in the state during the last 15 years. They include the following.

Timing matters

The high impact of the 2004 PNAS study/*Choosing our Future* report on California policy was likely, in part, because it was released at a point in time when the general public and key policy-makers were prepared for enacting regulations for AB 1493 and the Governor's Executive Order S-3-05. Similarly, findings from the Scenarios Project were released in time to have an impact upon deliberations associated with AB 32.

Since 1991 there have been several attempts to convey to the legislature and the governor the importance of climate change to California (Hanemann, 2007). Some of these attempts have been briefly mentioned in this chapter. Climate change has always had its champions in the legislature; but the major stumbling block has been the governor's office. An unexpected opportunity occurred with the election of Governor Schwarzenegger, who has been willing to hear what 'his' scientists have to say about climate change and has been convinced of the high vulnerability of his state to climate change and, therefore, the need for urgent action.

Credibility must be attained within and outside the scientific community

Credibility must begin with scientific peer review. Beyond this, it is important to engage local scientific and community leaders.

Findings must be accessible

Assessment findings must be presented in a manner that is accessible to non-technical audiences. This includes condensed 'user-friendly' summaries and simple graphics and images that communicate key messages. Essential to the accessibility of scientific results is framing them in a manner that presents climate impacts together with solutions. The impacts reported in the Scenarios Project highlighted those impacts that could be avoided through climate action.

Scientific networks can provide a foundation

Research activities funded at the national and international levels on climate change are extremely important, but are not designed to address regionally specific issues. At the same time, the California-specific studies being funded by PIER and others cannot take place in the absence of national and international research efforts. Thus, a coordinated linkage between national and international climate change studies and regional climate research efforts is critical in order to provide decision-makers with better information. Experience here suggests that these networks can be formal or informal, but are more likely to be effective if they are sustained over a period of several years.

In addition, as discussed above, the dissemination of relevant scientific findings to multiple audiences with different interests required the presence of California-based scientists with in-depth understanding of the local conditions. The extensive scientific infrastructure in California allowed for this to happen. In situations when local scientific resources are not adequate, it might be necessary to create regional partnerships involving multiple states.

Institutional memory must be maintained

Changes in administration in state government, in this case in Sacramento in both the legislature and in the governor's office, can result in a complete overhaul of top-level officials in the different state agencies. This can result in very little or no memory of past activities on climate change. Because of this turnover, and because climate science is not static but is still evolving, it is imperative that there is a process to periodically review and update key science findings for agency and administrative officials and their staff. Furthermore, having an ongoing linkage with government staff is crucial since in many cases they are responsible for advising elected officials and other high-level managers. Government staff having an extended involvement in the climate change science and policy arena are extremely valuable. This involvement is, in our experience, the only way to attain a working familiarity and sense of the growing volume of knowledge and literature, an extensive community of science and policy experts, and a complex set of entangled climate impacts. Thus, close coordination between researchers and technical staff at the different state agencies has multiple benefits, which include building a repository for long institutional memory and enhancing the expertise at these agencies.

Outreach is essential

Outreach efforts are essential in order to disseminate science information to decision-makers and to provide crucial feedback to researchers on climate impact applications. However, in order to maintain scientific credibility, extreme care should be taken to 'speak' with a clear non-advocacy voice.

Funding and research management programmes such as PIER are, by necessity, constrained in their outreach activities given the fact that they are part of state government. Outreach activities such as the ones organized by the UCS in the state legislature and the governor's office would have been impossible to undertake by a state-sponsored programme. PIER has attempted to broadcast the results of PIER-sponsored research and other relevant research projects via its annual research conferences. These conferences have been very successful in engaging and linking the technical and scientific communities, but only modestly successful in linking them to state government. They cannot replace direct meetings with legislature and the governor's office.

Conclusions

Concerted interest in climate change at different levels of the California state government started nearly two decades ago and recognition in Sacramento has grown since then to make it a key long-term issue. A number of political factors have helped to enable state policy-makers to take climate action. In particular, federal inaction on environmental issues has allowed California to build on its legacy of successful energy and air quality policies to develop a climate-specific policy agenda that has drawn national and international support. A body of global and regional scientific studies has motivated and reinforced this process. Regional climate change assessments have clarified global climate research findings, and together these have helped to convince state decision-makers of the reality of climate change and the need for state action.

Other jurisdictions may lack public and/or governmental support to adopt similar policy actions and legislation to California on climate change. This does not diminish the need for climate assessments and the dissemination of scientific findings to the public and decision-makers. These studies and education efforts will likely be needed when conditions are ripe for acceptance.

A long path of scientific investigations has been established and will surely continue. From the Californian experience, a crucial element, not traditionally a part of the science process, is the exchange of ideas and information among and between scientists and decision-makers. This exchange has proven to be instrumental in instigating and informing policies that aim to better prepare California for the serious challenges posed by a changing climate. The Scenarios Project has taken some steps in that direction; but there needs to be a continuing integrated assessment of climate change and its impacts in California. To this end, under funding mainly from PIER and CalEPA, the state is already preparing the 2008 Scenarios Project report.

Acknowledgements

We would like to thank all the participants in the Scenario Project and especially Eileen Tutt from the California Environmental Protection Agency. We thank two anonymous reviewers for their extremely valuable comments; however, we remain responsible for the content of this chapter and for any errors or omissions that may have escaped our attention. This chapter reflects the views of the authors and does not necessarily reflect the views of the California Energy Commission or the State of California. Dan Cayan was partially supported by NOAA through its RISA programme element.

Notes

1. This chapter is adapted from Franco, G., Cayan, D., Luers, A., Hanemann, M. and Croes, B. (2008) 'Linking climate change science with policy in California', *Climatic Change*, vol 87, supplement 1, ppS7–S20.
2. See www.cses.washington.edu/cig/pnwc/cc.shtml.
3. See www.metrokc.gov/exec/speeches/20060522stateofthecounty.aspx.
4. See www.rockymountainclimate.org/index.htm.
5. See www.rockymountainclimate.org/index.htm.
6. See www.climatechange.ca.gov/research/index.html.
7. See www.climatechange.ca.gov/climate_action_team/reports/index.html.
8. The Union of Concerned Scientists (UCS) played a major role in fostering this collating of scientists, as discussed in other parts of this chapter.
9. Governor Schwarzenegger took office in November 2003, replacing Governor Gray Davis, who had mostly pursued an extremely cautious and middle-of-the-

road approach to environmental policy. Prior to Davis, Governor Wilson had opposed initiatives on climate change (Hanemann, 2007).
10 See www.meteora.ucsd.edu/cap/.
11 See www.climatechange.ca.gov/research/index.html.
12 See www.climatechange.ca.gov/westcoast/documents/index.html.

References

Barnett, T. P. and Pennell, W. (eds) (2004) 'Impact of global warming on Western U.S. water supplies', *Climatic Change*, vol 62 (special volume)

Boyd, J. (2006) 'California Energy Commissioner, Statement at the Climate Change Science Meeting', Sacramento, 13 September 2006, www.climatechange.ca.gov/events/2006_conference/presentations/index.html

California Climate Change Center (2006) *Our Changing Climate: Assessing the Risks to California*, Summary report from the California Climate Change Center, CEC-500-2006-077

CAT (Climate Action Team) (2006) *Climate Action Team Report to Governor Schwarzenegger and the Legislature*, March, www.climatechange.ca.gov/climate_action_team/reports/index.html

Cayan, D., Luers, A., Hanemann, M., Franco, G. and Croes, B. (2006) *Scenarios of Climate Change in California: An Overview*, Final report from California Energy Commission, Public Interest Energy Research (PIER) programme, California Climate Change Center, Publication CEC-500-2005-186-SF, posted 27 February 2006, www.climatechange.ca.gov/biennial_reports/2006report/index.html

CEC (California Energy Commission) (1989) *The Impacts of Global Warming on California*, P500-89-004, CEC, California

CEC (1991) *Global Climate Change: Potential Impacts and Policy Recommendations*, CEC, California

CEC (1998a) *Historical and Forecasted Greenhouse Gas Emissions Inventories for California*, January, P500-98-001V3, CEC, California

CEC (1998b) *Greenhouse Gas Emissions Reduction Strategies for California*, January, P500-98-001V1, CEC, California

CEC (2002) *Inventory of California Greenhouse Gas Emissions and Sinks: 1990–1999*, November, P600-02-001F, CEC, California

CEC (2005a) *Inventory of California Greenhouse Gas Emissions and Sinks: 1990 to 2002 Update*, June, CEC-600-2005-025, CEC, California

CEC (2005b) *Integrated Energy Policy Report*, November, CEC-100-2005-007CMF, CEC, California

CEC (2007) *In The Public Interest: Developing Affordable, Clean, and Smart Energy for the 21st Century California*, March, CEC-500-2007-020-SD, CEC, California

DWR (Department of Water Resources) (2006) *California Water Plan, Update 2005*, Department of Water Resources, Sacramento, CA

Field, C. et al (1999) *Confronting Climate Change in California: Ecological Impacts on the Golden State*, Union of Concerned Scientists, Cambridge, MA, and Ecological Society of America, Washington, DC

Gleick, P. H. (1986) 'Methods for evaluating the regional hydrologic impacts of global climatic changes', *Journal of Hydrology*, vol 88, pp97–116

Gleick, P. H. (1987) 'Regional hydrologic consequences of increases in atmospheric carbon dioxide and other trace gases', *Climatic Change*, vol 10, no 2, pp137–161

Hanemann, W. M. (2007) 'How California came to pass AB32, the Global Warming Solutions Act of 2006', ARE Working Paper, University of California, Berkeley, CA

Hayhoe, K., Cayan, D., Field, C. B., Frumhoff, P. C., Maurer, E. P., Miller, N. L., Moser, S. C., Schneider, S. H., Cahill, K. N., Cleland, E. E., Dale, L., Drapecki, R., Hanemann, R. M., Kalkstein, L. S., Lenihan, J., Lunch C. K., Neilson, R. P., Sheridan, S. C. and Verville, J. H. (2004) 'Emissions pathways, climate change, and impacts on California', *Proceedings of the National Academy of Sciences*, vol 101, no 34, pp12422–12427

Knowles, N., and Cayan, D. R. (2002) 'Potential effects of global warming on the Sacramento/San Joaquin watershed and the San Francisco estuary', *Geophysical Research Letters*, vol 29, no 18, p1891

Morgan, M. G., Cantor, R., Clark, W. C., Fisher, A., Jacoby, H. D., Janetos, A. C., Kinzig, A. P., Melillo, J., Street, R. B. and Wilbanks, T. J. (2005) 'Learning from the U.S. National Assessment of Climate Change Impacts', *Environmental Science & Technology (Policy Analysis)*, vol 39, no 23, pp9023–9032

PPIC (Public Policy Institute of California) (2000) *Statewide Survey: Special Survey on the Environment*, Public Policy Institute of California, June 2000

PPIC (2006) *Statewide Survey: Special Survey on the Environment*, Public Policy Institute of California, July 2006

Roos, M. (1991) 'A trend of decreasing snowmelt runoff in northern California', *Proceedings of the 59th Western Snow Conference*, Juneau, AK, pp29–36

USGCRP (US Global Change Research Program) (2001) *Climate Change Impacts on the United States: The Potential Consequences of Climate Variability and Change*, Report of the National Assessment Synthesis Team, US Global Change Research Program, www.usgcrp.gov/usgcrp/nacc/allreports.htm

USGCRP (2002) *The Potential Consequences of Climate Variability and Change for California: The California Regional Assessment*, www.usgcrp.gov/usgcrp/nacc/california.htm

Wilson, T., Williams, L., Smith, J. and Mendelsohn, R. (2003) *Global Climate Change and California: Potential Implications for Ecosystems, Health, and the Economy*, Publication 500-03-058CF, www.energy.ca.gov/pier/final_project_reports/500-03-058cf.html

7

Russia's Climate Policy and the Kyoto Ratification Deal: Assessing the Science/Practice Interface

Elena Nikitina and Vladimir Kotov

Russia: Decision-making within the Kyoto Protocol ratification process

At the end of 2004 Russia finally declared its decision to ratify the Kyoto Protocol and entered into this international agreement. It was the result of a controversial process of domestic policy-making, accompanied by a great deal of debate, conflicts of interest and uncertainties related to the national obligations involved in this treaty. This process has also been accompanied by hot polemics between the protocol proponents and opponents about the vulnerability of Russia to global climate change and possible policy response options, including mitigation and adaptation measures. To a large extent these uncertainties were reinforced by contradictory declarations of Russian officials within the ratification decision-making – promises of imminent ratification were coupled with critiques of the Kyoto regime, including comments that it was discriminatory in its character and unacceptable for Russia. In general, the decision-making process on ratification in Russia has not addressed many uncertainties. The public and other stakeholders have also not actively participated in the decision-making.

It is remarkable that many analysts agree that the ratification decision was not only based on the results of scientific and expert assessments of the possible implications of the protocol for Russia and evaluations of Russia's vulnerabilities to the risks of global warming. The rationale for such a decision was slightly different. A powerful driver has been external to the climate change and climate policy domain. Experts suggest that 'a deal' was made by the European Union (EU) and Russia: the EU agreed to back Russia's bid to join the World Trade Organization (WTO) in exchange for Russia ratifying the protocol.

Russia's promise to ratify the protocol has been one of the most important results of the European Union/Russia Summit held in Moscow in May 2004 (i.e. before the ratification decision was undertaken). It demonstrated a serious shift in Russia's position within the ratification process. At that time important signals regarding ratification came from Russian Federation President Vladimir Putin. Attention was attracted by his three main statements at the EU–Russian Federation summit press conference:

1 'We support the Kyoto process.'
2 'The EU has met us halfway in talks over the WTO and that cannot but affect positively our position on the Kyoto Protocol.'
3 'We will speed up Russian Federation movement towards the Kyoto Protocol ratification' (Denisov, 2004; Kolesnikov, 2004).

These statements reduced, to a certain extent, the existing uncertainties about ratification prospects. The final decision on ratification in Russia was taken at the highest level.

Ratification of the Kyoto Protocol turned into a matter of bargaining a deal between the EU and Russia (Kotov, 2004). Both Russia and the EU demonstrated through this deal that they had common interests, and that they were seeking ways to realize them. It was important for the EU to secure ratification of the Kyoto Protocol and simultaneously to get access to Russia's domestic markets, which absorbed a substantial volume of EU exports, and also to provide substantial and sustainable energy flows from Russia to the EU countries. As for Russia, it was important to ensure access to WTO markets and to remove discriminatory barriers. For Vladimir Putin, membership of the WTO would also not only help to entrench transformation of institutional structures that already had occurred over the last decade, but to programme their implementation in the future, beyond 2008, when his term in office as the Russian Federation president ended. The entry of Russia into the WTO would allow it to embed the economic structures that had been established during the 1990s, but weakened by monopoly and corruption. The WTO provided a competitive space to oust the criminal and monopolistic structures from domestic economic sectors. WTO membership would also permit the linkage of these institutional changes with an acceleration of economic growth. The package of WTO entry and protocol ratification radically changed the context of ratification: it was no longer connected with impeding growth; on the contrary, its aim was to open up the way to accelerated economic growth.

This decision corresponded to national interests; the choice of the term 'national interests' as a key decision criterion changed the approach to the debates between the opponents and proponents of the protocol ratification inside the country. Ratifying the Kyoto Protocol was moved up to the agenda of 'big national policy'. Certain analogies can be drawn between this case and the Soviet entry into the United Nations' Convention on Long-Range Transboundary Air Pollution regime at the end of the 1970s, when the ratification decision was governed to a high extent by political considerations of *détente* between West and East, and not only for any ecological rationale. The protocol ratification issue has been one where various interests were concentrated and collided – not only the interests of climate policy as such, but interests rooted in the top priority of economic growth, social policy, energy and structural policies, as well as foreign policy (Moe et al, 2001). Similar to the Convention on Long-Range Transboundary Air Pollution case, the ratification decision was attributed primarily to economic and political issues rather than environmental interests and the rationale for a national climate policy. The latter, in any event, had been insufficiently developed by the time of the ratification decision.

Decision-making within the ratification process was accompanied by a variety of contradictions and uncertainties, including those relating to the scientific perception of climate change impacts and vulnerabilities. The new impetus for the Russian climate policy formation was acquired later – immediately after the official assessments of climate change and its future possible trends had been published in 2008 (Roshydromet, 2008). Its clear messages about possible impacts on socio-economic systems stimulated dynamic policy-making during the following period.

Russia's vulnerability to global warming: Facts and perceptions

The role of Russia in global warming

Russia ranks third in the world for anthropogenic greenhouse gas (GHG) emissions, after the US and China, accounting for 5 per cent of the world total in 2005 (Kokorin et al, 2008). During the 1990s, its GHG emissions declined by more than one third from their level in 1990 as a result of the national economic crisis. Emissions were expected to grow in the future and to follow the upward trend of economic growth which started in 2000. But despite high rates of economic development during the last decade, Russia reported that GHG emissions were decoupling from economic growth. Although there was an increase of emissions during 1999 to 2007, they remain below their 1990 base level, accounting for 2.2 billion tonnes CO_2e in 2007, which was about 34 per cent lower than in the base year (UNFCCC Secretariat, 2010). Energy efficiency improvements, changes in energy demand and structural changes in economy contribute to this. The structure of GHG emissions is typical for an industrialized country: the share of CO_2 is 72 per cent of total emissions; methane, 22 per cent; nitrous oxides, 4 per cent; and hydrofluorocarbons (HFCs) and perfluorated carbons (PFCs) combined, 2 per cent. The energy sector is responsible for about 82 per cent of GHG emissions; industry 9 per cent; agriculture 6 per cent; and wastes 3 per cent.

Russia has undertaken an obligation to keep its annual emissions within the limits of the 1990 base level during the next five-year period. The debates on whether and when GHG emissions might reach their pre-crisis level of the 1990s have been in the core of decision-making processes related to the Kyoto Protocol ratification. Today, existing assessments suggest that GHG emissions will remain below the 1990 level not only within the Kyoto period, but in a longer-term perspective up to 2020. The growing view about a considerable sequestration potential in Russia is an important factor in global warming policy formulation in Russia.

Evolution of domestic perceptions

While assessing the evolution of the social learning process and of global climate change perceptions in Russia and, formerly, in the USSR, it is worthwhile to note that traditionally global warming has never been a subject of broad attention in Russia until recently. Until the 1990s, when the international climate change regime entered the scene, the interest shown was mainly by scientists who made the Russian school of climatology famous around the world, or by some government officials directly involved in weather-related activities, or within international environmental debates. On the other hand, vulnerability to climate change, and particularly regional and seasonal variations and their impacts upon economic welfare, has always been an important topic for such a vast country with its different geographic zones and 18 climate types (one should not forget that the heating season in Russia is much longer than in other countries) (Social Learning Group, 2001). Until the beginning of the last decade, global warming has not been of a high priority in the national environmental agenda. Mitigation or adaptation strategies have also never been clearly formulated.

As to the evolution of official perceptions of the issue of Russia's vulnerability to global warming risk, it indicates significant modifications during the last decade in contrast to the previous period. There is a combination of polar approaches to the issue: from confirmation of general benefits and positive impacts of global warming upon the national economy and social sector to acknowledgement of a wide range of negative effects for the country, in general, and to human health, in particular. Such dualism was clearly reflected within protocol ratification debates and, according to opponents of protocol ratification, they were almost sure until the last moment that Russia would never enter into the Kyoto process. Their major stance has been that Russia will benefit from global warming. Uncertainties in knowledge regarding national vulnerability to global climate change and impacts upon various economic sectors and populations were widely used within ratification debates, both by the opponents and proponents of the protocol ratification.

During the following period after the assessments of climate change impacts by Roshydromet became public in 2008 (Roshydromet, 2008), the perception about the vulnerability of Russia to global warming was clarified. It indicates that during the last century the temperature rise in Russia was faster than global warming and that it would accelerate in the future. Negative effects are considerable and predominant across most regions of the country, especially in the northern parts.

The ratification process was aggravated by controversies and debates about the possible impacts of the protocol regime on Russia. Of particular concern was the uncertainty in forecasts about the possible time when national GHG emissions would surpass their baseline 1990 level due to rapid changes in the economic parameters of national development. In 2005 the annual gross domestic product (GDP) increase in rates was 6.4 per cent. Most of the opponents to the protocol ratification based their objections on the argument that Russia might not meet its national obligations under the Kyoto Protocol due to high economic growth. However, practice later showed that this did not become true.

National vulnerability to climate change

First assessments of the possible negative impacts of climate change were officially presented during the mid 1990s. They were formulated in the 1996 federal programme on climate change and in the Russian Federation's First National Communication to the United Nations Framework Convention on Climate Change (UNFCCC). They were based on scientific evaluations performed for the large-scale multifactor assessment for different sectors of the national economy.

During the ratification period there were signals that there was no consensus in the scientific community in Russia regarding perceptions of vulnerabilities to climate change, particularly regarding possible responses to them. Within the protocol ratification debates, variations in approaches were obvious. These controversies were used by the Kyoto opponents, including a group of scientists in the Academy of Sciences, underlining that the protocol ratification would have a negative impact upon Russia and that requirements of the protocol were of a discriminatory character with mechanisms that involved economic risks for Russia. Debates emphasized the absence of scientific substantiation of the protocol and its low effectiveness for reducing concentrations of GHG in the atmosphere as envisaged by the UNFCCC. Of course, this opinion was formulated as a result of wrestling among various scientific groups. It was not unanimous; there was a great deal of support for the Kyoto Protocol among academics.

Before the mid 1990s, the assessment of the extent of national vulnerability and, hence, adaptation and mitigation policy options, was not clearly formulated in government documents. Earlier official perceptions of climate change impacts underlined potential positive impacts upon livelihoods and upon the national economy (particularly upon agriculture) that would benefit from global warming. The notion of uncertainty of climate change impacts – trends, main causes and, especially, uncertainties in anthropogenic versus natural factors in global warming – had been at the core of official approaches. Still, numerous groups of scholars supported such a notion. It is remarkable that during the Soviet period the underlying notion for such an approach (i.e. 'It's cheaper to do nothing and benefit from global warming rather than to undertake mitigation measures') was quite strong and was actively exploited by policy-makers.

According to recent perceptions, Russia appears to be more vulnerable to global climate change than other regions of the world. Climate change and its impacts have already been registered in many regions during the last century and further changes are forecast for the major part of Russian territory. Due to the vastness of the country and differences in geographical zones and climates, vulnerability to global warming differs across regions and territories. Unified assessment of possible climate change and its impacts is not valid for the entire country because of such considerable regional variations and, hence, differences in policy response options.

Impacts of climate change are predominantly negative for most of the regions and sectors; but in some, positive effects might also occur. Among the most vulnerable sectors are water resources (from 2010 to 2039 water resources are expected to increase an average of 8 to 10 per cent) and agriculture. Only some regions of agricultural production in the country might benefit from changing climate, while in others expensive restructuring of production patterns would be needed. Destructive effects of global warming might be experienced in permafrost areas covering 67 per cent of the country's territory: in the northeast of European Russia, across most of Eastern Siberia and the Far East, and in the northern half of west Siberia. Destruction under climate change has already been registered;[1] further effects on infrastructure and settlements will occur. Sea-level rise is acknowledged to negatively impact highly populated coastal areas, such as the coast of the Gulf of Finland and the St Petersburg area. It might also have negative consequences for human health due to the spread of infectious diseases.

Negative impacts of climate change include a variety of water-related risks, such as an increase in intensity and length of droughts in some areas, and extreme floods and precipitation in others, as well as danger of forest fires and destruction of permafrost. Observed changes in climate and its impacts are already significant but spatially variable due to the vast territory of the country and high variability of geographical zones, with 17 types of different climatic belts; the forecast is that they will spread in the future across the major part of the country. According to observations during the last 20 to 25 years, for example, annual river flows in Siberia have increased – in the Lena, Irtysh and Tobol by as much as 20 to 30 per cent. At the same time, others such as the

Upper Ob in Siberia, and the Don and Dneper in European Russia have seen seasonal declines. Recently, the annual numbers of high and catastrophic floods in Russia increased by 15 per cent from the last decade of the 20th century, particularly in the rivers of the North Caucasus and in southern regions of the Far East.[2] They are expected to increase further across the major part of Russian territory (UNFCCC Secretariat, 2010). Since over one half of national annual river water inflow is formed by snow melting, which is usually the main reason for spring freshet floods, the risk of high freshet floods will grow, particularly in rivers where ice jams usually accompany river floods – in Archengelsk Oblast and Komi in European Russia, in the Urals, in East Siberia and in the northeast of Asian territory.

However, there are places and sectors for which some influence of climate change might be positive, including a decline in energy use for heating,[3] opportunities for increased river transportation, access to remote Arctic areas, and enhanced structure and productivity in some regions for planting and animal breeding.

National climate policy

Peculiarities in domestic policies

There are several phases in climate policy formation in Russia (Nikitina, 2001; Kotov, 2002). First, during the ratification period there was no comprehensive national policy on climate change: its institutional framework was not thoroughly developed and only some of its elements were established. Such a situation appeared to be a significant barrier for undertaking effective mitigation and adaptation measures for climate change. During the 1990s the emphasis had been on designing a possible means for GHG emissions regulation and an application of the protocol instruments; but efforts to develop and implement adaptation strategies and measures were quite poor. This was attributed largely to the existing modest level of knowledge of Russia's vulnerability to climate change, and particularly to significant lacunae in it.

Special climate change legislation had not been enacted in Russia, and neither a national programme nor an action plan was in place. Even after the protocol ratification, which was expected to speed up the climate change institutional regime formation and climate policy development, the process continued to progress quite slowly for a while. Under the direct impact of Kyoto ratification, the establishments for its implementation began to emerge – the national register for emissions quotas, the national system for the inventory of emissions and sinks, and the regime for joint implementation projects – and corresponding government regulations were introduced (Russian Federation Government, 2006a, 2006b, 2007). These were supplemented by the more general requirements for Russia as a UNFCCC member, including national climate policy formulation; assessment of vulnerabilities and risks associated with climate change; support of research; monitoring; education; dissemination of information; and so on. All of these elements required domestic capacity-building. Domestic climate policy regime formation was advancing very slowly. Critics argued that a slow regime formation process might result in 'lost opportunities' as potential partners for application of the protocol mechanisms might lose their patience under such legislative uncertainties and turn to other markets (Popov et al, 2003).

There was a problem in coordinating approaches within the government. No common opinion in government existed about the domestic design of the climate regime, and this influenced the development of national legislation. A group of government agencies headed by the Ministry for Economic Development backed the development of a package of a number of legal documents regulating joint implementation activities. They believed that all issues relating to the protocol could be regulated by existing national legislation, including laws in the energy sector, in environmental protection and in forestry. Other interest groups in the government, meanwhile, supported the adoption of a comprehensive framework law on climate change and Kyoto Protocol implementation. Even more uncertain was the evaluation of adaptation strategies and measures. Coordination of sectoral efforts remained quite poor.

There was a great deal of uncertainty about the vision of Russian climate change strategy development. The 1996 federal programme of climate change had expired, and it had not been extended. Its effectiveness was limited, as it had been too formal and not concrete. It was not supported by declared financial resources and it appeared that many of its components turned into poor implementation and dead letters. Many elements of climate policy and its coordination were characterized by high uncertainty. In Russia it is well known that unclear division of functions and responsibilities among ministries is an effective means of avoiding any responsibilities.

The new period of climate policy-making was shaped immediately after the official perceptions on climate change and impacts for the country were presented in 2008. An additional factor in this development process was the external driver of the post-Kyoto international negotiations. During the last few years, climate policy formation has acquired new dynamics (see Table 7.1). Its major innovations include adoption of the national strategy on climate change (Russian Federation Government, 2009), which undertakes climate policy-making at a high level, incorporation of climate change within the national security agenda, and placing it among priorities of sustainable development. New perceptions include the need to take into account in decision-making the combination of possible advantages and risks associated with climate change. The package of adaptation and mitigation action plans and concrete instruments and tools to realize them is in its core. For example, the new national *Climate Doctrine of the Russian Federation* (Russian Federation Government, 2009) is a strategic political document setting the framework for climate policy formation and implementation. It defines major goals, including:

Table 7.1 *Russian climate policy: Major recent milestones*

2008	
Assessment Report on Climate Change and its Impacts in the Russian Federation by Roshydromet	
'On some measures to enhance energy and ecological efficiency of the Russian economy', Decree of the Russian Federation president, N 889, 4 June	
2009	
'On energy-savings and enhancing energy efficiency', Federal Law, N 261-ФЗ, 23 November	
Climate Doctrine of the Russian Federation, 17 December	
President D. Medvedev participates in COP 15 and presents national emission reduction targets.	
2010	
President D. Medvedev convened a high-level meeting on climate change, 18 February.	
Russian Federation Security Council Meeting 'On measures to prevent Russian Federation national security threats related to global climate change', 17 March	
Fifth Russian Federation National Communication to the United Nations Framework Convention on Climate Change (UNFCCC)	

Source: chapter authors

- enhancing scientific and technical capacity for the assessment of climate change, its impacts and related risks;
- realization of near- and long-term mitigation measures; and
- adaptation to climate change (Russian Federation Government, 2009).

Adaptation

Following worldwide trends, only recently have the scientific community and policy-makers in Russia paid high attention to assessing national vulnerabilities to climate change and possible adaptation options.

As a result of recent research, an assessment of vulnerabilities for different regions as well as for a number of economic sectors in Russia has been performed (Kokorin et al, 2006). Among the immediate conclusions is a growing recognition that climate change is becoming an additional stress in many regions, particularly in the context of other existing social and environmental problems. The result is the view that adaptation strategies need to be developed, information about climate change impacts needs to be widely disseminated in society, and concrete measures need to be designed. But the detailed understanding of particular responses is still quite vague.

Adaptation options are now discussed more often, while during the last decade the major focus was primarily on mitigation strategies. Discussion of adaptation focuses both on the problems of ecosystem adaptation and on adaptation measures that can be undertaken to reduce negative effects on society. There is a growing understanding that climate change impacts are progressing more rapidly than the ability of natural ecosystems to adapt to them. This requires the introduction of a regular monitoring system of the changes in nature that will serve as a basis for decision-making. Such a comprehensive information system is now absent. A dialogue between the scientific community and decision-makers about vulnerabilities and adaptation responses in water use and water protection, in agriculture and in human health protection is starting. However, this dialogue is still at initial phases and substantial research and formulation of concrete actions are urgently required.

Adaptation to climate change in Russia is now incorporated within the national climate strategy. It entered the national agenda only recently, and today it becomes equally important with mitigation policies. There is even an impression that adaptation has drawn a comparatively stronger focus. Adaptation strategy is the focus of the new national *Climate Doctrine*. The doctrine targets the increase of Russia's resilience to climate change, and it phases further research as a basic groundwork for adaptation. Particular emphasis is given to the evaluation of climate change impacts upon various sectors, households and ecosystems (Article I.4) because comprehensive assessments are still lacking. The so-called 'preventive adaptation' to climate change impacts is described as a major climate policy priority (Article 22). It is yet to be seen what particular adaptation measures and tools it incorporates and how it will be implemented in practice.

Regions and provinces are encouraged to develop their adaptation and mitigation programmes and plans. The design of adaptation measures in various Russian regions is based on integrated assessments of regional vulnerabilities and risks, possible opportunities and disadvantages related to climate change impacts, costs/benefits, and adaptation potential based on economic, social and other considerations (Russian Federation Government, 2009). Adaptation strategy secured support at a high political level: at the recent meeting of the Security Council President Medvedev emphasized that climate change and adaptation to it were on the list of national security issues, and he set concrete deadlines for development of action plans by 1 October 2010 to implement national climate strategy. An action plan up to 2020 was developed in July 2010 and sent for coordination among responsible government agencies.

Despite recent progress in shaping a strategic vision for adaptation, more active measures are urgently needed. Climatology assessment is a first important step, but it is not sufficient. The next steps, including detailed evaluation of economic and social impacts supplemented with cost-benefit analysis, and then outlining concrete adaptation measures for mid- and long-term perspectives, are essential. One of the problems is that today many decision-makers suggest as a priority the enhancement of monitoring and forecasting of climate change and further research. Independent experts suggest that other urgent actions are required, such as increasing preparedness for floods and droughts, strengthening epidemiological control over infectious diseases, controlling water resources in water-deficit areas, planting trees, starting reconstruction of infrastructure in permafrost areas, assisting indigenous groups in their adaptation activities, introducing new designs for marine vessels and ice-breakers, and modernizing coastal infrastructure (Kokorin et al, 2009).

Interests and approaches of actors

During the 1990s, when the international climate change regime entered the scene, several major groups of actors in climate policy in Russia could be identified. These included policy-makers in the government, regions (federation subjects), business, the scientific community, the public, non-governmental organizations (NGOs) and 'external' experts. Domestic actors sometimes had a variety of interests, sometimes none, in relation to the issue of global warming and the associated risks and opportunities. Although their perceptions and knowledge on the issue were well advanced, in general, the major common feature was that, given their positive stance towards the Kyoto Protocol, they would like to use the climate regime and its mechanisms to meet their economic interests. In particular, this related to the business community and to the regions of Russia and its federation subjects. The public, in general, still had a more neutral attitude to global climate change and to the protocol as such.

Public

The structure of priorities, perceptions and knowledge/practice interfaces in Russian society has some specific features that seem different today from those typical in the West. One of the key characteristics is the low ranking given to ecological concerns in society. Within the current national public environmental agenda, the role of global environmental change, in general, and global climate change, in particular, is extremely low. Domestic ecological concerns or, to a greater extent, local ones dominate the public agenda. Global warming mitigation/adaptation continues to appear at the bottom of the public agenda in Russia.

The reasons for this are apparent. Public polls indicate that the population of Russia, particularly during the 1990s, has been involved primarily in solving the problems of survival, and has been interested, first of all, in addressing low salaries and pensions (about one third of the population has been below the subsistence level), medical services, criminalization, corruption and insecurity. In addition, the public was still heavily relying on the paternalism of the state in the environmental field, which was, to a great extent, a heritage of the Soviet past; although concerned with ecological problems, the public tended not be heavily involved in environmental actions. Knowledge about societal vulnerabilities to climate change was also not widely disseminated and popularized; the knowledge base was not sufficient for the public to make decisions. Because the level of environmental culture and public ecological education was far from what was needed, it was most necessary to raise environmental awareness, extend public education, and disseminate user-friendly information about mitigation and adaptation to climate change.

Public perceptions related to vulnerabilities to climate change were mixed. Many believed that warming might be beneficial to their locales and regions. Most of Russia is a territory of severe natural and climatic conditions, so most of the public was not inclined to strongly oppose global warming. The expectation was that global warming might be associated with more comfortable living conditions and higher productivity of agricultural crops.

Until recently the climate change problem also ranked comparatively low in the programmes of political parties and among politicians, especially those who were not involved directly in decision-making related to it. Today there is a certain activation of environmental NGOs. According to many experts, however, the activities of NGOs in Russia are unfortunately still of a decorative character, and the public does not take any real part in their efforts, which are mostly limited to the activities of a narrow group of functionaries. If mobilized, however, the environmental NGOs could easily support more climate change mitigation and adaptation.

However, opponents to the Kyoto Protocol who were trying to influence public attitudes usually indicated that the protocol was not able to prevent an increase in global emissions, which were growing in developing countries, while the US had quit the regime so as not to limit its national economic development. Opponents noted that the protocol, although creating an impression of problem-solving, was, in reality, without real solutions, while economic growth and poverty alleviation were the top public priorities in Russia.

Under these conditions, climate policy can easily be pressured and influenced by interest groups. In this context, climate policy might be implemented not for the sake of environmental goals, but for the realization of certain economic benefits, including, for example, the creation of new jobs and for economic development.

Regions

In Russia, the main features of regional participation in climate policy formation, decision-making and possible roles in application of protocol tools can be summarized as follows:

- active domestic discussion about defining roles of the regions in the process;
- expression of interest in taking part in the application of Kyoto Protocol tools at regional levels;
- consolidation of regional interests and formulation of regional approaches;
- the predominance of some regions over others in taking practical steps, such as the development of regional legislation, administrative structures and the compilation of regional GHG inventories.

Interests

Since the 1990s, as a result of real federalism in Russia, regions (89 federation subjects) began to play an increasingly important role in climate and environmental policy. This was a new phenomenon for Russia because under the Soviet regime environmental interests of the regions were restrained. This development opens additional options in the application of protocol tools.

Following decentralization and the establishment of new institutional patterns of vertical subsidiarity, some regions expressed particular interest in participating in joint implementation activities and in emissions trading (Nikitina and Kotov, 2003). Many of them supported Russia's ratification of the protocol. Usually, in a set of regional priorities, pragmatic economic goals (along with political ambitions) had a much higher standing than ecological ones; regional assessments of global warming implications were frequently not advanced.

Regional interests can be summarized as:

- use of Kyoto Protocol tools for attracting foreign investments to energy, industrial, forestry and municipal sectors of the regions;
- use of protocol tools for enhancing energy efficiency and energy savings through refurbishment of obsolete facilities and their modernization;
- application of green reinvestment schemes;
- generating additional financial revenues for regional budgets from international emissions trading that might contribute to social and economic problem-solving;
- establishing institutional schemes for regional participation in regulations and control over emissions trading and joint implementation, as well as division of competences between the federation and the regions;
- air quality amelioration as a co-benefit from GHG mitigation.

Regional approaches to the application of Kyoto Protocol tools have been actively formulated. About 30 federation subjects participated in the discussion of this issue. Identification and consolidation of the regional interests were actively supported by domestic and international groups lobbying Kyoto Protocol ratification. To a large extent, the latter were serious pushers for the formation of regional approaches. Among the main principles of a regional approach was the delegation of broader responsibilities and control functions in climate policy performance from the federation centre to the regions. Proponents of the regional approach formulated the view that obligations of the country are the sum of obligations of the regions. It meant that together with responsibilities, the idea of the dissemination of broader authorities in climate policies to the regions was maintained. Regions were attempting to influence the formation of a national institutional framework and vertical division of power among different levels of government administration in the application of protocol mechanisms. For example, regions strongly supported the idea that the central government should transfer GHG emission quotas to the regions and their administrations, and, along with them, the corresponding management functions and major levers of control over allowance allocations.

The regional approach usually emphasized that existing institutional structures of governmental management of environmental protection in Russia were based on territorial principles, and, accordingly, it would be much easier to incorporate elements of climate policy within the existing institutional frameworks of environmental management. This process is characterized by strong debates between opponents and proponents of a high role for the regions, both in institutional schemes and in decision-making versus balance of interests between the federation and its subjects. They were accompanied, in short, by a clash between centralized and decentralized approaches.

Conclusions: New dynamics

During the years since Russia joined the Kyoto Protocol, only part of the Kyoto ratification deal has been met. Russia has not entered the World Trade Organization. But its climate policies have shown significant dynamics, especially during the last couple of years. They include the formation and diversification of domestic and international climate policy, evolution of official perceptions about climate change and its impacts, and, finally, adoption of a national climate change strategy and the entry of climate change into the national security agenda.

The new dynamics illustrate perfectly well that a science/policy interface is crucial. As soon as a detailed scientific assessment of climate change and its impacts was presented by the scientific community with clear and user-friendly messages, the policy-making process was propelled towards further action. The effect of the assessment for the development of climate policy was even stronger than the Kyoto impact. Previously, the major decisions about the Kyoto Protocol ratification by Russia were undertaken by policy-makers in a context of high uncertainty about scientific assessments, which appeared to be a significant barrier for policy advances, both domestic and international. The result was controversies in decision-making, contradictions and conflicts among interest groups.

The ten key messages about the new dynamics in the national climate policy are as follows:

1 Global climate change gradually turns into an issue of high-level political decision-making. Under the new presidency in Russia, much more attention is paid to the design of climate strategy and concrete tools and measures for its implementation.
2 Climate change and its impacts turn into national security issues and are included in the agenda of the Russian Federation Security Council. The new national *Climate Doctrine* has become the major strategy that defines the framework for climate policy development.
3 During the last century, observed climate change in Russia has progressed faster than global warming; observed changes in climate are significant but spatially variable across the country; the forecast is that climate change will increase in the future in major regions.

4   Negative socio-economic impacts of climate change are officially acknowledged after the publication of Russia's first climate change assessment report. They include a variety of risks associated with the increased possibility of floods and droughts, forest fires, destruction of permafrost, spread of infectious diseases, intrusion of alien fauna and flora species, etc. Some positive impacts will occur as well.
5   Adaptation to climate change entered the national climate change agenda only recently, and it is now incorporated as a focus within its climate strategy. Emphasis is given to further evaluation of climate change impacts upon economic sectors, households and ecosystems. Adaptation becomes equally as important as mitigation measures; there is an impression that adaptation has drawn a comparatively stronger focus in policy decision-making.
6   Growing attention is paid to adaptation to climate change impacts in the Arctic and in the Russian northern provinces, which are among the areas most severely affected by global warming. Together with new opportunities related to economic development of these regions, negative impacts are rapid. International cooperation on Arctic climate change issues is suggested as a key tool for solving potential conflicts among states.
7   New targets for domestic emission reduction have recently been set. President Dmitry Medvedev declared the goal of 25 per cent GHG emissions reductions in 2020 from their 1990 levels. Ongoing structural changes in economy significantly contribute to meeting this target.
8   A set of mitigation policies focuses on the increase of energy efficiency in all sectors and households; on enhancing renewable and alternative energy sources; on the introduction of financial tools, incentives and taxation instruments to support GHG emissions reductions by various actors; on the creation of green jobs; and on enhancing sinks' capacities through sustainable forest management as a part of emission control strategy.
9   Introduction of energy-efficient and low-emission technologies is a part of a new national innovation strategy that allows shifting towards 'green' economic development. Building government/business partnerships and interactions is at the core. This strategy is planned to be followed independently from the progress in international regime formation.
10 Further decentralization of climate policies is under way. Regions and provinces are encouraged to develop their adaptation and mitigation programmes and action plans based on regional integrated assessment of regional vulnerabilities and risks, opportunities and disadvantages of climate change impacts, cost/benefit of adaptation and mitigation potential based on socio-economic considerations, and introduction of new technologies for GHG emissions reduction and their absorption.

Acknowledgements

The authors acknowledge the contribution of the Russian State Fund on Humanities Research, *Russia's Interests in the Arctic: Sustainable Development and Energy Security*, N 09-02-00600a/p. The research leading to these results has received funding from the European Community's Seventh Framework Programme (FP7/2007-2013) under grant agreement no. 226571.

Notes

1   For example, in Western Siberia about 12 per cent of annual accidents at oil and gas pipeline systems is caused by mechanical factors associated with deformation of basement and maintenance constructions (Roshydromet, 2008, vol 2, p59).
2   Russia is among moderately flood-affected countries, with about 2.5 per cent of its territory prone to floods. Flood-prone areas account for 400,000 square kilometres and about one eighth are annually flooded; about 746 cities and 40,000 towns are vulnerable. Major flood-prone regions are the Far East provinces (Amur and Sakhalin Oblasts, and Primorsk and Khabarovsk regions), Baikal regions, the Urals, the Lower Volga, west and east Siberia, and the North Caucasus.
3   In many regions the length of the heating season might be reduced by 5 per cent by 2025, and by 5 to 10 per cent by 2050 against the 1961 to 1990 period.

References

Denisov, A. (2004) 'Mastera kompromissa', *Vremia Novostey*, 24 May
Kokorin, A., Kuraev, S., Minin, A. and Stecenko, A. (2006) *Problems of Adaptation to Climate Change*, Working Paper, Environmental Defence (in Russian), Moscow
Kokorin, A., Garnak, A., Gritsevich, I. and Safonov, G. (2008) *Economic Development and Solving the Problem of Climate Change*, DEA, Moscow
Kokorin A., Kuraev S. and Ulkin. M. (2009) 'The economics of climate change: The Stern Review', WWF–GOF, WWF Russia, Moscow
Kolesnikov, A. (2004) 'Rossia sdala normy VTO', *Kommersant*, 22 May
Kotov, V. (2002) 'Policy in transition: New framework for Russia's climate policy', *Nota di Lavorno*, FEEM, Milan
Kotov, V. (2004) 'The EU-Russia ratification deal: The risks and advantages of an informal agreement', *International Review for Environmental Strategies*, vol 5, no 1, pp157–168
Moe, A., Tangen, K., Stern, J., Grubb, M., Berdin, V. and Korpoo, A. (2001) *A Green Investment Scheme: Achieving Environmental Benefits from Trading with Surplus Quotas*, Briefing Paper, Climate Strategies, University of Cambridge, Cambridge, UK
Nikitina, E. (2001) 'Russia: Climate policy formation and implementation during the 1990s', *Climate Policy*, vol 1, pp289–308
Nikitina, E. and Kotov, V. (2003) 'Russia: National framework for GHG emission trading', Paper presented to the OECD Global Forum on Sustainable Development, 14 March, CCNM/GF/SD/ENV(2003)8
Popov, A., Pluiznikov, O. and Gavrilov, V. (2003) 'Kiotsky protocol: Perspektivy, vygody I zatraty dlia Rossii', in *Kyotsky Protocol: Otvetstvennost i perspectivy dlia biznesa*, ICEP, Moscow

Roshydromet (2008) *Assessment Report on Climate Change and its Impacts in the Russian Federation*, vols 1–2, Roshydromet, Moscow

Russian Federation Government (2006a) *O sozdanii v tseliah realizacii obiazatelstv vytekauishih iz Kiotskogo protokola Rosiiskogo reestra uglerodnyh edinits*, Ordinance of the Russian Federation Government, N 215-p, 20 February

Russian Federation Government (2006b) *O sozdanii v tseliah realizacii obiazatelstv vytekauishih iz Kiotskogo protokola rosiiskoy systemy otsenki antropogennyh vybrosov iz istochnikov i adsorbcii poglotiteliamy parnikovyh gazov ne reguliryemych Monrealskym protokolom po veshestvam razruishauishim ozonovy sloy*, Ordinance of the Russian Federation Government, N 278-p, 1 March

Russian Federation Government (2007) *O poriadke utverzhdenya I proverki hoda realizatsii proektov osuishestvliaemyh v sootvetstvii so st.6 Kiotskogo protokola k ramochnoi konvencii OON ob izmenenii klimata*, Resolution of the Russian Federation Government, N 332, 28 May

Russian Federation Government (2009) *Climate Doctrine of the Russian Federation*, Ordinance of the Russian Federation President, N 864-рп, 17 December

Social Learning Group (2001) *Learning to Manage Global Environmental Risks*, vol 1, W. C. Clark, J. Jaeger, J. van Eijndhoven and N. Dickson (eds) MIT Press, Cambridge, MA

UNFCCC (United Nations Framework Convention on Climate Change) Secretariat (2010) *Fifth Russian Federation National Communication to the UNFCCC*, UNFCCC Secretariat, Bonn

8

Urban and Social Vulnerability to Climate Variability in Tijuana, Mexico

Roberto Sánchez-Rodríguez

Introduction

Attention to climate change and its negative consequences for society emphasizes the need for reflection about current patterns of growth. Two interrelated elements are important in this reflection: the growing disparities among and within societies leading to impressive levels of segregation, marginalization and injustice; and the imbalance in the society/nature relationship with a remarkable rate of depletion of natural resources. These two combined elements present a shady future for societies in the 21st century. Calls for new patterns of growth have been shaped in different forms, particularly under the broad umbrella of sustainable development. Unfortunately, sustainable development remains a rather vague and rhetorical concept and there is an urgent need to expand the discussion of alternatives to transform it into an operational concept capable of addressing the complex realities of the 21st century.

The expectation that science will take a leading role in these efforts requires rethinking traditional conceptions about the creation of knowledge. Scientific knowledge has advanced our understanding of the problems that societies face in the 21st century and the challenges for sustainability. However, experience shows major obstacles and challenges in transforming scientific knowledge into practice. Part of the problem is that traditional conceptions about science as an unquestionable truth, and the role of scientists as generators of objective and useful knowledge, are deeply rooted in disciplinary cultures and in our societies. These approaches provide few incentives and opportunities to transform the divide between science and practice. A growing number of scholars and organizations are now recognizing the importance of transforming the traditional roles of scientists as generators of knowledge and practitioners as the end users of that knowledge. A more balanced participation of scientists and practitioners in the generation and use of knowledge and its conversion into practical actions to benefit society is receiving attention.

One area where efforts to bridge the science and practice gap could yield positive results is in attempting to reduce and adapt to the negative consequences of climate variability and change in urban areas. Urban areas have several advantages compared to other sectors in the study of the impacts of climate change: there is a long tradition in urban studies that enhances our understanding of social and physical processes; most urban areas have a broad range of generators of practical knowledge (planners, decision-makers and other stakeholders); there are international and national efforts to expand capacity-building among planners and local urban decision-makers. Evidence of the negative consequences of climate variability and change are easier to identify and are closer related to pressing urban and environmental problems that are also connected with opportunities leading to sustainable development. It is worth remembering that these issues will likely affect millions of people and create extensive economic and social costs if not addressed.

Despite these advantages, there is still much to be learned in the search for new ways to transform scientific innovations into practical actions in urban areas. Recent contributions towards interdisciplinary and transdisciplinary approaches to ecological and environmental issues are relevant to the study of vulnerability, resilience and adaptation, but they need to overcome disciplinary obstacles among natural and social sciences (Fry, 2001; Schoenberger, 2001; Tress et al, 2001; Petts et al, 2006; Luks and Siebenhuner, 2007; Tress et al, 2007). These approaches are important in creating the integrated perspectives needed to better understand the interactions between social and biophysical systems. At the same time, there is a risk of overestimating the capacities and interest of practitioners to participate in new pathways of knowledge generation to address vulnerability and adaptation problems related to climate change in urban areas. Some authors highlight the deficiencies of public institutions to solve development problems at the national and local levels. Pritchett and Woolcock (2004) point out that, despite some successes, the constraints and technical responses of those institutions under a one-size-fits-all model for rich and poor countries have transformed them into part of the problem of development. Development – or, more specifically, sustainable development – is a fundamental framework to address vulnerability and adaptation to climate variability and change.

This chapter discusses lessons learned from the study of social and urban vulnerability to climate variability in Tijuana, Mexico. Tijuana is an interesting case study. It provides an example of conditions frequently found in poor countries where fast urban growth extends into areas prone to flooding and landslides. The fast pace of urban growth, the type of urbanization created, the growing social inequality and marginalization, and the modifications to the landscape are factors behind social and urban vulnerability to the negative consequences of climate variability in Tijuana.

Although the study of urbanization in poor countries has been well documented and analysed during the last 30 years, little attention has been given to the way in which the driving forces of urbanization affect urban and social vulnerability to the negative consequences of climate variability and change. The science/practice divide in urban areas is particularly evident in poor countries where scientific contributions on urban climate, ecological services, climate forecast and other topics useful to reduce urban and social vulnerability to climate variability are seldom used by practitioners.

Tijuana illustrates the shortcoming of the dominant perspectives that orient growth in poor countries. These policies consider a strong urban economy as a prerequisite for development. Tijuana has a vibrant and diversified urban economy, but its economic success has been a source of imbalance rather than development. Tijuana's urban economy is strongly affected by rapid industrialization associated with socio-economic and transboundary geopolitical processes taking place at different geographical scales with few local benefits. Rapid industrialization has aggravated urban deficiencies caused by fast urban and population growth. These are important elements in understanding urban and social vulnerability to climate variability in Tijuana.

Study object and argument

This study of social and urban vulnerability in Tijuana was carried out by a multidisciplinary research team (biologists, climatologists, geographers, geologists, sociologists, urban and regional planners) who sought to create an interdisciplinary perspective. The project included four institutions, three of them academic – the University of California, El Colegio de la Frontera Norte (COLEF) and Centro de Investigación y Ensenanza Superior de Ensenada (CICESE) – and the Municipality of Tijuana (Dirección de Protección Civil and Comité de Planeación Municipal). It evolved as an attempt to assist Tijuana to prepare better for the negative consequences of climate variability. The participation of local authorities in the development of our project from its early stages created a bidirectional learning process.

The context: Social and urban vulnerability to climate variability in Tijuana, Mexico

Tijuana has experienced impressive growth because of its location on the US/Mexico border. It is a centre of trade and services, the last Mexican stopping point before crossing the international border, and a site of rapid industrialization to supply the US and other foreign markets. Tijuana's accelerated growth over recent decades has made it Mexico's largest city on the US/Mexico border and one of the fastest growing cities in Mexico. Its population grew from 240,000 in 1970 to 1,148,681 in the year 2000 (INEGI 2004). Tijuana's urban economy and its location adjacent to the international border with California continue to be a strong pole of attraction for migration from inland Mexico.

Tijuana is marked by two important characteristics. The first is its inability to keep pace with the demands of its fast-growing population and accelerated urban expansion. These shortfalls have resulted in large areas of incomplete urbanization which have generated severe social and environmental problems. The second is the city's rapid industrialization during the last three decades, which has diversified its urban economy but also modified its urban structure – its daily urban life – and introduced a new set of social and environmental problems. These characteristics have created a peculiar situation. The same factors that give rise to opportunities for economic growth also present obstacles to a balanced development. They have also created fragmented spaces with high spatial segregation that aggravates the social exclusion characteristic of Mexico and other poor countries (Lopes de Souza, 2001; Beall, 2002; Pirez, 2002). Urban space in Tijuana is a mosaic of contrast with a clear division between the formal and informal, the legal and illegal, the rich and poor. Important components of Tijuana's urban growth are irregular claims to property, with irregular settlements accounting for 43 per cent of Tijuana's land area, concentrating 53 per cent of its population and containing 52 per cent of its family dwellings (Alegriá and Ordoñez, 2005, p121). These areas have grown outside any urban planning regulations, often in risk-prone areas.

Rapid and often chaotic growth in Tijuana has aggravated its vulnerability to periodic floods and landslides during the rainy season (winter). The most severe floods have been associated with El Niño. During the 1992 to 1993 El Niño, 36 people in Tijuana died in flooding and landslides. Thousands of people were evacuated and close to 2000ha (10 per cent of the urban area) suffered significant damage from flooding, erosion and landslides (Bocco et al, 1993). There were also immeasurable economic losses. Most of the city was paralysed for two weeks as transport routes were severely damaged or destroyed, leaving parts of the population immobilized or isolated in a fragmented city. An unanticipated outcome of the disaster was the discovery that local officials were totally unprepared to manage this type of emergency and that there was no forecasting capacity that might have helped to prevent the loss of human life. Local authorities had advance warning for the next big El Niño event of 1997 to 1998. They took preventive actions to avoid disasters similar to 1993. Despite those preparations, 14 people died in flooding and landslide-related problems.

Research has shown that climate-related floods in Tijuana also occur during normal years when total daily rainfall causes flooding and, sometimes, landslides. Our compilation of historical records identified 349 significant rainfall-related events in Tijuana between 1970 and 2001. In 2000, for example, flooding affected over 1000ha and nearly 14,000 homes in Tijuana, and another 4600 homes suffered some sort of landslide damage. Most of this damage occurred in areas ranked by our project as having medium and high marginality, underscoring the close relationship among poverty, social exclusion and vulnerability to the negative consequences of climate variability.

The problems that rainfall brings to Tijuana are compounded by the city's urban growth pattern, which includes housing construction in low-lying areas along riverbeds and in canyons and ravines. Flooding results because of topographic conditions in the area and the city's inadequate and inefficient network of storm drains, which covers only 60 per cent of the urban area (mostly in middle- and high-income well-urbanized areas). Tijuana's urban area in 1972 was 7789ha, which grew to 26,046ha in 2000. This explosive growth was dispersed and unplanned, extending into flood-prone lowlands and up slopes with inclines of up to 40 per cent. Although urban growth first occupied lowlands, by 1980 there was a progressive pattern in urban growth by slope incline. That is, from 1972 to 1989, most growth took place on inclines of 20 to 30 per cent; from 1989 to 1994, urban growth began to occupy 30 to 40 per cent inclines; and from 1994 to 2000, new settlements were established on inclines of between 40 to 50 per cent. The low and central parts of Tijuana are occupied by consolidated urban areas, while areas of recent and incomplete urbanization are located in the peripheries with difficult topography for urban growth.

This project used entitlement theory as its conceptual framework. Our working definition considered vulnerability as the exposure of individuals, groups or communities to harmful perturbations from nature at a specific time and in a specific space, where the level of vulnerability is defined by the dynamic balance between the external side of events or perturbations and the internal side, which is created by a wide range of social, economic and political relationships that govern access to the resources needed to cope with minimum damage loss. This working definition is borrowed from contributions of scholars working on natural hazards (Blaikie et al, 1994; Adger, 1999; Kasperson and Kasperson, 2001), food security (Bohle et al, 1994) and urban poverty (Moser, 1998).

The model used for the project reflects major aspects of the conceptual framework. One section focused on the external side of vulnerability and studied seasonal climate variability and its impact upon extreme daily precipitation (these records were available only from 1950 to 2000). Reports of the state's major newspapers were also consulted to document major damages caused by climatic events between 1970 and 2001. Data was compiled into four databases: events, damages, consequences and responses by local authorities.

The characteristics of the landscape play an important role in the study of urban vulnerability in Tijuana. Local conditions with little land suitable for urbanization at low cost have pushed urban growth into hazardous areas. The project identified landscape units (geology, soils, slopes, vegetation and hydrology) with a particular focus on the stability of slopes and the natural drainage of the terrain. These units were used to identify hazardous areas (flooding and landslides) within the urban area and correlated with historical records of climate-related damages during the last 30 years in Tijuana.

On the external side of vulnerability, the project confirmed initial hypotheses that extreme daily precipitation is associated with El Niño years. Results obtained from the study of extreme winter precipitation in Tijuana showed that intense wet years correlate better with strong El Niño than dry years with strong La Niña. The most extreme wet years and the larger number of floods per year took place after 1976 to 1977, during constructive phases of El Niño and the Pacific Decadal Oscillation (PDO) that occurred in 1978, 1983, 1993 and 1998. But floods also occurred during weak El Niños (e.g. January 1980 and February 1995) and even during non-El Niño years (e.g. January 1967 and January 1991 and 2000). Intra-seasonal phenomena such as the Madden/Julian Oscillation and synoptic activity from the mid-latitudes also play an important role during flood events. Unfortunately, intra-seasonal climate anomalies are more difficult to predict. These results are relevant to the study of vulnerability in Tijuana since much less attention is given and less preventive actions are taken by local authorities during weak El Niño or non-El Niño years. The correlation with damage in the urban area during those years is an indicator of the uncoordinated responses to those hazards. Indeed, local authorities have learned from past disasters and have expanded the drainage system in some areas affected by flooding. They have also cleaned up canals in the municipal drainage system, ordered evacuation of some hazardous areas and prepared emergency response actions before strong El Niño events. Unfortunately, no attention has been given to identifying urban and social vulnerability and adaptation to reduce the negative consequences of extreme climatic events.

The study of the internal side of vulnerability addressed the driving forces for urban growth, the type of urbanization created, the characteristics of the population and the modifications to the landscape. Using remote sensing, the project reconstructed urban growth and land use in Tijuana from 1972 to 2000. Updated urban data was incorporated (public services, transportation, paved roads, schools and healthcare centres, and legal and illegal urban developments). A geographic information system (GIS) was created to manage data and facilitate their analysis and visualization. Population data were added for 1990 and 2000 by census tract (not available at this level of analysis for previous years).

The project documented poverty and social exclusion in the urban area. Poverty is a multidimensional problem and it is difficult to document through a single index. In order to avoid distortions caused by a single variable (income), we used a marginality index developed by Mexico's National Population Council based on eight variables – income; age; gender; education; access to drinking water; sewage; electricity; and healthcare – as well as time of residence in Tijuana. Data were compiled by census tract for 1990 and 2000. This information was complemented by 150 interviews in hazardous areas (flooding and landslides) identified by the project. The interviews provided a profile of the inhabitants, their perception of risk, an idea of their assets, and a notion about their willingness to relocate to secure areas. Results from the interviews showed that close to 40 per cent of the households would not evacuate their homes in the event of exposure to an extreme climatic event, even if demanded by local authorities. These inhabitants feared that they would not be allowed to return to their homes. Many of them were recent migrants with less than five years in Tijuana and without previous experience of extreme flooding and/or landslides. Lack of awareness about climate-related hazards among these inhabitants is only part of the explanation for their attitudes. Inhabitants with a longer residence time in Tijuana who suffered from the 1993 to 1994 or the 1997 to 1998 El Niño events would also have resisted forced evacuations for fear of not being allowed to return to their homes. Additional data about social organizations in each neighbourhood was later added in order to have complementary parameters of social assets.

Another 27 additional interviews were conducted with local and state officials, including current and previous mayors of Tijuana during the 1993 and 1997 El Niño events. These officials were responsible for emergency response to environmental occurrences. The interviews sought to learn about the perception of hazards by local and state authorities and the use of scientific data (mainly climate forecasts) in decision-making. The interviews revealed a lack of knowledge about Tijuana's vulnerability and the potential role of scientific analysis in reducing and managing climate-related hazards.

Results from the project showed a complex web of social processes behind social and urban vulnerability in Tijuana. The study of urban vulnerability focused particularly on understanding modifications to the landscape by urban growth (type of urbanization and land use) between 1972 and 2000. The new landscape resulting from these modifications affects the type of interactions between climate and urbanization. A good illustration of the effects of these modifications is changes in precipitation runoff. The pattern of runoff is important in the study of flooding hazards. Runoff has diverse characteristics and consequences according to the characteristics of the landscape, the amount of precipitation, and the type of urbanization and land use within the urban area. Urbanization also introduces significant changes in the stability of slopes. Modifications of the landscape by urban growth, for example, dramatically reduce rainwater infiltration and accelerate runoff downhill, increasing flooding hazards in low areas. Problems with slope stability are also common in areas with poor or incomplete urbanization and even in well-urbanized areas.

Based on the modification of the landscape, we identified different paths of urbanization responsible for those modifications. We found two main processes of urbanization. On the one hand, legal and illegal large-scale urban developments have created massive movements of soil, fracturing the stability of slopes and affecting the natural drainage of the terrain. The high demands for housing and urban land are factors driving these developments. On the other hand, the constant influx of low-income migrants to the city and the lack of land suitable for urban growth are driving forces for modifications to the landscape. Individuals extend to outlying parts of the urban area, particularly the peripheries, and settle on flood-prone areas or high slopes to meet their housing needs.

Key aspects for understanding alternatives to reduce vulnerability in Tijuana are the social processes that drive urban and population growth at the local, national and transnational levels. The study of vulnerability emphasizes local processes (vulnerability is time and site specific); but the alternatives to address and reduce that vulnerability need also to consider social processes at different scales (national, regional and international), operating as driving forces of local processes. Among factors driving urban and population growth in Tijuana are domestic affairs in Mexico that include recurrent economic, social and political crises, and the redefinition of the role of the state, which since the 1980s has adopted a strong market-economy ideology. Those occurrences have been translated into massive unemployment and underemployment in large parts of the country, a regional social and economic divide between the north and south, a strong increase in violence, particularly in urban areas, and severe cuts in social investment. Thus, the impact of financial crises upon the construction of housing and operation, upon maintenance of urban public services, and upon access to education and healthcare was strongly aggravated.

Processes operating at a regional level include the North American Free Trade Agreement (NAFTA), implemented in 1994. The evaluation of NAFTA yields a negative balance for Mexico, particularly in terms of the loss of domestic economic activities and jobs (both in agriculture and industry), and the aggravation of the regional divide between North and South Mexico. Processes at the international level include the relocation of industry associated with transnational corporations (subsidiaries or subcontractors) to Tijuana. Rapid industrial growth in Tijuana since the late 1970s has created a constant demand for jobs. The opportunity to find a job in the maquiladora industry operating there has been a strong attraction factor for a large number of unemployed in Mexico. The maquiladora has become an important pole of attraction for migration, it should be noted, and has rapidly modified the urban economy and urban structure, aggravating urban fragmentation and social exclusion in Tijuana. Rapid expansion of maquiladora plants throughout the city has increased the competition for access to urban land to the detriment of other social groups. Tijuana is an urban space that results from constant confrontations between the structural conditions of exclusion and resistance by low-income groups (Harvey, 1973; Castells, 1974). The combined effect of these multi-scale factors helps to explain rapid but unbalanced urban and population growth in Tijuana. These factors also make it clear that the control of urban growth is an almost impossible enterprise for local authorities.

In summary, social and urban vulnerability in Tijuana is created by structural conditions of social inequality and marginalization, fast uncontrolled population and urban growth, deficiencies in urban planning, severe modifications to the landscape by urban growth, corruption of local authorities, lack of financial resources for the development and maintenance of housing and public services, and lack of information and knowledge about hazardous areas. As mentioned above, in reality, local authorities have little control of small- and large-scale legal and illegal urban growth. The major driving forces of that growth are associated with transnational, national and local socio-economic and geopolitical processes. There is also a lack of political will in Tijuana to tackle social exclusion, poverty and major urban and environmental problems by local decision-makers. The three-year term limit for municipal administrations in Mexico, with no possibility for immediate re-election, provides little incentive and time for a strong commitment to address structural problems. Current geographies of exclusion limit the opportunities to reduce social and urban vulnerability to extreme climatic events in Tijuana.

The state of science

There has been impressive attention to vulnerability during recent years. Vulnerability is often considered to comprise a number of components, including exposure to impacts, sensitivity and the capacity to adapt (Turner et al, 2003; O'Brien et al, 2004; Adger and Vincent, 2005). Adger's (2006) comprehensive review of research traditions of vulnerability to environmental change is a good reference point of the state of science. This section has a much more modest task. It reflects on potential obstacles and opportunities to bridge the science/practice divide in the study of vulnerability. A point of departure is the difference among research traditions in the study of vulnerability. Different conceptual traditions create different research results. It is unwise to assume that all scientific studies of vulnerability can yield positive practical actions and policies leading to its reduction and enhancing opportunities for adaptation.

One size fits all

One of the obstacles to reducing the science/practice divide in vulnerability is the trend to create models that would apply in diverse sets of conditions and societies. This might be a problem that has not received enough attention. Scientific research, like other creative processes, is prone to seek unique and universal contributions. While there is a clear value in these efforts, there is also a risk that such contributions can become detached from a specific reality, a problem in the context of efforts oriented to understand and reduce vulnerability within the science/practice interface. Vulnerability is time and site specific (Blaikie et al, 1994; Adger, 2006), and its study cannot be transformed into a blueprint to be used in different societies and conditions. One-size-fits-all approaches could have unexpected and unintended negative, practical and policy results. Part of the problem is precisely the science/practice divide. Scientific contributions are often conceived and developed in isolation from their potential policy and practice implications. Since the translation of research results is deferred to other actors, scientists rarely consider themselves liable for the unintended negative consequences in the event that those contributions achieve a practical application. In fact, policy-makers and practitioners are also rarely liable for the consequences. This lack of liability aggravates the science/practice divide. Scientists do not feel compelled to address the structural factors in society behind vulnerability, thus creating more distance between research (objectives, conceptual frameworks, methods and results) and reality in terms of the alternatives to reduce vulnerability. For some scholars, the study of vulnerability has become a fascinating effort to understand the interactions between the social and the biophysical systems (also called the coupling of a social–ecological system), but not necessarily as part of an effort to reduce it. It is often expected that research results will eventually, somehow, be translated into useful practices and policies. Practitioners and policy-makers are generally not attracted to scientific contributions that do not address the specific problems that they confront in their communities.

The risk of becoming a popular but vague concept

Vulnerability, like other popular concepts addressing social processes (development, sustainable development, decentralization), has been subject to diverse interpretations. The appeal of vulnerability is particularly strong in the current context of attention to global environmental problems (particularly climate change) and their connection to local environmental problems, social inequality, failing institutions, and the search of new paradigms to orient growth. Vulnerability may face a similar fate to development and sustainable development; it can become used in all possible situations and overextended. Ambiguity in the use of analytical concepts transforms them into rhetoric rather than operational tools. As we have said, vulnerability is time and site specific and its study should reflect and address local conditions.

The trap of technical approaches

Scholars and practitioners tend to fragment or simplify complex problems in order to address them. Although these approaches have practical benefits, they involve the risk of oversimplifying complex problems and losing key elements for building integrated perspectives that lead to better understanding. Many of these approaches are presented as 'technical solutions'. Vulnerability is a multidimensional phenomenon that requires interdisciplinary and transdisciplinary approaches to address the interactions among its dimensions, which operate in multiple scales. However, it is not uncommon to find only technical approaches to vulnerability among practitioners or decision-makers, often presented as emergency responses to environmental contingencies (flooding, landslides and other climate-related extreme events). Thus, there is risk of reducing the study of vulnerability to methodological and somewhat mechanical approaches, to the detriment of the conceptual analysis of interactions between social and biophysical processes – the root causes of vulnerability. Vulnerability assessment could become a technical rather than an analytical task.

Schröter et al (2005) and Patt et al (2005) suggest methodological guidelines for the study of vulnerability. For them, vulnerability is not so much the development of new conceptual domains but integration across three distinct domains: natural hazards, famine relief and climate change. These authors do not recognize that the differences among studies of vulnerability are not methodological (how studies consider single or multiple stressors), but conceptual. Why is this important? Different conceptual frameworks in the study of vulnerability produce different research results. The emphasis of methods over concepts can create fragmented perspectives that are common in environmental management, where the environment is detached from the socio-political and cultural dimensions from which the problems emerge (Redclift, 1994; Bryant and Wilson, 1998; Gibbs and Jonas, 2000). Attention to the conceptual elements of vulnerability is particularly important in poor countries, where drivers for vulnerability have different characteristics than those in rich countries (although Hurricane Katrina and New Orleans have shown that these characteristics are perhaps more frequently neglected than non-existent).

The study of vulnerability requires addressing the balance between structural constraints that limit assets and access to resources by individuals, social groups and their agency to cope with the exposure to extreme events. The same can be said for the design of adaptation strategies. It is worth stressing the need for a better understanding of social processes in vulnerability. Approaches that neglect the understanding of social processes could lead to incomplete or incorrect conclusions.

The search for alternatives: Connecting vulnerability and adaptation

Vulnerability is a useful analytical concept for identifying the structural conditions of susceptibility to harm and risk, but perhaps has limitations in guiding actions and policies to reduce both harm and risk. These limitations are due to the subject of the analysis rather than deficiencies in the concept itself. Actions to reduce vulnerability need to address social and economic equity issues outside the priorities of local and national decision-makers. Vulnerable people become visible to decision-makers when they become victims of extreme events, particularly, but not exclusively, in poor countries. They are usually excluded from decision-making and from access to power (Pelling, 2003; Adger, 2006). Efforts and actions to reduce vulnerability need to identify and address its root causes, often associated with social inequality, poverty and marginalization.

As we know, social inequality is rooted in structural contradictions in societies and is present throughout human history. What has changed in this era of globalization is the aggravation and extension of social inequality. The reduction of traditional physical barriers in the interaction among societies during the last century has had impressive benefits (communications, opportunities for economic growth, exchange of knowledge and ideas, among many others); but it has also expanded the reach of social inequality, marginalization, exploitation and poverty. Global environmental change should be added to the list of factors aggravating social inequality by global, regional and local socio-economic and geopolitical processes. O'Brien and Leichenko (2000) used the term 'double exposure' to call attention to the interactions of climate change with economic globalization. Adger et al (2005) argue that risks from climate change are imposed upon present-day society as a result of previous actions in perturbing the climate system and highlight the key role of underlying distributions of power within the institutions that manage resources that often create vulnerabilities. For them, present-day adaptation actions reinforce existing inequalities and do little to alleviate underlying vulnerabilities. They suggest that measures to reduce poverty and increase access to resources could reduce present-day vulnerabilities, as well as vulnerability to climate variability and climate change. Thomas and Twyman (2005) highlight the fact that climate change does not occur independently of other processes affecting poor societies and call attention to how the interface of climate change and development processes can enhance existing inequalities.

The discussion of the science/practice interface in the context of climate variability and change would benefit from connecting the concept of vulnerability with other concepts that address actions to reduce harm and risk and open opportunities for social well-being. Adaptation is a natural choice for that purpose. Adaptation is associated with actions needed to reduce harm and risk to an extreme event or to evolve to a higher stage of social well-being in societies. It provides an umbrella for the design and enforcement of actions intended to reduce vulnerability (Adger and Vincent, 2005; Adger et al, 2005). Both concepts would also benefit from a connection to specific issues, problems and opportunities for local development in the agendas of stakeholders, decision-makers and practitioners (Burton, 1997; Beg et al, 2002). The multidimensional perspective of vulnerability and adaptation is central to establish those connections, and equity should be an important component.

Several scholars have emphasized the importance of equity concerns of the consequences of climate change and how they are connected to development challenges, particularly in poor countries (O'Brien and Leichenko, 2000; Mirza, 2003; O'Brien et al, 2004; Tol et al, 2004; Thomas and Twyman, 2005; Paavola and Adger, 2006; Reid and Vogel, 2006). The close association of vulnerability and equity is highlighted by contributions drawing attention to factors influencing individuals' or groups' capacity to anticipate, cope, resist and recover from the impact of a natural hazard (Bohle et al, 1994; Wisner, 2004). These factors (assets, sources of livelihood, class, race, ethnicity, gender and poverty) are also part of the discussion of social justice presented in approaches seeking higher states of social well-being (development, sustainable development and decentralization).

Scholars have also raised awareness about equity issues in adaptation to climate change. Adger et al (2005) call attention to distributional issues of adaptation and to the balance between private and public costs and benefits of adaptation action. Their efforts to identify the criteria for success in adaptation recognize possible externalities at other geographical and temporal scales and the risk that actions effective for the adapting agent may produce negative externalities and spatial spill-overs, potentially increasing impacts upon others or reducing their capacity to adapt. Dow et al (2006) argue that a moral imperative exists for the most vulnerable people in adaptation efforts if social justice is to be achieved.

Similar concerns have been expressed before about development and sustainable development seeking to promote equity and new patterns of growth in balance with nature. Initiatives for sustainable development in a community can be to the detriment of the social well-being of other communities, even aggravating local social inequality, poverty and marginalization. The importance of equity underlines the importance of including contradictions, conflicts and imbalance within and among societies in the discussion of sustainable development, as well as in vulnerability and adaptation to climate change.

Tompkins and Adger (2005) highlight this issue, proposing that any response to climate change must be cognizant of wider development pressures as well as goals, instead of focusing solely on single-system management. They point to the relationships among assets, institutions and society, the role of cultural and regional differences among societies, and the importance of public policy in responses to different hazards and different types of climate change.

The relationships among vulnerability, adaptation and development can be an effective way to broaden the scope of policy and actions to reduce vulnerability and facilitate adaptation to climate variability and change. But that framework is prone to incorporate actions within the scope of local decision-makers and practitioners and is not necessarily responsive to individual and group agency. The reduction of vulnerability and opening opportunities for adaptation, development or sustainable development depend upon the balance between structural constraints and agency. Structural constraints to higher stages of social well-being for larger parts of society require institutional change and depend upon complex politically negotiated processes. But it would be unrealistic to expect that such processes could make a difference by themselves. Agency (individual or social groups) is an important component of social change needed to reduce vulnerability and open opportunities for adaptation. The relationship mentioned above would benefit from incorporating a concept like sustainable livelihoods, capable of addressing agency at a larger scale. It is also an efficient way to present the discussion of vulnerability and adaptation within the broader context of development and sustainable development that can enhance their visibility, attention and priority among decision-makers and stakeholders in society. In fact, to some extent it is a logical connection since approaches to vulnerability and sustainable livelihoods share a common conceptual basis stemming from the contributions of Amartya Sen (1981) on entitlement theory.

The sustainable livelihoods concept can be understood as individual or household projects that engage actors and are landscaped across scales as they seek to make a living and make it meaningful (Bebbington, 1999). It is the challenge of securing a viable way of guaranteeing the material basis of their livelihood and, at the same time, building something of their own. The livelihood analysis seeks to understand how political economic structures at different scales shape marginalized households and landscapes (Bebbington, 1999). There are several useful connections between this approach and vulnerability, adaptation and sustainable development. Livelihood projects such as vulnerability, adaptation and sustainable development engage multiple realities across scales. Individuals or households are continuously renegotiating their role as part of interlocking projects with their economic activities, landscapes, social networks and institutions. The livelihoods approach links individual or household strategies and landscapes (the environment). It identifies the type of relationship established between the economic and social dimensions of these strategies and the environment. It also links actors and practices, providing a detailed characterization of the livelihood strategies. Finally, it links agency at the individual or household level with structure that conditions and shapes access to resources. All of these aspects are part of the discussion of vulnerability and adaptation and are central to the examination of sustainable development.

Reid and Vogel (2006) support the sustainable livelihoods framework as a multidimensional concept that enables the understanding of some of the interacting factors that shape the way in which communities respond and interact with climate variability and other stresses. They highlight the sustainable livelihoods framework focus on people and how their assets in the form of various capitals enable them to achieve positive livelihood outcomes. They also seek to connect factors that appear to enhance local livelihoods with options for future development and adaptation, and enhance resilience with interventions to reduce vulnerability.

Figure 8.1 illustrates the connections among these concepts. Sustainable livelihoods play an interface role among vulnerability, adaptation and sustainable development. Reducing vulnerability, enhancing adaptation to climate change and improving sustainable livelihoods can be considered as a continuum with different stages and as contributing to sustainable development. The focus of sustainable livelihoods on people provides the concept of a central role in the continuum. Seeking new patterns for science in addressing and eventually solving society's problems in the 21st century should foster contributions from science along different stages of the continuum, but also from other forms of knowledge (practitioners). It is worth restating the relevance of linking actions to reduce vulnerability and enhance adaptation to climate variability with broader development strategies. The science/practice interface can play a key role in this direction. Knowledge generated in the study of vulnerability

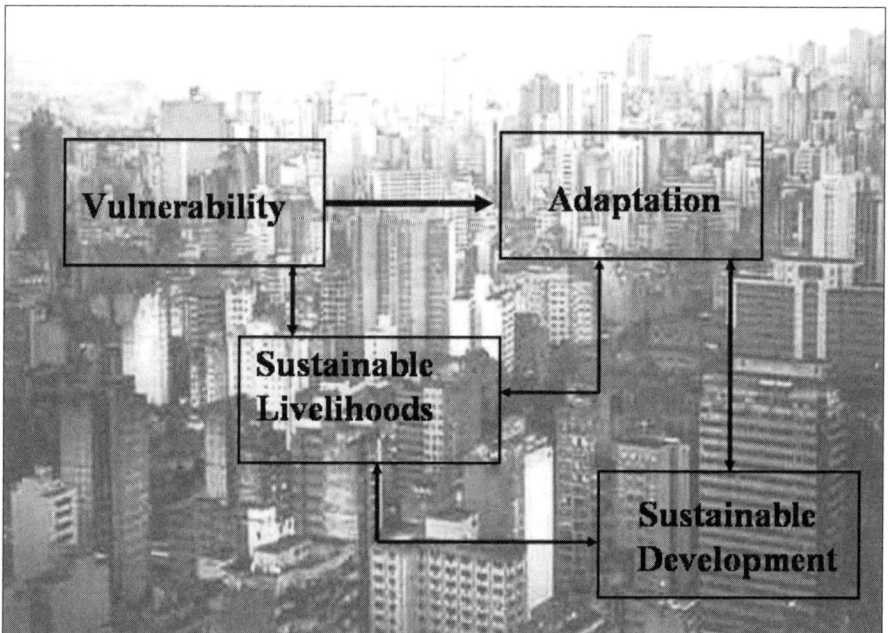

Figure 8.1 *A continuum from vulnerability to sustainable development*
Source: chapter author

and adaptation to climate change under the approach suggested above perhaps has better opportunities to become appropriated by stakeholders and transformed into concrete actions.

The science/practice interface

Concerns about the role of science in the study of global environmental change have fostered fresh ideas to reduce the divide between science and practice. These efforts recognize the need to transcend disciplinary boundaries and reductionist and positivist approaches to science into a new role in the generation of knowledge that includes a balance and dynamic participation of stakeholders (Robertson and Hull, 2003; Bammer, 2005; Naess et al, 2006; Roux et al, 2006; Turnhout et al, 2007). Robertson and Hull (2003) put forward the notion of public ecology, aimed at building common ground among competing beliefs and values for the environment, where science is produced in collaboration with a wide variety of stakeholders in order to construct a body of knowledge that will reflect the pluralistic and pragmatic context of its use, while continuing to maintain the rigor and accountability of scholarly knowledge.

Scholars have suggested a number of methods to facilitate collaboration among these groups. Brown (2003) and Bammer (2005) list a number of participatory methods that can enable practitioners and researchers to learn together about problems of common interest and provide reciprocal benefits. This is, however, a broad area of exploration that requires further research in order to find methods capable of concealing differences in disciplinary cultures, ideologies, interests and expectations among scientists and between scientists and practitioners.

A number of authors have highlighted the importance of facilitating understanding and communication between scientists and practitioners. Brown (2003) underscores the value of understanding the needs, tools, values, language and process functions of practitioners by scientists as an effective way to bridge the divide between these two communities. Roux et al (2006) highlight differences in operational cultures and working philosophies between scientists and practitioners and put forward the notion of bidirectional knowledge flows between them, recognizing that highly relevant knowledge is also created in the policy, management, societal and traditional knowledge domains. Turnhout et al (2007) conceptualize an interface area where science and policy meet. At this interface, knowledge is translated into usable knowledge, policy questions are translated into research questions, and knowledge translation and use take place as well.

It is worth addressing the discussion above in the light of our experience in Tijuana. The main practitioners from Tijuana's municipal authority participating in the project were Protección Civil, responsible for emergency response to natural and man-made disasters in coordination with the police and fire departments, and Comité de Planeación Municipal (Municipal Planning Committee), a coordinating body for municipal development with the participation of stakeholders and local authorities. Each of these two departments had a particular motivation for their involvement in the project. For Protección Civil, the project provided a unique opportunity to carry out a detailed study of vulnerability, as well as the creation of databases and a GIS to improve their operation and response capacity to emergencies. Comité de Planeación was interested in assisting the municipal authority and the city in expanding and updating information about the urban area. The project offered these two offices the opportunity to overcome resource limitations that constrained their minimal research capacities. These public officials shared with project researchers a common interest in understanding and reducing vulnerability to climate variability in Tijuana. The project created a middle space mentioned by Brown (2003) and Turnhout et al (2007), where scientists and practitioners met to bridge the divide between the two domains of knowledge.

We had advantages in the project that facilitated the creation of this shared space. A number of the researchers had a good understanding of the needs, tools and thought processes of practitioners in Tijuana obtained from previous collaborations with Mexican municipal, state and federal authorities. They also had a good knowledge of the city of Tijuana and its urban problems, and the socio-demographic characteristics of its inhabitants. Previous knowledge of urban issues in Tijuana created a pool of shared knowledge with practitioners that facilitated dialogue and communication, essential for the development of trust among the research team and ownership of the project. An important element in the collaboration between scientists and practitioners was the presence of leadership with transdisciplinary skills. Scholars highlight the critical role of transdisciplinary skills to lead and facilitate the convergence of the knowledge domains in science and practice (Roux et al, 2006). It should be clear that although transdisciplinary skills are an asset when studying complex and multidimensional issues such as vulnerability and bridging knowledge domains in science and practice, leadership in research is also a learning process and it cannot be taken for granted. Little attention has been given to this issue in the literature.

Lessons from the project demonstrate that differences between the two knowledge domains (science and practice) do not disappear but become subordinated to their common elements. Differences in the perception of knowledge to be extracted from the research process represent a clear illustration of this point. Practitioners were interested in knowledge that could help them to better carry out their work: identification of hazardous areas, reduction of vulnerability and how to better prepare in case of an environmental emergency. Researchers sought a broader multidimensional understanding of vulnerability and its driving forces. In the praxis of the project, these two perspectives turned out to be complementary. Researchers benefited from the practical approach by practitioners as well as from a broader perspective of vulnerability that expanded uses of the knowledge generated by the project. It might also be worth suggesting that key researchers in the project played a facilitator role in reaching understanding between the two groups. This mediating role of scientists has been suggested by some authors (Brown, 2003; Roux et al, 2006; Hordijk and Baud, 2007; Turnhout et al, 2007); but perhaps it has not received enough attention in the literature. The dynamic role of key researchers as mediators in the project did not have an equivalent counterpart among practitioners. There was, however, a flexible attitude and strong interest in the collaboration among practitioners in key positions within the municipal government in Tijuana. This implies that there was not a balanced distribution of tasks and responsibilities (most of the research was carried out by the scholars in the project).

Our experience in this project highlights the importance of realistic expectations in terms of the involvement of practitioners in the research process, particularly in the context of poor countries. Practitioners are subject to a number of demands and pressures (time and resources) from their jobs and have limited availability to participate in research. A selective and efficient use of their time in that process might yield better results.

Other actors

Our project could have benefited from the participation of other stakeholders. Unfortunately, it was not designed with this objective in mind. It is worth highlighting the difficulty of incorporating stakeholders within the research process. Community groups in hazardous areas are not well organized; their input is important, but our project was not set up to facilitate their participation. Also of interest is the participation of actors who can provide continuity in the transfer of knowledge. In the case of Tijuana, non-governmental organizations (NGOs) and community organizations (COs) can play such a role. Experience from past projects in the area have shown that NGOs and COs can operate as an interface among stakeholders and assist in securing the continuation of projects beyond the involvement of researchers and public officials. Problems in securing political support for vulnerability reduction in Tijuana within the mayor's office (past and present administrations) stress the need to secure governance processes capable of surviving beyond the three-year terms of municipal administrations. NGOs, COs and other actors are capable of maintaining involvement in those processes.

There are difficulties, however, in securing the involvement of NGOs working on environmental issues in Tijuana. These groups are going through a tricky time. Their number has declined dramatically for several years due to financial difficulties. Support for these organizations during the 1990s flourished in light of the attention to border environmental problems generated by the debate around the North American Free Trade Agreement (NAFTA). That support has gradually disappeared due to changes in interests and priorities among foundations and other funding organizations. Ironically, the environmental problems remain; but the support to address them has diminished significantly. Environmental groups and other NGOs currently operating in Tijuana give no attention to vulnerability and adaptation to climate variability and change at this time.

There are also questions about the role of NGOs in governance. Most of the literature on NGOs in development issues is optimistic about their role in international, national and local arenas. But some scholars stress the need for a better assessment of the role of NGOs in governance. Jordan and Van Tuijl (2000) consider the appropriate role of NGOs' advocacy as influencing reality directly, rather than through policy. This is a role beyond the delivery of a narrow service (helping people to access information or providing tools to reach out to decision-makers), which is a role supported by the World Bank, international agencies and national governments. For others, NGO actions relate to influencing policy, especially public policy (Edwards and Hulme, 1995). However, in the case of environmental NGOs, much less attention has been given to their role in governance.

The following are questions for further research in this area:

- Is the role of these organizations limited to a narrow service mentioned above (raising the profile of issues)?
- Are these organizations seeking to change reality and formulate actions oriented to directly controlling the negative consequences of growth and, in our case, of climate variability and change?

- Is the role of these groups in governance limited only to influence the decision-making process of policies, leaving their implementation to other actors (governments, international organizations, the business community, individuals or community groups)?

There are no clear answers to these questions and they will depend upon the particular circumstances of each case. In the case of the project in Tijuana, the involvement of NGOs in creating and using knowledge on vulnerability and adaptation to climate variability could have been an effective vehicle to reach stakeholders, provide continuity in those actions, and enhance the visibility and political relevance of addressing the problems.

Other actors could also play a role in creating and using knowledge about vulnerability and adaptation in Tijuana. These are professional organizations and religious and business groups. Some professional organizations (engineers, architects, medical associations, teachers, etc.), business organizations (Chamber of Commerce, Chamber of Manufacturing Industry, Chamber of the Construction Industry) and religious groups (Christian) have shown interest in urban and social problems in Tijuana and their willingness to participate in seeking solutions to some of them. As mentioned above, connecting the discussion of vulnerability and adaptation to climate variability with a broader discussion of livelihoods and sustainable development could enhance the interest of those groups. Their participation could also become an effective tool in raising the relevance of these issues in Tijuana.

Obstacles

Scholars have paid attention to the obstacles bridging the science and practice domains of knowledge (Robertson and Hull, 2003; Brown, 2003; Bammer, 2005; Naess et al, 2006; Roux et al, 2006; Hordijk and Baud, 2007; Turnhout et al, 2007). The project encountered other sets of obstacles not mentioned often in the literature. Access to reliable data at the local level was one obstacle; but it is also common in studies at the urban level in poor countries. Data and information on social, biophysical and urban variables relevant to the study of vulnerability were often not easily available and required significant amounts of fieldwork and collaboration with local authorities. In the case of Tijuana, it took our research team over two years to complete databases and systematize information needed for the research. We encountered all types of difficulties in this regard: parts of the data were gathered in diverse offices within the municipal government; data were not easily available or in different formats; responses from offices took much longer than expected due to inexplicable causes, etc. The participation of local officials provided significant assistance in collecting data from offices of the municipal government. Without their collaboration and direct involvement in the request for data, we could have not completed our databases. Our research team also carried out a significant amount of fieldwork to verify existing data and to collect new data.

The lengthy process of data collection caused unexpected delays in the development of the project which, in turn, created another unexpected obstacle. There was a change in the administration in the municipal government of Tijuana before we completed our research. An opposition party won the local election and a complete turnover of public officials occurred. Practitioners involved in the project were removed from their positions. The short terms of municipal administrations in Mexico (three years), together with the lack of tenure of public officials are obstacles for middle- and long-term collaborations between scientists and practitioners at the local level. This is also an issue to which the literature has provided little attention.

The transition to a new municipal administration created obstacles in transferring knowledge and research results to decision-makers and to individuals and groups in hazardous areas. Local officials from Protección Civil were important players in communicating with individual and community groups in hazardous areas, given the relationships that existed in some of them. One obstacle was establishing a reliable, trusty and efficient route of communication with those individuals and social groups. Transferring knowledge is a delicate matter. Lessons from previous projects on sensitive environmental issues underscore the controversial nature of communicating environmental hazards, even if there is clear scientific evidence backing up research results. There are ample references on risk communication; but its practice depends upon the structures and characteristics of local society. In the case of Tijuana, Protección Civil had already established a relationship with inhabitants in hazardous areas and we had planned on working together in communicating knowledge on vulnerability to those communities. The change of municipal administration before the project ended hindered that communication.

Additional obstacles in transferring knowledge from the project were due to the lack of attention to vulnerability to climate variability within the municipal government in Tijuana. Despite historical evidence of social, economic and environmental damages caused by flooding and landslides there, vulnerability reduction to climate variability has not been a priority for its past or current mayors. The former director of Protección Civil complained about his difficulties in attracting attention to these issues from the mayor's office. He believed the project could be instrumental in communicating the urgency for reducing vulnerability and enhancing adaptation to high-ranked officials in Tijuana.

The discussion above highlights the need for institutional changes in Tijuana and Mexico, more generally, in order to increase attention and priority and to expand the scope of environmental issues in that country. The environment continues to be considered as an afterthought of economic growth. It is also detached from its socio-political and cultural dimensions. Despite annual climate-related disasters in several parts of Mexico with significant economic, social, environmental and political cost, attention to vulnerability and adaptation to climate variability and change is more rhetorical than real.

Scientific contributions to sustainable development challenges (including socio-economic, cultural and political aspects traditionally considered under the umbrella of development, as well as environmental issues embedded in biophysical processes at local and global scales) play an important role in our understanding. It is worth remembering, however, that there is a difference between understanding and acting (van den Hove, 2007), and institutional change should target both aspects.

Institutional changes can be a lengthy process in Mexico; but better understanding of vulnerability and adaptation and the way in which those issues are introduced to stakeholders and decision-makers could increase attention to them. Explicit connections between vulnerability and adaptation with urban issues can be an effective method to achieve this goal. The advantage of this approach is to present a vision of vulnerability to which stakeholders and decision-makers can relate in their daily lives and areas of concern.

It is worth returning here to a point introduced in the first part of this chapter – namely, considering vulnerability and adaptation in conjunction with sustainable livelihoods and sustainable development as a continuum (see Figure 8.1). Some scholars have suggested connecting vulnerability and adaptation. Tompkins and Adger (2005) suggest a framework establishing relationships among assets, institutions and society that could be extended to facilitate connecting vulnerability and adaptation. Naess et al (2006) suggest vulnerability assessment as an instrument for local-level adaptation and recognize that existing municipal institutional arrangements provide few incentives for proactive adaptation at the local level in Norway. They also stress the importance of addressing climate vulnerability in a way that the information fits existing structures. Adger and Vincent (2005) use adaptive capacity, a component of vulnerability, to connect it with adaptation. They suggest that adaptive capacity is a vector of resources and assets that represent the asset base from which adaptation actions and investments can be made. For them, adaptive capacity has diverse elements encompassing the capacity to modify exposure to risks associated with climate change, absorb and recover from losses stemming from climate impacts, and exploit new opportunities that arise in the process of adaptation. The pool of resources and assets that comprise adaptive capacity are also part of sustainable livelihoods and can become components of operational strategies in sustainable development. Taking advantage of these connections can be an efficient strategy to illustrate the broader benefits of vulnerability reduction and enhance opportunities for adaptation to climate variability and change in Tijuana.

Conclusions

The project in Tijuana fosters some reflections about the science/practice interface in climate variability and climate change. One major conclusion is the benefit of collaboration between scientists and practitioners from the early stages of research. This collaboration is important not only for the generation of knowledge but also for its use. Lessons from the project have also shown a need for the participation of other actors in the transfer of knowledge and its translation in concrete tangible actions that lead to reductions in vulnerability and enhance adaptation to climate vulnerability and change.

The growing attention to the science/practice interface in the study of climate change and other processes in global environmental change highlights obstacles in bridging the divide between the domains of knowledge in science and practice. The case study in Tijuana encountered obstacles not frequently addressed by the literature on this topic. Those obstacles underscore the importance of institutional change and governance in order to address the challenges introduced by climate variability and change, but also to assist societies in improving their relationship with the environment.

The experience in Tijuana illustrates that there is a broad range of potential uses of knowledge on vulnerability and adaptation to climate variability. However, that knowledge should integrate multidimensional perspectives of vulnerability and adaptation. It is worth stressing the importance of not assuming that knowledge generated in these types of studies would necessarily be translated into actions leading to vulnerability reduction and adaptation for the marginalized individual and social groups in Tijuana. There are important implications of equity in vulnerability and adaptation. The case of Tijuana highlights the need to present the knowledge obtained within the broader context of development in the urban area. The multidimensional perspective of vulnerability facilitates the connection to development issues. Additional support in this regard can be obtained from connecting vulnerability and adaptation with sustainable livelihoods and sustainable development, as suggested above. This continuum also facilitates addressing issues of equity embedded in actions for the reduction of vulnerability and opening opportunities for adaptation to climate variability and change in Tijuana.

In fact, social equity seems to be one of the biggest challenges for efforts seeking vulnerability reduction, adaptation, sustainable livelihoods or sustainable development, as well as other efforts seeking to improve social well-being within and among societies. Humankind has proved incapable of reaching higher states of social well-being equitably throughout history. But the consequences of inequity and exclusion in current times have reached dramatic extension in terms of the number of people, the geographies and the manifestations.

The resilience and resistance of humankind to inequality is a source of hope for social change. Agency for resistance has maintained a constant confrontation with the structural forces of exclusion. Addressing these processes through multidimensional and multiscale perspectives in creating new paths of knowledge is a significant step forward, compared to the way in which decisions are being made today in Tijuana and other parts of the world. It is certainly an alternative worth developing compared to the fragmented perspectives used to orient growth under the market ideology that prioritized economic growth over other dimensions, or the limited technical perspective of environmental management and emergency response to the impacts of climate variability and climate change. Tijuana's situation seems difficult; but there are remarkable examples of social resilience and agency at the city, neighbourhood and household levels that give hope for positive changes in the future. Narrowing the gaps in the science and practice interface could help to catalyse efforts in this direction.

References

Adger, W. N. (1999) 'Social vulnerability to climate change and extremes in coastal Vietnam', *World Development*, vol 27, no 2, pp249–269

Adger, W. N. (2006) 'Vulnerability', *Global Environmental Change*, vol 16, no 3, pp268–281

Adger, W. N. and Vincent, K. (2005) 'Uncertainty in adaptive capacity', *Comptes Rendus Geoscience*, vol 337, no 4, pp399–410

Adger, W. N., Arnell, N. and Tompkins, E. (2005) 'Successful adaptation to climate change across scales, *Global Environmental Change*, vol 15, no 2, pp77–86

Alegría, T. and Ordoñez, G. (2005) *Legalizando la ciudad: Asentamientos informales y procesos de regularización en Tijuana*, El Colegio de la Frontera Norte, Tijuana

Bammer, G. (2005) 'Integration and implementation sciences: Building a new specialization', *Ecology and Society*, vol 10, no 2, article 6, www.ecologyandsociety.org/vol10/iss2/art6/

Beall, J. (2002) 'Globalization and social exclusion in cities: Framing the debate with lessons from Africa and Asia', *Environment and Urbanization*, vol 14, no 1, pp41–51

Bebbington, A. (1999) 'Capital and capabilities: A framework for analyzing peasant viability, rural livelihoods, and poverty', *World Development*, vol 27, no 12, pp2021–2044

Beg, N., Morlot, J. C., Davidson, O. and Afrane, Y. (2002) 'Linkages between climate change and sustainable development', *Climate Policy*, vol 2, pp129–144

Blaikie, P., Cannon, T., Davis, I. and Wisner, B. (1994) *At Risk: Natural Hazards, People's Vulnerability, and Disasters*, Routledge, London

Bocco G., Sánchez, R. and Riemann, H. (1993) 'Evaluación del impacto de las inundaciones en Tijuana (enero de 1993)', *Frontera Norte*, vol 5, no 10, pp53–84

Bohle, H., Downing, T. and Watts, M. (1994) 'Climate change and social vulnerability: Towards a sociology and geography of food insecurity', *Global Environmental Change*, vol 4, no 1, pp37–48

Brown, A. L. (2003) 'Increasing the utility of urban environmental quality information', *Landscape and Urban Planning*, vol 65, pp85–93

Bryant, R., and Wilson, G. (1998) 'Rethinking environmental management', *Progress in Human Geography*, vol 22, no 3, pp321–343

Burton I. (1997) 'Vulnerability and adaptive response in the context of climate and climate change', *Climatic Change*, vol 36, pp185–196

Castells, M. (1974) *La Cuestión Urbana*, Siglo Veintiuno Editores, Mexico City

Dow, K., Kasperson, R. E. and Bohn, M. (2006) 'Exploring the social justice implications of adaptation and vulnerability', in W. N. Adger, J. Paavola, S. Huq and M. J. Mace (eds) *Fairness in Adaptation to Climate Change*, MIT Press, Cambridge, pp79–96

Edwards, M. and Hulme, D. (eds) (1995) *Nongovernmental Organisations: Performance and Accountability*, Earthscan, London

Fry, G. (2001) 'Multifunctional landscapes: Towards transdisciplinary research', *Landscape and Urban Planning*, vol 57, pp159–168

Gibbs, D. and Jonas, A. (2000) 'Governance and regulation in local environmental policy: The utility of a regime approach', *Geoforum*, vol 31, pp299–313

Harvey, D. (1973) *Social Justice and the City*, John Hopkins University Press, Baltimore, MD

Hordijk, M. and Baud, I. (2007) 'The role of research and knowledge generation in collective action and urban governance: How can researchers act as catalysts?', *Habitat International*, vol 30, no 3, pp668–689

INEGI (2004) *XII Censo Nacional de Población y Vivienda*, Instituto Nacional Estadistica y Geografia, Mexico, DF, Mexico

Jordan, L. and Van Tuijl, P. (2000) 'Political responsibility in transnational NGO advocacy', *World Development*, vol 28, no 12, pp2051–2065

Kasperson, J. and Kasperson, R. (eds) (2001) *Global Environmental Risk*, United Nations University Press, Tokyo, and Earthscan, London

Lopes de Souza, M. (2001) 'Metropolitan decentralization, socio-political fragmentation and extended suburbanization: Brazilian urbanization in the 1980s and 1990s', *Geoforum*, vol 32, pp437–447

Luks, F. and Siebenhuner, B. (2007) 'Transdisciplinarity for social learning? The contributions of the German socio-ecological research initiative to sustainability governance', *Ecological Economics*, doi10.1016/j.ecoloecon.2006.11.007

Mirza, M. (2003) 'Climate change and extreme weather events: Can developing countries adapt?', *Climate Policy*, vol 3, pp233–248

Moser, C. (1998) 'The asset vulnerability framework: Reassessing urban poverty reduction strategies', *World Development*, vol 26, no 1, pp1–19

Naess, L. O., Norland, I. T., Lafferty, W. M. and Aall, C. (2006) 'Data and processes linking vulnerability assessment to adaptation decision-making on climate change in Norway', *Global Environmental Change*, vol 16, no 2, pp221–233

O'Brien, K., and Leichenko, R. (2000) 'Double exposure: Assessing the impacts of climate change within the context of economic globalization', *Global Environmental Change*, vol 10, pp221–232

O'Brien, K., Leichenko, R., Kelkar, U., Venema, H., Aandahl G., Tomkins, H., Javed, A., Bhadwal, S., Barg, S., Nygaard, L. and West, J. (2004) 'Mapping vulnerability to multiple stressors: Climate change and economic globalization in India', *Global Environmental Change*, vol 14, pp303–313

Paavola, J. and Adger, N. (2006) 'Fair adaptation to climate change', *Ecological Economics*, vol 56, pp594–609

Patt, A., Klein, R. and de la Vega-Leinert, A. (2005) 'Taking the uncertainty in climate-change vulnerability assessment seriously', *Comptes Rendus Geoscience*, vol 337, pp411–424

Pelling, M. (2003) *The Vulnerability of Cities: Natural Disasters and Social Resilience*, Earthscan, London

Petts, J., Owens, S. and Bulkeley, H. (2006) 'Crossing boundaries: Interdisciplinarity in the context of urban environments', *Geoforum*, doi:10.1016/j.geoforum.2006.02.008

Pirez, P. (2002) 'Buenos Aires: Fragmentation and privatization of the metropolitan city', *Environment and Urbanization*, vol 14, no 1, pp145–158

Pritchett, L. and Woolcock, M. (2004) 'Solutions when the solution is the problem: Arraying the disarray in development', *World Development*, vol 32, no 2, pp191–212

Redclift, M. (1994) 'Development and the environment: Managing the contradictions?' in L. Sklair (ed) *Capitalism and Development*, Routledge, London, pp123–137

Reid, P. and Vogel, C. (2006) 'Living and responding to multiple stressors in South Africa: Glimpses from KwaZulu-Natal', *Global Environmental Change*, vol 16, pp195–206

Robertson, D. and Hull, B. (2003) 'Public ecology: An environmental science and policy for global society', *Environmental Science and Policy*, vol 6, pp399–410

Roux, D., Roggers, K., Biggs, H., Ashton, P. and Sergeant, A. (2006) 'Bridging the science–management divide: Moving from unidirectional knowledge transfer to knowledge interfacing and sharing', *Ecology and Society*, vol 11, no 1, article 4, www.ecologyandsociety.org/vol11/iss1/art4/

Schoenberger, E. (2001) 'Interdisciplinarity and social power', *Progress in Human Geography*, vol 25, no 3, pp365–382

Schröter, D., Polsky, C. and Patt, A. (2005) 'Assessing vulnerabilities to the effects of global change: An eight-step approach', *Mitigation and Adaptation Strategies for Global Change*, vol 10, pp573–596

Sen, A. (1981) *Poverty and Famines: An Essay on Entitlements and Deprivation*, Oxford University Press, Oxford

Thomas, D. and Twyman, C. (2005) 'Equity and justice in climate change adaptation amongst natural resources-dependent societies', *Global Environmental Change*, vol 15, pp115–124

Tol, R., Downing, T., Kuik, O. and Smith, J. (2004) 'Distribution aspects of climate change impacts', *Global Environmental Change*, vol 14, pp259–272

Tompkins, E. and Adger, N. (2005) 'Defining response capacity to enhance climate change policy', *Environmental Science and Policy*, vol 8, pp562–571

Tress, B., Tress, G., Decamps, H. and d'Hauteserre, A. (2001) 'Bridging human and natural sciences in landscape research', *Landscape and Urban Planning*, vol 57, pp137–141

Tress, G, Tress, B. and Fry, G. (2007) 'Analysis of the barriers to integration in landscape research projects', *Land Use Policy*, vol 24, pp374–385

Turner, B. L. II, Kasperson, R., Matson, P., McCarthy, J., Corell, R., Christensen, L., Eckley, N., Kasperson, J., Luers, A., Martello, M., Polsky, C., Pulsipher, A. and Schiller, A. (2003) 'A framework for vulnerability analysis in sustainable science', *Proceedings of the National Academy of Sciences of the United States of America*, vol 100, no 14, pp8074–8079

Turnhout, E., Hisschemöller, M. and Eijsackers, H. (2007) 'Ecological indicators: Between the two fires of science and policy', *Ecological Indicators*, vol 7, pp215–228

van den Hove, S. (2007) 'A rationale for science–policy interfaces', *Futures*, doi:10.1016/j.futures.2006.12.004

Wisner, B. (2004) 'Assessment of capability and vulnerability', in G. Bankoff, G. Freks and T. Holhorst (eds) *Vulnerability, Disasters, Development and People*, Earthscan, London, pp183–194

Part IV

The Science/ Practice Gap: Global Perspectives

9

Food Insecurity in South Africa

Scott Drimie and Gina Ziervogel

Introduction

Despite a strong government commitment to addressing development issues in South Africa, tremendous disparities in food security exist between communities and households across the country, reflecting continuing social and economic inequalities. Estimates suggest that approximately 14 million people are food insecure and 1.5 million children suffer from malnutrition (HSRC–FIVIMS, 2003; HSRC, 2004). Food insecurity is prevalent particularly in areas that have been historically disadvantaged during the colonial and apartheid eras and in the growing urban centres. The cause of hunger and malnutrition is not due to a shortage of food, but, rather, inadequate *access* to food by certain categories of individuals and households in the population (Vogel and Smith, 2002). The growing need to address South Africa's food insecurity has therefore emerged as a priority for both scientists and policy-makers.

In response to improving the understanding and monitoring of food insecurity in South Africa, a pilot study was implemented during 2004 to 2005 to establish a Food Insecurity and Vulnerability Information Management System (FIVIMS-ZA). The project was initially driven by policy-makers and supported by scientists in order to improve knowledge around food insecurity issues. Its goal was to establish a system that would provide an interface between scientists and policy-makers and practitioners engaged in such questions. This interface would enable information to be provided in an accessible format and permit improved food security programming. The pilot study is therefore a good case of examining how science engages with policy and practice, revealing some of the challenges of creating a usable interface between different users of knowledge.

Study objectives and argument

The authors will demonstrate an attempt to establish a 'boundary organization' in South Africa – namely, FIVIMS-ZA, which attempted to facilitate linkages between science and practice in food security issues. Although very ambitious, FIVIMS-ZA performed a 'brokership' role between divergent groups in the science and policy communities and provided an important contribution to confronting food insecurity in South Africa.

This chapter outlines the process of developing the pilot system and the roles of the scientists and policy-makers. It assesses how the project can be seen as a success in some ways, although there were numerous limitations. The disparities will be explored to expand further what it means to have a meaningful integration of science into policy or whether this challenge will always result in limited success from one (or more) party's point of view.

Food insecurity is a concern to policy-makers, practitioners and academics alike. Yet, the understanding of food insecurity and the process of identifying priority areas for intervention tend to differ among groups, increasing the challenges of addressing the problem. The most vulnerable groups are difficult to target and monitor. The FIVIMS-ZA process has helped different stakeholders to analyse food insecurity in South Africa, hopefully enabling future interventions to be better grounded by the realities of the affected population, limited resources and institutional challenges.

Boundary organizations

Scholarship in the social sciences has argued convincingly that what demarcates science from non-science is not some set of essential or transcendent characteristics or methods, but rather an array of contingent circumstances and strategic behaviour known as 'boundary work' (Guston, 2001, p399). The concept of boundary work has been applied in studying the strategic demarcation between political and scientific tasks, arguing that the blurring of boundaries between science and politics, rather than the intentional separation often advocated and practised, can lead to more productive policy and decision-making.

Recognizing both that there is no unbridgeable chasm between science and non-science and that the flexibility of boundary work may threaten some important values and interests, scholars have discussed some possible factors that contribute to linking the two domains and stabilizing some boundary work – this includes the identification of boundary objects and boundary organizations.

Boundary objects sit between two different social worlds, such as science and non-science, and can be used by individuals within each world for specific purposes without losing their own identity. Guston (2001, p400) gives the example of a patent on research results that can be used by scientists to establish priority – and be used simultaneously by politicians to measure the productivity of research. Boundary organizations involve the participation of actors from both sides of the boundary, as well as professionals who serve a mediating role, and they exist at the frontier of the two relatively different social worlds of politics and science. Nonetheless, they have distinct lines of accountability to each. In other words, boundary organizations perform tasks that are useful to both sides and involve people from both communities in their work, but play a distinctive role that would be difficult or impossible for organizations in either community to play.

A successful boundary organization will thus manage to accommodate two sets of principals and remain stable to external forces astride the internal instability at the actual boundary (Guston, 2001). The success of the organization in performing these tasks can then be taken as the stability of the boundary, while in practice the boundary continues to be negotiated at the lowest level and the greatest nuance within the confines of the organization. This dual agency has been labelled 'co-production': the simultaneous production of knowledge and social order (Jasanoff, 1996). Boundary organizations are involved in co-production in two ways: they facilitate collaboration between scientists and non-scientists, and they create the combined scientific and social order through generating boundary objects, which are something either party can use for its own purposes.

In order to play this role, a boundary organization must have credibility through strategies of engagement and inclusion of the interested parties, rather than insulation from them. By appealing to, and balancing between, multiple principals, a boundary organization may become an arbiter of the quality of policy-relevant research. In addition, a boundary organization (e.g. an agricultural extension service) can facilitate a more effective flow of information by augmenting the creation and transfer of usable knowledge and coordinating science and decision-making across lines.

The boundary organization is able to project authority by showing its responsive face to either audience. To the scientific principal it says: 'I will do your bidding by demonstrating to the politicians that you are contributing to their goals, and I will help facilitate some research goals besides.' To the consumer of knowledge, who is also a principal, it says: 'I will do your bidding by [ensuring] that researchers are contributing to the goals you have for the integrity and productivity of research' (Guston, 2001, p405). The boundary organization thus gives both the producers and consumers of research an opportunity to construct the boundary between them in a way favourable to their own perspectives.

As will be demonstrated below, the intention of FIVIMS-ZA was to become a 'broker' between the two communities, working as a 'facilitator' or a 'trusted third party'. FIVIMS-ZA has appeal to both the science and practitioner communities in terms of playing a direct role in brokering or facilitating a greater exchange between scientists and practitioners when working with food insecurity issues. As such, it provides an important case study in terms of establishing a boundary organization in South Africa to serve the needs of both scientists and policy-makers and practitioners.

Understanding food insecurity in South Africa and the role of FIVIMS-ZA

Although the concept of the Food Insecurity and Vulnerability Information Mapping System (FIVIMS) emerged from the Food and Agricultural Organization of the United Nations (FAO), the specific South African initiative resulted from an identified need to provide better information for a range of requirements across government in relation to food security. These requirements included reporting on the commitments made to the World Food Summit of 2002, the Millennium Development Goals (MDGs) and the Southern African Development Community's Regional Indicative Social Development Plan (SADC RISDP). Information requirements were also identified at local and municipal levels for food security programming that complemented poverty alleviation processes.

Combating food insecurity in South Africa involved actors of various types that operated at different levels. As one example, government officials operated at various levels. At the national level, there were the policy-makers concerned with the development of the monitoring system and how it might be used. At the provincial level, there were actors who mediated information dissemination between the national and district levels and oversaw the implementation of various projects. The district-level structure was relatively new and coordinated a number of municipalities. These municipalities were interested in a range of data for their planning and operational use. Information sources were limited and dissemination had numerous challenges in flowing from both the local to the national level, and from the national to the local. An improved monitoring and dissemination system in relation to food insecurity and vulnerability was therefore of interest to multiple government users.

FIVIMS-ZA was established to consolidate existing scientific knowledge around food security and to generate new information to supplement gaps. This information was to be made available through a system accessible to a wide range of users. FIVIMS-ZA can be described as a tool, process and product to assist with South African national and sub-national food security interventions. The intention was to provide this aid through a:

- structural food insecurity and vulnerability baseline;
- food security model;
- reporting capability (human capacity to interpret information which included maps, short analysis and the understanding of impacts upon particular areas and population groups); and

- monitoring system that complemented existing early warning systems (e.g. agro-meteorological monitoring) and accounts for a community's predominant risk factors to food insecurity and its capacity to cope in a rural or urban environment.

As there was some uncertainty about the state of scientific knowledge around food insecurity, it was decided that a FIVIMS should be piloted in one district. The pilot study focused on the Greater Sekhukhune District Municipality, Limpopo Province. It was intended that the project would be tested by defining the framework, implementing methods for the collection of necessary data, and developing an appropriate food insecurity and vulnerability model and a decision support system that included a reporting system for the South African Department of Agriculture (DoA). It was envisaged that once this project was completed and a successful model established, it would then be scaled up to 13 integrated sustainable rural development nodes, of which Sekhukhune would be one, before being implemented across the country.

The aim of FIVIMS-ZA was therefore to help identify the nature and location of groups that were vulnerable to food insecurity. This necessitated a definition of food security that moved away from the notion of its being linked to agricultural production to a more holistic understanding, as well as a shift in considering how to evaluate and react to food insecurity. Because the project was driven by the DoA, the investigators had tended to focus on the agricultural aspects of food security; but they were open to a broader definition.

The research concentrated on developing a chronic food insecurity baseline through creating a repository of useful national datasets around food availability, access and utilization. A detailed livelihoods survey was conducted to help understand the complex and multifaceted nature of food insecurity and vulnerability that people face and to verify the quality of the repository of national datasets.

Actual food insecurity in South Africa

The cause of hunger and malnutrition in South Africa is not due to an overall shortage of food, but to access to food by certain groups. Statistics South Africa showed in 2000 that food insecurity is not an exceptional short-term event in the lives of many South Africans, but a continuous threat for more than one third of the population (Shabalala and Mosima, 2002). Since the vast majority of people in South Africa buy their staple foods from commercial suppliers, rather than growing it themselves, access to food is largely dependent upon (direct or indirect) access to cash.

Among the poor, who by definition suffer the brunt of the lack of jobs in the South African economy, the main sources of cash are insecure piece-jobs; the government social welfare safety net, primarily in the form of old-age pensions and child support grants; and private transfers from working relatives and neighbours. In addition to cash, the 'bundle of entitlements', which enables individuals and households to feed themselves, also includes access to land (especially in rural areas) for supplementary food production, as well as access to family and community networks for sharing food that is available. As such, many households in South Africa are not in a position to address their food needs through household-level food production, as production levels are not sufficient (Shabalala and Mosima, 2002).

As will be seen below, the pilot study conducted by FIVIMS-ZA confirmed much of this and concluded that food insecurity in Sekhukhune was largely driven by:

- a relatively great reliance on purchased food;
- a relatively high dependence upon wages and remittances;
- poor people's lack of access to land and other assets essential for economically sustainable food production;
- the meagre contribution of subsistence agriculture to household food needs; and
- a disproportionately higher exposure to inflation and price shocks.

From the survey, it was clear that the majority of households interviewed (94 per cent) had the largest portion (42 per cent) of a total monthly income being spent on food. Other shares of total monthly expenditure were on clothing (9 per cent), education (8 per cent), fuel for energy (7 per cent) and burials (4 per cent). Although over 40 per cent of households indicated that they grew their own crops, this was largely for supplementary purposes through a fruit tree near the homestead, a vegetable garden or a maize plot. Major constraints to agricultural production for household food security (for consumption or sale) were a lack of inputs such as seed, fertilizer, money and water. This reiterated the importance of purchasing food for household food requirements and the related necessity of having income sources for food security.

It was also apparent from the survey that there were four common sources of household income in Sekhukhune: government-provided old age and child support grants (each being received by one third of households), in addition to remitted income from migrant labourers (31 per cent) and income from regular wage employment (27 per cent). The remaining types of social assistance (foster care grant, disability grant, care dependency grant and compensation funds) all had a limited coverage in the survey area, with none being present in more than 5 per cent of households. The same was observed for other income sources, including pension funds from work, selling of production and non-production-related assets, and the receipt of gifts in kind. Four per cent of households reported that they had received no form of income during the month prior to the survey. It was revealing that over half of the respondents (54 per cent) indicated that the household ran out of money to buy food.

In terms of household food security, questions around food consumption indicated that household members often skipped meals because of a lack of food (53 per cent), while children ate less than they should because of insufficient food (51 per cent) and sometimes went to bed hungry because of a lack of money to buy food (36 per cent). Anthropometric measurements taken during the survey confirmed provincial statistics that the prevalence of stunting was high (36 per cent) for different age groups of children between 13 and 215 months. Similarly, almost one fifth (19 per cent) of children between the ages of 13 and 215 months were underweight.

The science/practice interface

This information was intended to give substance and impetus to food security policies and programmes in the country, and, in particular, a response in Sekhukhune. The knowledge gap, especially at the local level, had been identified by the DoA and, although a range of policy instruments existed, practical information about how to deal with food insecurity was lacking. In order to better understand the problem of South Africa's food insecurity, a nationally developed FIVIMS was envisioned as a necessary part of the solution. The requirements from the Directorate for Food Security of the Department of Agriculture, South Africa, therefore identified the need for a FIVIMS that would take into consideration national policies, the Integrated Food Security Strategy for South Africa, and international best practice in the development and implementation of a pilot model that integrated international indicators on food insecurity and vulnerability.

Policy-makers and practitioners

A policy document relating to food security existed before FIVIMS-ZA began. The Integrated Food Security and Nutrition Programme (IFSNP) was designed to enable the coordination of different government line departments to meet the food security needs of the indigent population. Although the programme lacked a legislative basis, the South African Cabinet believed that the government's Social Cluster provided enough impetus to meet the political imperative of halving hunger in South Africa by 2015.

The Social Cluster was comprised of two tiers: the directors-general or their deputies leading on the political level, and assigned senior civil servants leading on the technical aspects. The interaction between the two was intended to give real impetus to integrated planning and programming. The reality was that line departments continued to be governed by budget lines and accountability to senior managers dealing with overstretched staff, who had enough on their hands without being pulled into extra-departmental activity.

Essentially, the IFSNP was comprised of the following sub-programmes:

- Food Production and Trade Programme managed by the Department of Agriculture and the Department of Health;
- Community Development Programme managed by the Department of Public Works and the Department of Social Development;

- Nutrition and Food Safety Programme managed by the Department of Health and the Department of Agriculture;
- Safety Nets and Food Emergencies Programme managed by the Department of Social Development, and provincial and local government (DoA, 2002).

The objective of the Food Production and Trade Programme was to improve resource productivity, food production and trade through strategies that included the improvement of soil and water management and control, crop intensification and marketing, and better agricultural information and communication networks. Community Development set out to create employment, income generation and the development of skills necessary for economic empowerment. Nutrition and Food Safety was intended to promote disease-specific nutritional support, to prevent malnutrition, and to promote knowledge and awareness of food nutrition information. Safety Nets and Food Emergencies provided very short-term relief from food crises through food and other aid and rehabilitation.

The institutional arrangements around the IFSNP are illustrated in Figure 9.1.

The DoA, as the lead department in the Social Cluster charged with implementing the IFSNP, clearly identified the need for information to underpin action. Taking the lead on meeting this need, the DoA contracted with a consortium to undertake the development of an information system in the pilot project.

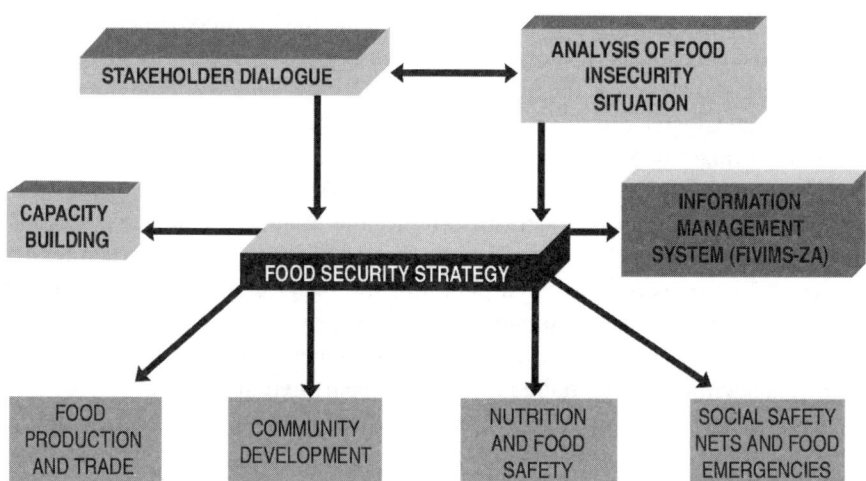

Figure 9.1 *Integrated Food Security and Nutrition Programme (IFSNP) institutional structure*

Source: chapter authors

Scientists and the creation of knowledge

The consortium that was awarded the tender was led by the Human Sciences Research Council (HSRC) and involved a range of stakeholders from both academic, research and practice backgrounds.[1] There were a number of parallel processes that fed into the pilot FIVIMS-ZA, including an advisory panel that consulted government and other stakeholders during the process. Background documents were prepared for the consortium members and government stakeholders. These explored existing material on Sekhukhune and the theoretical components of food security and vulnerability that helped to identify the complex nature of the problem and highlighted the need to be cautious in trying to develop indicators to monitor food insecurity.

The research process was sensitive to the complexities of the problem and therefore a range of methods was used during the fieldwork, including the livelihood survey, focus group discussions, household food diaries, key informant interviews and feedback sessions. This was important from the scientists' point of view, as it helped to capture both the quantitative nature of food insecurity among individuals and households and the qualitative nature of food insecurity and vulnerability that related to entitlements, livelihood activities, social capital and the dynamic nature of the problem. An institutional analysis was also undertaken to ensure that the relationship and roles of district and local stakeholders were identified. This was crucial in understanding both the dynamics of local food insecurity and the opportunities and constraints to implementing a FIVIMS-ZA at various levels.

A survey was undertaken with 597 households contributing to an in-depth understanding of food insecurity and vulnerability in Sekhukhune. The data was integrated with secondary sources and the qualitative data to produce a model. This essentially integrated spatial data from a geographic information system (GIS) with location-specific information in order to identify areas vulnerable to hunger. The various outputs represent 'boundary objects' that were created throughout the process, providing value for both scientists and policy-makers.

Towards the end of the project, training on the use of this model was undertaken with government officials from various levels and departments. Although this was an identified requirement of the process, training at the end of the pilot phase was essentially limited in effect. The research consortium argued from the inception of the project that 'capacity-building' could not be achieved by training at the end of the pilot, but that dedicated officials from a range of departments needed to be engaged throughout the process. This was achieved at the highest level in the steering committee meetings, where senior agricultural officials began to engage in a highly effective manner as their understanding of the issues increased. Little dedicated interaction occurred at other levels, partly due to the lack of real engagement in the IFSNP.

However, as the pilot unfolded – and more regular interaction with government stakeholders occurred in an interactive process – the 'ownership' of the system was embraced by the DoA. In fact, upon realizing the dangers of maps being taken at face value, the senior manager of the food security directorate insisted that the 'M' in FIVIMS-ZA be changed from the original term 'mapping' to 'management'. This relatively insignificant turn of phrase represented a significant moment in the pilot in that the department took firm ownership of the system, recognizing one of the major challenges raised by the scientists.

Establishing FIVIMS-ZA as a boundary organization: Institutional challenges[2]

At the close of the pilot phase, the consortium felt that it was not yet possible to specify detailed institutional arrangements that would support a FIVIMS because these depended upon detailed forms of stakeholder consultation and user involvement, particularly between the Department of Agriculture and other possible end users, which simply had not happened or were at an early stage. This created some tension between the department and the consortium, but was ultimately accepted as the reality after several discussions with various government stakeholders. In the institutional analysis undertaken by the consortium, certain government stakeholders outside of the DoA had never heard of FIVIMS. This was despite an awareness-raising process and some discussion among departments about the FIVIMS-ZA process. This created the danger that the FIVIMS-ZA would be judged to be irrelevant, or that there would be unrealistic expectations of the pilot system, which might not be possible in practice.

Complementing this analysis, the DoA commissioned another report on the effective coordination and implementation of the IFSNP, which looked critically at the institutional arrangements of the programme (Hamid, 2005). The unreleased report, authored by the FAO, identified a major impediment to the effective coordination of the IFSNP (and the FIVIMS) as an 'absence of legislation that clearly defines its authority and responsibility, organizational structure and working procedures' of the inter-ministerial committee (Social Cluster) responsible for the programme. This issue had major implications for a future FIVIMS as the system was intended to facilitate the information needs of the IFSNP as a multi-stakeholder forum of government departments.

Essentially, the policy-makers who commissioned the scientists to undertake the research had expectations about the FIVIMS that could not be met without serious engagement with difficult institutional questions. The fact that there was no engagement with other stakeholders within policy-making processes meant that there was no co-management of the process or challenges as they arose, and no co-creation of knowledge, which was essential if FIVIMS-ZA was to become a reality and to be sustainable. This raised the question as to whether the science/policy interface created in this context was the solution.

An independent review of the Food Emergency Scheme in Bohlabela, which neighbours Sekhukhune, echoed some other concerns raised in the FAO

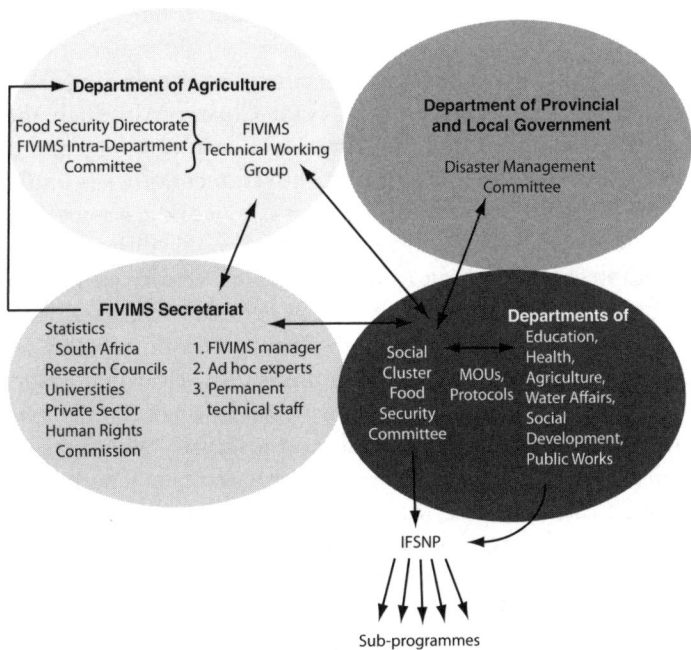

Figure 9.2 *Proposed Food Insecurity and Vulnerability Information Mapping System (FIVIMS) institutional arrangements*

Source: chapter authors

report and by the FIVIMS-ZA consortium (Polzer and Schuring, 2003). Highly complex institutional questions arose during the delivery of the food parcels, which, according to both Department of Agriculture and Department of Social Development officials, a FIVIMS would support as part of a short-term response to a food shortage in conjunction with longer-term monitoring of food insecurity. Institutional issues included communication and coordination issues among the wide range of stakeholders involved in planning and implementing the programme. Interprovincial, interdepartmental and provincial-to-district communication and coordination all proved to be problematic. Inconsistent information at higher planning levels, especially concerning the continuation of food parcel distributions after the three-month pilot, created great insecurity and distrust among local implementers who had to communicate the details of the programme to communities.

Institutional arrangements and effective utilization of information cannot be discussed without reference to the core characteristics of FIVIMS-ZA as a system itself. Research undertaken during the FIVIMS process revealed that until the overarching design, intention, identification of stakeholders and limitations involved in designing various types of systems were clarified, there was little that the consortium could do than 'tinker' with a food security 'setup'. In an attempt to help the DoA policy group think through these questions – particularly where a 'FIVIMS' might reside – Figure 9.2 was devised.

Thus, the actual concept of 'food insecurity' as a multidimensional issue became crucial for the DoA and the Social Cluster. Although theoretical constructs of the concept of food insecurity had been compiled and discussed by the scientists, this had little meaning for government officials until the DoA was faced with the reality of how difficult it was to work in a multidimensional way. One of the key ideas behind FIVIMS was to enable different departments to work in an integrated fashion to meet the food security needs of all South Africans, a major responsibility of the state as defined in the overarching social contract governing the country: the South African Constitution. As depicted in the preceding analysis of hunger in Sekhukhune, it is clear that ensuring food security went well beyond enabling effective production of food at a household level. Many of the interventions around food insecurity would reside outside the DoA; hence, the importance of greater interdepartmental needs assessments and user assessments. The FIVIMS was therefore envisaged by hard-pressed officials, battling to enforce coordination and integration, to 'facilitate' and 'enable' departments to work together to execute the IFSNP through an *integrated* food security strategy.

Another major reason for the piloting of the FIVIMS was the need for information for decision-making, packaged in an accessible manner to enable different levels of officials (and civil society, which became a theme later in the process) to respond effectively to the situation 'on the ground'. However, the dangers of overlaying information from diverse sources such as Statistics South Africa and university research became clear when viewing maps of Sekhukhune. These were characterized by a mosaic of primary colours that were alarmingly misleading in some instances. For example, the maps of the human immunodeficiency virus (HIV) prevalence statistics were confusing. In 2005, Sekhukhune was defined as a cross-border district council as it was comprised of municipalities from two provinces. The areas from Limpopo to the north displayed a lower HIV prevalence of 0 to 13.1 per cent, compared to the southern areas of Mpumalanga, with a prevalence of between 17.6 and 27.1 per cent. With the political decision to incorporate the southern Sekhukhune municipalities within Limpopo, the official HIV prevalence for Sekhukhune automatically became that of Limpopo, thus dropping significantly in terms of the colour coding.

What this revealed was that government officials not only require information but also an analysis and interpretation of the information. As Andries du Toit (2005) pointed out, it is highly instructive to listen to the kinds of things officials need to know in order to guide interventions on the ground: they need to know who the poor people in a district are; they need to know what the reasons are for their poverty; they need to know what kind of intervention will be most appropriate.

The narrative account of what was working and why in a particular municipality raised another dimension for the science/policy interface. The articulated desire for FIVIMS-ZA to provide a one-stop information system was based on the need to synthesize information from a wide variety of sources and to put it together in a way that was accessible and enabled action in a particular context. This need, however, could not be met by an information management system. It could only be met by developing the capacity for understanding, synthesizing and working with information at local, provincial and national levels.

At the close of the pilot phase the DoA undertook to inform and debate with the Social Cluster the future role of FIVIMS-ZA. Discussions took place within the Food Security Directorate of the DoA, within the technical committee of the Social Cluster, and between the DoA and the World Food Programme largely around possible institutional arrangements that would best facilitate the effective development and roll-out of the system. Some of these discussions focused on building capacity to use information – provided by a system such as FIVIMS, but also through other sources of existing information in an effective and smart manner. In order to support these discussions, the DoA visited the Mozambique Technical Secretariat for Food Security and Nutrition (SETSAN) in early 2005 to see the institutional arrangements laid out to facilitate monitoring systems in that country. These opportunities and discussions initiated a process of focused stakeholder consultation and user involvement that was necessary for FIVIMS-ZA to develop further.

As a final contribution to these deliberations, the consortium proposed options about how the FIVIMS-ZA might complement the DoA, the IFSNP and the Disaster Management Committee based in the Department of Provincial and Local Government while utilizing the analytical capability found within government, civil society, academia and the private sector. Essentially, these proposals hinged on arrangements that would provide the information and analysis required by decision-making by creating an effective interface between science and policy hinged on the concept – rather than the instrument – of a FIVIMS.

Conclusions

FIVIMS-ZA provided an opportunity to set up a science/policy interface for consultative and collaborative problem-solving. This interface can be seen as a boundary organization with the aim of integrating science and non-science to address food insecurity in South Africa. The process of piloting the FIVIMS-ZA achieved some important outcomes that lay beyond the technical products of the information baseline, the food security model, the monitoring system and numerous research outputs. Human capacity development had occurred beyond the training of officials to use the 'system'; for example, some senior officials had begun to grapple with difficult questions around vulnerability to hunger and to engage in more complex debates with the scientists. It was clear that part of the effectiveness of FIVIMS-ZA – as a boundary organization – was underpinned by the development of personal relations between representatives of science and policy. This was exemplified in the relationships established within the steering committee and particularly between the senior manager of the food security directorate and the consortium's managers. However, the momentum from these relationships grew slowly, as some institutional knowledge was lost by a high turnover of staff. The implications for future interfaces of this sort suggest that social relations among policy-makers, practitioners and academics should be explicitly prioritized and fostered.

Reflecting on the FIVIMS-ZA process highlights a number of issues about how science interfaces with policy and vice versa. For instance, the nature of the pilot meant that the boundary organization under construction could take into account a range of perspectives and grapple constructively with the multidimensional challenge of understanding and monitoring food insecurity. Thus, some scientists and government officials developed a greater understanding of the issues that they faced collectively. The project was initially conceptualized and contracted by the Department of Agriculture, but was given important flexibility to allow different approaches to be tested. This led to a move away from agriculture as a dominant issue to focus on a range of issues that underpinned vulnerability to hunger and the challenges associated with an integrated approach to food security. The boundary organization thus allowed a more comprehensive understanding of the issue to be developed, although this also revealed the inadequacies of the state in implementing such an approach.

FIVIMS-ZA was a project that built the capacity of both scientists and government officials through learning by doing. This is a critical task for boundary organizations, particularly in developing countries where stakeholders have to assume multiple roles due to limited capacity. Although capacity strengthening should have been better integrated within the process from the inception of the pilot, the involvement of government stakeholders as well as scientists enabled a degree of cross-fertilization to occur. The science/policy interface that FIVIMS-ZA facilitated provided a forum for an iterative process where there was space for academics to voice their views and for practitioners to demand clarity beyond the language of science, while strengthening capacity on both sides. Scientists were held accountable and were forced to make their science digestible. Government, in turn, was encouraged to recognize the validity of science while seeking manageable solutions to food insecurity.

FIVIMS-ZA was not driven by a 'blueprint' from elsewhere or from examples of similar processes but was 'home-grown', as intended by the DoA. Boundary organizations cannot follow a formula. The characteristics of institutions, people and resources need to be carefully evaluated before determining how the boundary organization should be structured. In this process, the regional and international perspectives were brought to bear where appropriate. For example, FIVIMS-ZA drew on examples of good practice in institutional arrangements, such as from the Mozambique Technical Secretariat for Food Security and Nutrition.

There were, however, a number of limitations in the pilot process. For example, there was an overriding expectation from government officials, particularly within the DoA, that FIVIMS-ZA would solve a range of problems confronting them in executing their core mandate. In reflecting on the process, one of the key scientists argued that:

> ... in spite of the recognition of the role of local history, power relations and in spite of the acknowledgement of the importance of practical local knowledge embedded in institutions on the ground, very little could be done to shift the perception on the part of the officials involved that, ultimately, what was practical was a GIS-based system that would provide information about 'indicators' of 'structural vulnerability' in unambiguous, mapable, quantifiable terms. (du Toit, 2005, p13)

Ultimately, these unrealistic expectations created fissures in the science/policy interface. Food security provides challenges for scientists, both conceptually and methodologically in terms of measuring and monitoring. Translating this complexity into a simplified, accessible 'system' was therefore bound to have limitations. Scientists were hesitant to promote the final product of the pilot and the extent to which it would be able to monitor and manage the multidimensional nature of food insecurity and vulnerability. Defining the aim and goals of FIVIMS-ZA as a boundary organization at the inception may have helped to anticipate the final limitations.

As mentioned above, an important lesson was that the nature of the boundary organization ensured that scientists were held accountable for ensuring that their practice was grounded in an accessible form relevant to a wide range of primary and secondary stakeholders. However, keeping the science accessible was a major challenge, despite the recognition that real participation 'oils the mechanisms' of the interface. In practice, scientists and practitioners are often impatient, driven by budget lines and the need to meet deadlines for deliverables. An effective interface is dependent upon adequate 'time to engage' and to co-learn, raising the need for success to be partly judged by process rather than output. It is likely that an ongoing challenge of integrating science and policy will be the different timelines on which different groups operate. Again, if this is factored into the design of a boundary organization, there may be compromises that can be met earlier on in the process.

Another key lesson was that although FIVIMS-ZA originated out of the need to monitor local food insecurity and to help find solutions by using existing frameworks such as the IFSNP, the lack of legislation behind this programme meant that there was very little interdepartmental coordination and commitment. The institutional realities of how line departments function undermined what the science was calling for in terms of addressing food insecurity. This was a major stumbling block in the ongoing development of FIVIMS-ZA beyond the pilot phase.

The challenges of both process and content diluted the impact of FIVIMS-ZA. The system, regardless of challenges around integrating data, required personnel from a range of government departments and other organizations to work collectively. The real institutional challenges of integrated development raised by the scientists led many in government to regard the FIVIMS-ZA pilot phase as an expensive academic exercise. This was largely because there was no delivery of a blueprint that could be systematically rolled out across the country. This was a primary reason why a 'second phase' of the exercise was commissioned, which was predicated upon the delivery of a 'hard system with lots of outputs ready to use'. Thus, the need to adapt the goals and activities of boundary organizations to the most important challenges, as agreed upon by a range of stakeholders, becomes evident. It is then also critical to be explicit about which goals and activities are not addressed and the implications of this.

Notes

1   The consortium was comprised of the Human Sciences Research Council (HSRC); Agricultural Research Council (ARC); Agrista (Pty) Ltd; Council for Scientific and Industrial Research (CSIR); GeoTerraImage (Pty) Ltd; Kayamandi Development Services (Pty) Ltd; Melinda Potgieter (Private Consultant); University of Cape Town, Department of Environmental and Geographical Sciences; University of the Western Cape, Programme for Land and Agrarian Studies, School of Government; and University of the Witwatersrand, School of Geography, Archaeology and Environmental Studies.
2   The views of Dr Andries du Toit of the University of the Western Cape and Professor Coleen Vogel of the University of the Witwatersrand are acknowledged as being key to this section.

References

DoA (Department of Agriculture) (2002) *The Integrated Food Security Strategy (IFSS) for South Africa*, Department of Agriculture, Pretoria, South Africa
du Toit, A. (2005) 'Poverty measurement blues: Some reflections on the space for understanding "chronic" and "structural" poverty in South Africa', www.plaas.org.za/publications, accessed 15 March 2007
Guston, D. H. (2001) 'Boundary organizations in environmental policy and science: An introduction', *Science, Technology, & Human Values,* vol 26, no 4 (Special Issue), pp399–408
Hamid, G. M. (2005) *The Effective Coordination and Implementation of the IFSNP*, Draft issue paper, Food and Agriculture Organization of the United Nations, Rome
HSRC (Human Sciences Research Council) (2004) *Food Security in South Africa: Key Policy Issues for the Medium Term*, Position paper, Integrated Rural and Regional Development, HSRC, Pretoria
HSRC–FIVIMS (Human Sciences Research Council–Food Insecurity and Vulnerability Mapping Systems) (2003) Tender document, November, Tender no 5/3/1-29/03, Appointment of a contractor to assist the Department of Agriculture with the analysis, design and development of a model for a food security and vulnerability information and mapping system (FIVIMS) for South Africa

Jasanoff, S (1996) 'Beyond epistemology: Relativism and engagement in the politics of science', *Social Studies of Science*, vol 26, pp393–418

Polzer, T. and Schuring, E (2003) 'To eat is an everlasting thing: Evaluation of the food emergency scheme in Bohlabela District, Limpopo Province, South Africa', Acornhoek Advice Centre, Forced Migration Studies Programme, University of the Witwatersrand, Johannesburg

Shabalala, N. and Mosima, B. (2002) *Report on the Survey of Large and Small Scale Agriculture*, Statistics South Africa, Pretoria, South Africa

Vogel, C. and Smith, J. (2002) 'The politics of scarcity: Conceptualising the current food security crisis in southern Africa', *South African Journal of Science*, vol 98, pp315–317

10

Science and Vulnerability Reduction in Taiwan after the 1999 Chi-Chi Earthquake

Kuan-Hui Elaine Lin, Huei-Min Tsai and Chang-Yi David Chang

Introduction

Sustainability science and research on global environmental change increasingly recognize the importance of addressing vulnerability. The issue has been addressed and researched by scientists and applied by policy-makers and practitioners of diverse interests. Strategies for both short-term and long-term planning to reduce human vulnerability to environmental disturbances have become a focus of contemporary scientific and political debate.

The emerging concept of vulnerability has evolved from early work by Timmerman in 1981 and Chambers in 1989 (for an explicit review, see Cutter, 1996), subsequently developing a political ecology approach to analysing famine (Downing, 1991; Watts and Bohle, 1993), studies of land degradation (Blaikie, 1985; Blaikie and Brookfield, 1987) and disaster risk reduction analysis (Blaikie et al, 1994; Wisner et al, 2004). As attention to the negative effects of climatic and global environmental change has grown since the late 1990s, vulnerability has rapidly become a core concept in the intellectual and academic fields, such as the Intergovernmental Panel on Climate Change (IPCC), the United Nations Framework Convention on Climate Change (UNFCCC) and the International Human Dimensions Programme (IHDP) on Global Environmental Change. Knowledge about vulnerability has also advanced through contributions from social, physical and ecological sciences. This multidisciplinary approach addresses the interaction of human and environmental systems (Turner et al, 2003). Some scholars use the term 'social–ecological system' (see especially Berkes and Folke, 1998) and have formed new discourses on the relationships among vulnerability, adaptation and resilience (Berkes and Folke, 1998; Adger et al, 2001; Clark et al, 2001; Berkes et al, 2003; Adger et al, 2005a; Kasperson and Kasperson, 2005).

The coupled human/environment systems framework emphasizes multiple interacting features, such as multiple stressors, social actors and place-based analysis viewed in the context of large-scale change. Vulnerability is conceptualized as exposure and sensitivity to perturbations or external stressors, as well as the capacity to adapt. It is generally portrayed, in negative terms, as susceptibility to harm and viewed as a function of exposure, sensitivity and adaptive capacity (McCarthy et al, 2001; Fraser et al, 2003; Turner et al, 2003). Conversely, the concept of social–ecological resilience, with its roots in ecology, refers to the magnitude of disturbance a system can absorb and still maintain its general state and function, as well as the capacity to self-organize and the capacity for learning and adaptation (Carpenter et al, 2001; Folke, 2006). Vulnerability is thus influenced by the build-up or erosion of the elements of social–ecological resilience (Adger, 2006).

The building of resilience and adaptive capacity can be broken down into different levels, from individual, household, community and sub-social units, to national- or international-level strategies. In this study, we focus on the institutional dimensions of vulnerability mitigation, emphasizing the role that science played in transforming adaptive governance as a response to environmental changes and disasters. Policy interventions and governance solutions for reducing vulnerability to environmental change have recently become key issues. However, the inclusion of vulnerable social groups and scientific intelligence within governance structures remains an under-researched, yet important, area of enquiry (Adger et al, 2005b; Adger, 2006).

In this discussion we examine the Taiwan experience as our case study for several reasons. First, Taiwan experienced a devastating earthquake on 21 September 1999. The earthquake scored 7.3 on the Richter scale, hit Chi-Chi township region in Central Taiwan, caused more than 2400 deaths and injured 11,000 people. After the earthquake, a large number of landslides occurred.

In combination with abundant precipitation in the typhoon season, floods and debris flows in Taiwan have become severe, constituting new threats to local communities and exposing them to greater risk. Second, in response to these multiple natural stressors, Taiwan has gone through a series of institutional transitions, establishing new laws for disaster mitigation that have resulted in polycentric disaster mitigation governance (see below). Finally, the scientific community has played a significant role in illuminating directions for institutional transition in Taiwan. Scholars have the opportunity to become involved in policy-making and to create a boundary organization as an intermediary to promote cross-scale interplay and dialogue across boundaries of science, policy-making and the public (for a fuller discussion see Chapters 2 and 9). Embedded within this socio-political context, scientists are an important social source for the construction of collaborative/adaptive disaster mitigation governance.

The purposes of this study are to analyse the potential effectiveness of Taiwan's emerging adaptive governance system, and to understand the social processes and mechanisms that make it possible for scientists and scientific knowledge to be effectively integrated within it. It will also examine the emergence of cross-boundary communication and knowledge co-production among multiple actors in an attempt to build a kind of governance which comprises scientific knowledge, governmental resources and local participation.

This chapter first reviews the literature on the implications of adaptive governance, science and boundary organizations for vulnerability. Second, we examine the background of the Taiwan case. Third, we delineate the institutional transition process and the roles of science, scientists and developing boundary organizations. We then study a village as an example of how adaptive governance is practised at a community level. Finally, we analyse the Taiwan experience of bringing science into vulnerability reduction practices in adaptive governance, as well as any obstacles or limits.

Theoretical review: Adaptive governance, science and boundary organizations

With the deepening phenomenon of global environmental change, the environmental crises and problems we confront now are much more complicated and unpredictable than before. In addition, traditional environmental management and solutions are ineffective in resolving today's dilemmas. Over the last two decades, there has been much literature analysing the factors that make conventional scientific knowledge and environmental management approaches fail. Different perspectives have been presented, including scientific theory that sees the ecosystem as stable and in equilibrium (Holling, 1986; Gunderson et al, 1995). Ignorance of scale dynamics (Cash et al, 2006), blocked communication among disciplines, which hinders knowledge integration, and the gap linking scientific knowledge with policy-making and practices are all involved.

We are now facing a dynamic, disequilibrium, cross-scale and coupled human/environment system. Many environmental problems are themselves inherently embedded in the system's complexity and uncertainty. Management practices, therefore, should more cautiously address the multi-scale environmental problems inherent in the biogeophysical and human systems and the interactions between them across scales (Berkes and Folke, 1998; Folke et al, 2005). Scholars suggest that by gaining a more synoptic understanding and integrating the different knowledge systems at all scales, problems that occur in matching natural systems to management systems through scales can be reduced (Cash and Moser, 2000). This theoretical reflection reveals a need for a paradigm shift that can support a higher degree of knowledge integration into more resilient management approaches that occur in top-down authoritative management. Bottom-up, localized, community-based experience, and a knowledge system that integrates scientific knowledge with practice-based knowledge are all needed (see Chapter 4).

Theories of adaptive management that have developed from natural resource management over the last decades have evolved as a potentially powerful framework for the dynamic linkages among knowledge, institutions and management practices across scales to address present-day environmental crises (Cash and Moser, 2000; Cash et al, 2006). Essentially, adaptive management involves perspective on a system's ability for learning, self-organizing and the transformability of a social–ecological system in responding to environmental changes (Walker et al, 2004; Berkes and Turner, 2006). C. S. Holling (1986) has proposed that hazards be viewed as ecological surprises that bring opportunities for society to reorganize. Gunderson and Holling (2001), in their classic book *Panarchy*, develop an adaptive cycle theory that emphasizes the possibility for system reorganization in times of abrupt change. Their concepts shed light on the positive meaning of hazards, pointing out the possibility for society's reorganization through adaptive adjustment. This work has attracted many scholars to study social processes that can help to increase a society's resilience and adaptive capacity in the face of change (Berkes and Folke, 1998; Berkes et al, 2003; Olsson et al, 2004; Tompkins and Adger, 2004; Adger et al, 2005b; Folke et al, 2005; Smit and Wandel, 2006). The system's adaptive management and its capacity to reorganize, learn and adapt to environmental threats or changes have now been taken as important institutional factors for society to reduce its vulnerability to external human/environment problems.

The central notion of this perspective is that crisis may trigger learning and knowledge generation and open up space for new management trajectories of resources and ecosystems (Folke et al, 2005). For the policy and management domain, this perspective indicates that adaptive policies involve an explicit learning-oriented policy experimentation that aims to understand human/environment system dynamics and infuse diversified knowledge bases from different scales and different types into management practices and to test the outcomes (Cash and Moser, 2000). It is a process by which institutional arrangement and knowledge are tested and revised in a dynamic, ongoing process of experimentation (Berkes and Turner, 2006). This characteristic makes the approach exceptional in linking knowledge, policy and practices. It offers social space for multi-stakeholder involvement, such as local people, scientists, governmental agencies and non-governmental organizations (NGOs), and allows feedback from scientists and decision-makers to respond to the outcomes of management experimentation and thus increase the effectiveness of policies and actions.

An adaptive management approach also addresses the scale issue. Many empirical studies have pointed out that, in responding to complex and multi-scale environmental problems, cross-scale cooperation or multi-scale institutional design is usually required. Take ecosystem management, for example (Hahn et al, 2006). Good management practice requires broad support and legitimacy among local residents, experts, local governments and state or national institutions. Vertical or horizontal institutional interplay (Young, 2002; Cash et al, 2006; Young, 2006) and polycentric institutional arrangements (Folke et al, 2005) that mediate between centralized and decentralized modes of management are emerging as important institutional designs to link multi-scale and cross-scale interactions. Co-management – which means the sharing of management power and responsibility among different levels and scales of governments and local people – is an important alternative to create cross-scale interactions and produce flexible and multilevel governance systems. By adaptive co-management, Cash et al (2006) stress that successful co-management often arises from the adaptive, self-organizing processes of learning by doing, rather than from optimal power-sharing across levels.

Thus, the meaning of adaptive co-management is redefined in the values of cross-scale interaction, learning and knowledge generation, and combines the dynamic learning characteristic of adaptive management with the linkage characteristic of cooperative management (Folke et al, 2005). Dietz et al (2003) further use the concept of adaptive governance to expand the category of adaptive management to embrace the broader social contexts that enable ecosystem-based management.

Governance is the structures and processes by which people in societies make decisions and share power (Lebel et al, 2005). Folke et al (2005) argue that the term governance has recently become a catchword for various alternatives to conventional top-down government control, including collaboration, partnerships and networks. It relies on networks that connect individuals, organizations, agencies and institutions at multiple organizational levels and provides for collaborative, flexible, learning-based approaches to managing social–ecological systems (Olsson et al, 2006).

The concepts of adaptive co-management and governance offer a theoretically robust basis to construct knowledge-based and multi-scale environmental management approaches as alternatives to conventional management practices. However, we should still ask by what kind of institutional designs or mechanisms, precisely, can we reach real cross-scale interactions and integrated knowledge. Occasionally, gathering people together in seminars or meetings to discuss the environmental problems that they confront without institutional support is apt to fall into inefficient communication and come out with vague conclusions lacking common views or any regulative enforcement. Some scholars have suggested that the setting of a boundary (or bridging) organization can facilitate an adaptive approach by accessing information and linking agents through its use (Cash and Moser, 2000). Empirical studies also provide some good illustrations. Hahn and colleagues, in their research on adaptive co-management of a wetland landscape in Kristianstad, Sweden, suggest that, as an integral part of adaptive governance of social–ecological systems, a bridging organization provides an arena for trust building, vertical and horizontal collaboration, learning, sense making, identification of common interests and conflict resolution (Hahn et al, 2006).

We argue that a boundary organization can provide a valuable institutional function in maintaining knowledge production links through multi-stakeholders' communication and interactions, and creating more effectively integrated information and decision systems that are able to retain scientific credibility while ensuring political saliency in adaptive governance (Jasanoff, 1990). When the term boundary organization was initially created during the 1990s, it meant institutions that mediated the shifting divides and tensions between science and policy (Guston, 2001), aiming to replace traditional approaches by incorporating scientific information within a policy process that Cash and Moser (2000) term a 'pipeline model'. This traditional model, which operates through producing scientific assessment reports and funnelling them to policy-makers, forms a unidirectional flow of information and drifts into the defects of ignoring the differential needs for knowledge at different scales (e.g. large-scale scientific knowledge that has little relevance to local decision-makers). It also overlooks the discrepancy between knowledge systems among disciplines and the usefulness and relevance of the knowledge to policy-making. As opposed to the pipeline model, the function of a boundary organization is to build a platform that helps to transform scientists from being the sole knowledge producers to being information co-producers (along with policy-makers) to create scientific information that is salient to specific needs, is scientifically credible and is legitimate to all stakeholders (Cash et al, 2006). Boundary organizations are thus able to facilitate the multidirectional flow of information by iterative communication and negotiation between scientists and decision-makers.

The category of boundary organization has since expanded to describe the boundaries between different scales or functional levels (Cash and Moser, 2000). When Cash studied the role of boundary organizations in American agricultural extension in the High Plains, the Cooperative State Research, Education and Extension Service of the US Department of Agriculture (CSREES) was treated as a boundary organization that moderated among county extension offices, state land-grant colleges, the US Department of Agriculture (USDA), the US Geological Survey (USGS), state regulatory agencies, local farmers and other local offices to manage aquifer depletion problems and their relationship to agriculture production (Cash, 2001). In Cash's study, the concept of boundary organization extended beyond the science/policy dimension to incorporate other levels of organization (such as local to state and national levels and local individual actors). Similarly, Westley (1995) uses the term 'bridging organization' to refer to inter-organizational collaboration (Westley, 1995). Folke et al (2005) propose that a bridging organization can provide opportunities by bringing in resources, knowledge and other incentives for ecosystem management. It encompasses the function of a boundary organization by communicating, translating and mediating scientific knowledge to make it relevant to policy and action. Thus, in this chapter we refer to an organization that plays an intermediary role between different arenas, levels or scales, and that is focused on this intermediary function as a boundary or bridging organization.

Cash et al (2003) propose that boundary organizations are more effectively organized when they contain the following three important institutional features:

1   They involve specialized roles within the organization for managing the boundary.
2   They have clear lines of responsibility and accountability to distinct social arenas on opposite sides of the boundary.
3   They provide a forum in which information can be co-produced by actors from different sides of the boundary through the use of boundary objects.

Boundary objects, as noted in earlier chapters, are important products of the collaborative efforts. The objects could be models, maps, reports or forecasts that are co-produced by actors on different sides of the boundary and carry information and knowledge that are salient, credible and legitimate to all actors. Such objects provide a medium for further discussion, revision and application with multiple parties regarding differences in perspectives, values and desired outcomes. Conclusively, a boundary organization moderates knowledge co-production and communication, mediating, translating and negotiating across scale-related boundaries to facilitate solutions to complex problems (Cash et al, 2006). The collaboration process is actually a broader process of social learning that not only empowers policy-makers and the public to develop adaptive expertise for managing uncertainty or risks, but enables scientists to learn local knowledge and to develop more policy-relevant information or technology.

Environmental surprise and crisis may create space for social reorganization and stimulate new forms of governance systems; however, the transformation of institutions and the emergence of adaptive governance also rely on the social capacity for resilience-building (Berkes et al, 2003). There are some important social factors underlying the operations of boundary organizations that influence the performance of adaptive governance. Scholars have pointed out that the collaborative process needs a good trust relationship for communication and negotiation. Trust grows over time in a jargon-free, non-condescending communication environment, as each side increases its mutual understanding of each other's language, institutional culture, standards and expectations of professional conduct and the larger decision-making context (see Chapter 2). Leadership is also a key function in the building of trust and in the development of a boundary organization (Olsson et al, 2006). Leaders can create a vision that frames and gives direction to the cross-scale interaction process (Cash et al, 2006), and can fabricate new and vital meanings, overcome conflicts or contradictions in creating a new synthesis, and forge alliances or partnerships between actors or organizations (Westley, 1995). Some key individuals in networks that develop through collaboration with other organizations and actors also play vital roles in managing boundaries in the context of learning, knowledge generation and negotiation (Folke et al, 2005). In addition, there are some institutionalized or contracted relationships among actors, governmental agencies, NGOs and boundary organizations. These social networks, whether formal or informal, form important partnerships that provide solid support for broader social learning processes. Trust, leadership, social networks and partnerships construct the most essential social sources or mechanisms that maintain the collaborative process. Nevertheless, some scholars have proposed the concept of 'social memory' to emphasize the importance of the accumulation of a diversity of experiences which concern management practices and rules. These are used at the collective level to influence future management policies (Berkes et al, 2003; Folke et al, 2005). The term 'social capital' is now widely used to explain and encompass all of these social mechanisms, demonstrating how they generate collaborative governance and strengthen people's resilience to deal with and adapt to environmental change and increasing crises (Adger, 2005b; Folke, 2006).

Background: The event and aftermath of the 1999 Chi-Chi Earthquake

An account of the 1999 Chi-Chi Earthquake

Situated on the Tropic of Cancer, Taiwan is an island located 150km off the southeast coast of mainland China, with cooler temperate Japan to the north, subtropical south China to its west, and the tropical Philippines and Indo-Malayan islands to the south. The island lies at the juncture of two plate boundaries, on the edge of a continental shelf. Given this location and its high mountain ranges – more than 200 peaks over 3000m in height and hilly topography – Taiwan is confronted with frequent earthquakes and typhoons, as well as the floods, landslides and debris flows that follow. During the last few decades, rapid economic and technological progress, population growth and urban sprawl have been associated with increased economic growth, on the one hand; on the other, they have led people to invest in property in potentially disaster-prone areas in ignorance of the power of nature.

Owing to geological conditions, between 1900 and 2005 there were 93 hazardous earthquakes with a magnitude of 5 or more on the Richter scale. The 1999 Chi-Chi Earthquake was a result of a major reverse fault, the Che-lung-pu Fault. The upward fault movement hit a relatively shallow point, just 7km to 10km below land surface in Central Taiwan. The movement produced a 100km long uplifted fault with a vertical height up to 2m high. Along the fault, roads, railroads and dams were damaged; thousands of structures – schools, community agencies, business offices, hospitals and residential buildings – collapsed.

The earthquake not only resulted in loss of life and property, but also shifted the mountains' structures, causing 25,845 landslides over a total area of 15,977ha, creating a higher risk of further disaster. It was one of the most severe earthquakes in Taiwan in the past 100 years.

Multiple hazards: Combined effects of nature and human systems

As a consequence of global climate change, the severity and frequency of typhoons have been increasing in the years since the Chi-Chi Earthquake. Figure 10.1 shows the number of typhoons hitting Taiwan from 1980 to 2005. The number has increased since 1999, and the intensity of storms and rainfall has increased as well. The combined effects of geological factors and changing meteorological conditions have reshaped the landscape of mountain areas, not only by landslides occurring after the earthquake, but also through severe flooding and mud and debris flows along rivers and mountainsides in the years that followed. Table 10.1 shows the casualties resulting from these typhoons and their related effects.

According to a scientific survey sponsored by the Taiwan Soil and Water Conservation Bureau (SWCB), rivers and streams with a high potential for debris flows increased threefold, up from 485 in 1996 to 721 after the 1999 earthquake and increasing to 1420 in 2003 (Lin et al, 2005; Soil and Water Conservation Bureau, 2005a, 2005b, 2005c). Landslide areas also expanded dramatically year after year, as is illustrated in Figure 10.2.

The unpredictability of nature is a great threat to people living in affected areas. We should note that during the Chi-Chi Earthquake, the most devas-

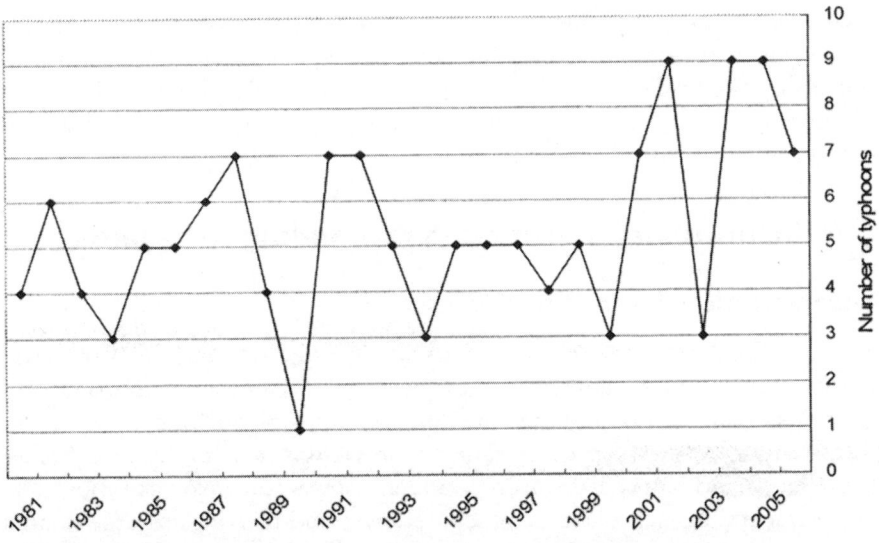

Figure 10.1 *Number of typhoons in Taiwan between 1980 and 2005*

Source: adapted from Taiwan Central Weather Bureau database, www.cwb.gov.tw

Table 10.1 *Casualties of the most serious typhoons in Taiwan, 1995–2005*

Date	Typhoon name	Maximum hourly rainfall (mm/hour)	Total accumulated rainfall (mm)	Casualties: deaths/missing	Houses destroyed/damaged
29 July 1996	Herb	114.5	1986	51 22 / 73	503 880 / 1383
21 August 2000	Bilis	51	937	14 7 / 21	434 1725 / 2159
29 October 2000	Xianshen	70.5	1054	64 25 / 89	*
28 July 2001	Toraji	146.5	757	111 103 / 214	645 1972 / 2617
15 September 2001	Nari	142	1462	94 10 / 104	*
28 June 2004/ 2 July 2004	Mindulle	166.5	2005	31 13 / 44	304 107 / 411
23 August 2004	Aere	61.5	2279	14 15 / 29	72 44 / 116
16 July 2005	Haitang	176	1670	13 2 / 15	*

Notes: From 1980 to 2005, the most serious casualties caused by typhoons happened in 1996 (Herb), 2000 (Bilis and Xianshen), 2001 (Toraji and Nari), 2004 (Mindulle and Aere) and 2005 (Haitang). Most victims were lost or injured in flooding, mud and debris flows.
* = no data.
Source: Taiwan Central Weather Bureau (2006, www.cwb.gov.tw/); National Disaster Prevention and Protection Commission (www.ndppc.nat.gov.tw/)

tated areas were in Central Taiwan, especially Nantou and Taichung counties. Villages along the fault and near the epicentre were reduced to ashes. Unfortunately, because of the mountainous topography of Central Taiwan, these counties also suffer most from the combined effects of multiple hazards. Furthermore, the diverse terrain and complicated micro-topographical variation also make predicting the precise location of new debris flows or the extent of flooding from typhoons challenging. It is hard to predict in advance which village will be in critical danger.

Institutional transition in responding to hazards

The institutional transition process

The Chi-Chi Earthquake was a major catastrophe. During the aftermath, rescue teams sent by governmental agencies, NGOs, humanitarian and religious organizations and the Taiwan public, in general, all contributed to disaster-relief efforts. For quick response to victims' needs to rebuild homes and villages and reduce harm from future disasters, academic institutions and governmental agencies held many conferences and meetings to urge related scientific research and motivate further actions by integrated task forces. The 921 Post-Earthquake Reconstruction Council was set up in 2000 with the

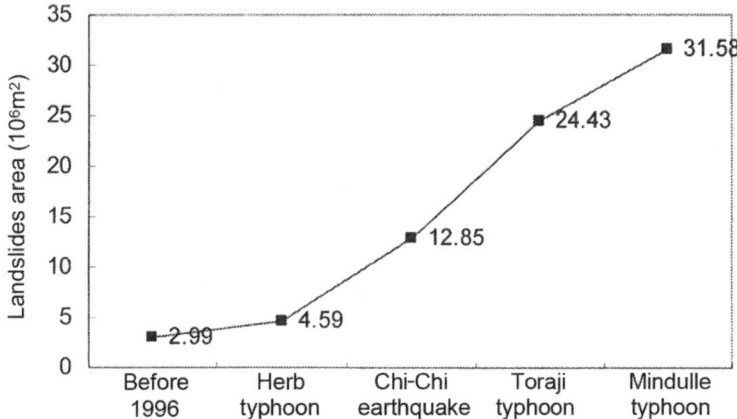

Figure 10.2 *Expanding area vulnerable to landslides on the Chen-Yo-Lan River watershed: Typhoons Herb (1996) to Mindulle (2004)*

Source: adapted from Lin (2005)

purpose of managing cross-level and cross-scale cooperation and coordination among several governmental agencies (such as the Soil and Water Conservation Bureau, the Council for Cultural Affairs, the Forestry Agency and Construction and Planning Agency) and to develop an integrative community reconstruction plan. In addition to governmental efforts, many social workers, local university scholars and students, and community associations all voluntarily contributed to reconstruction works. These voluntary teams worked closely with governmental agencies and local people during the following years to formulate an important social milieu that could be labelled as the original model of wide participation in community and public affairs.

This collaborative experience also accumulated a kind of social capital and social memory that governed the value change in influencing the institutional transition of Taiwan's disaster mitigation governance and regulation. Towards the end of 1999, the vice-premier of the Executive Yuan, who had been a professor of science before serving in government, authorized the vice minister of the National Science Council, who is also a professor of meteorology, to organize a committee for preparing new legislation for disaster mitigation. This committee comprised scholars from several different institutions and disciplines, such as the National Science and Technology Programme for Hazard Mitigation and the Central Police University, as well as senior officers of the Ministry of the Interior. Some of these individuals had been involved in community relief and reconstruction work, going into villages and collaborating with supporters. They applied the experiences of the Chi-Chi Earthquake, on the one hand, and referred also to other countries' disaster-mitigation experiences and institutional arrangements, such as those of the US Federal Emergency Management Agency and Japanese hazard reduction programmes. On the other hand, they developed a tentative framework for new legislation.

The new legislation, entitled the Disaster Prevention and Response Act (DPR Act), was issued in July 2000. It has three distinct characteristics. First, it treats disaster management as horizontal and vertical collaborative work, replacing the former uni-operational system, which only authorized the fire department as the main authority for disaster-related tasks. It confirms a three-tiered polycentric disaster management system from national to county (city) to township levels. Each level has quasi-autonomous authority and the responsibility to develop a Disaster Prevention and Response Basic Plan (DPR Plan) with regional specificity. Mayors have to convene the Disaster Prevention and Response Council (DPR Council) regularly with officers of local governmental agencies, as well as operate an Emergency Operation Centre (EOC) during emergency periods and organize a Reconstruction Implementation Council after disasters occur. Second, it defines the type of disaster that each authority must focus on. For example, the Soil and Water Conservation Bureau is the main agency in charge of typhoon-related hazards (such as debris flows, landslides and floods). Third, it emphasizes the importance of scientific assessment and public participation in disaster-management regimes.

This law is an important innovation in Taiwan's disaster regulation history and creates a new age for future disaster management. Before 1965, there were no official disaster-related laws. In 1965 a directive, Regulation for Natural Hazards and Response, was set up as the first regulation for responses to disaster (see Figure 10.3). The regulation, however, was confined only to natural hazards and was limited to response and recovery aspects. In 1994, the Disaster Prevention and Response Action Plan (DPRA Plan) was issued, replacing the former regulation. The plan expanded the scope of hazards to include human dimensions and began to examine aspects of disaster mitigation. The DPR Plan was succeeded by the Disaster Prevention and Response Act (DPR

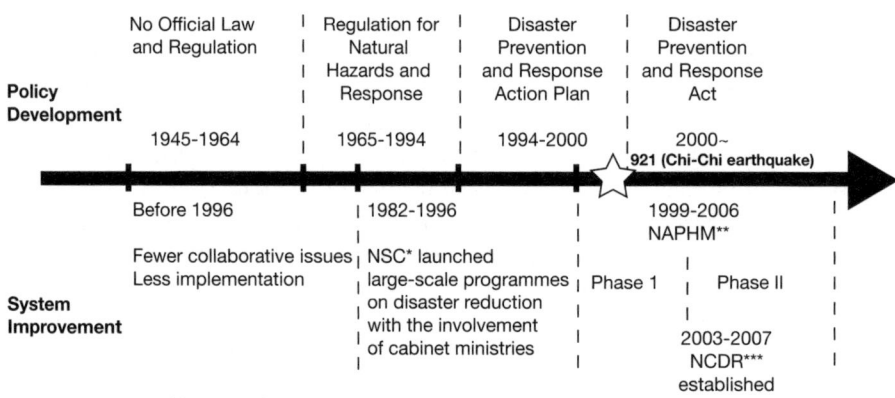

Figure 10.3 *History of system improvement and policy development in disaster reduction in Taiwan*

Source: adapted from Chen (2006a)

Act of 2000) and polycentric disaster management governance has been confirmed. The National Disaster Prevention and Protection Commission (NDPPC) has been established as the highest level of national administrative agency. It coordinates nationwide disaster management affairs across scales. By setting the vice-premier of Executive Yuan as the chair and the vice-minister of each ministry as committee men, the NDPPC has attained a solid position to deploy and moderate among governmental agencies.

The role of science and scientists, and the development of a boundary organization

Science and the role of scientists have occupied an influential place in disaster management affairs, although they have gone through different stages. In the political domain, inherent in the traditional socio-political context, Taiwan scientists used to have a higher accessibility to government officials. Eminent scientists tended to be recruited by government and to occupy important governmental positions, such as deputy chief executive, senior advisers to the president, ministers, etc. As mentioned earlier, they even played key roles in pushing the institutional transition of disaster management governance. In scientific research, in terms of the dramatically changing landscape of Central Taiwan after the Chi-Chi Earthquake, the importance of environmental research has been increasingly emphasized. The National Science Council funded and synergized 21 domestic universities to conduct environmental and hazard research (Lin et al, 2005), and many governmental agencies provided substantial financial support for science as well. Take typhoon-related hazards, such as debris flows, landslides, floods and mud, for example. The Soil and Water Conservation Bureau commends experts to conduct a great deal of scientific assessment involving the identification of locations and scales of the hazards, building hazard models to identify critical hazard-prone areas, setting risk levels and thresholds, and continuously monitoring environmental changes and updating the data. A nationwide geographic information system (GIS)-based typhoon-related hazards early warning system has been established in this way.

These scientific assessments for environmental monitoring are implemented by scientists alone, conventionally after producing the reports to the decision-makers. Scientists are then insulated from policy-making. This science/policy relationship condition, however, has changed since the DPR Act. As the DPR Act recognizes science as significant in the decision-making process, the National Science and Technology Programme for Hazard Mitigation, an integrated interdisciplinary science programme under the aegis of the National Science Council since 1973 (see Figure 10.3), was asked to expand its scope and was re-established as the National Science and Technology Centre for Disaster Reduction (NCDR) in 2003. The NCDR takes on the responsibility of the science and technology consulting unit to the NDPPC, and changes the role of scientists in responding to the need for new forms of collaboration among scientists and policy-makers in policy-making.

The NCDR is a transdisciplinary organization consisting of geologists, geographers, psychologists, sociologists and urban planners. Its three distinct functions are to:

1   promote disaster-related research and technology and enhance the integration of scientific knowledge by developing an information database;
2   use the information database for disaster analysis and provide the information to policy-makers at various levels;
3   improve communication among governmental agencies, academic institutions and the public on local and regional disaster issues and to promote a process of information sharing across levels.

The bureau collects environmental information from governmental agencies and promotes technology innovation for knowledge integration. When a hazard event occurs, it takes a place in the Central Emergency Operation Centre (CEOC), which is convened by the NDPPC and the cabinet ministers for risk assessment and disaster emergency management. Take typhoon-related hazards, for example (see Figure 10.4). The NCDR and several governmental agencies organize an assessment group, gather real-time data, develop scenario analysis and risk maps for floods and debris flows, forecast potential areas of these floods and debris flows according to the risk map, and provide the

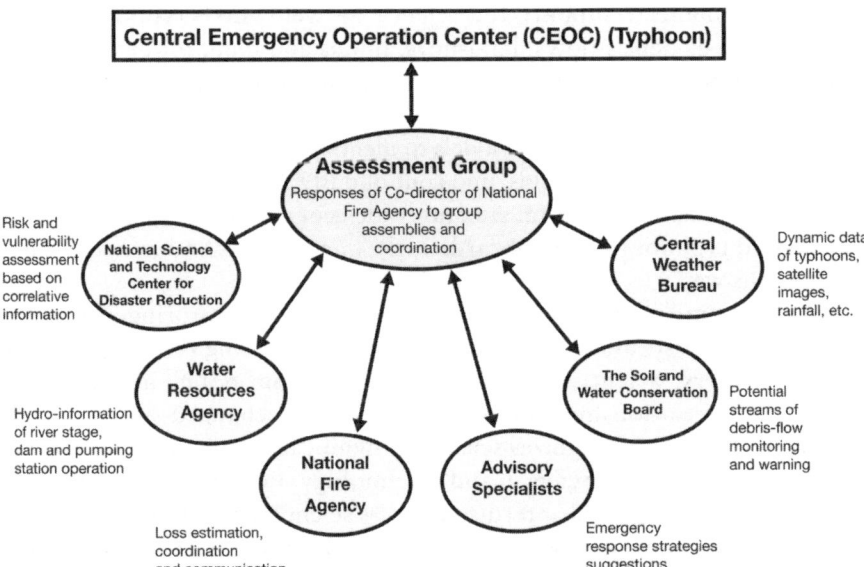

Figure 10.4 *Schematic diagram illustrating the functional role of the Central Emergency Operation Centre (CEOC) and its interactions with various typhoon disaster-related government agencies through the Assessment Group*

Source: adapted from Chen (2006a)

forecast results to the CEOC for decision-making. The whole process is governed by cross-level collaboration, transdisciplinary integration and technology application. The output of this risk assessment and the activation of a warning system is a cooperative product. Scientists are not the sole information producers, but work closely with governmental agencies to co-produce the information that is science-based and has policy relevance at the same time.

The Disaster Management Decision Support Project (DMDS)

As decreed by the DPR Act, Taiwan has formed a three-tiered polycentric management system in its disaster mitigation from national to county (city) to township scales. At the national scale, the NDPPC integrates the central governmental agencies that the NCDR has prepared in constructing a horizontal network for the knowledge-based decision-making system. However, at other scales of county (city) and township there is a considerable space to cover. In order to improve local governments' capacity for disaster management, the NDPPC mandated the NCDR to develop a multi-year Disaster Management Decision Support Project (DMDS) in 2003.

The NCDR constructs partnership networks by incorporating local university scientists in different fields. It acts as an intermediary in the networks by sharing its information database and technology, and by providing a forum, in the form of training seminars and workshops, for all stakeholders, including local and central government officials and local university scientists. The most important link in this project is the relationship between local university scientists and officials of local governmental agencies. They hold regular working meetings every two weeks, as well as some training workshops. These meetings and workshops not only educate officials about scientific knowledge on the hazards in their areas, but also allow university scientists to learn what information these local managers need. Through intensive discussions, critical environmental issues and their priorities regarding local specificity in disaster mitigation, prevention and response for their DPR plan are verified one by one. As the NCDR shares the entire information database and technology, local university scientists thus gain access to information to build local disaster models by modifying national models with more precise local data, such as population distribution, land-use types and large-scale topology and terrain maps.

The information incorporated within local disaster models is quite scale dependent and is in accordance with local characteristics. The cooperative approach and transparent process make the models salient to the information users, credible to the scientists and local managers who use the data for scientific analysis and modelling, and also maintain legitimacy in the process. These models become important boundary objects that facilitate discussion about application and revision of the county (city) DPR Plan and are also applied to the county (city) Emergency Operation Plan (EOP). During typhoon warnings, the county Emergency Operation Centre (EOC) is established simultaneously with the central EOC. The disaster models at that time provide a substantial boundary function in communicating data to different parties engaged in revising models that illustrate the potential threat and risk for certain disasters. Local university scientists are involved in the EOC, along with mayors and engineers and officials from local agencies. Models that are iteratively revised according to timely information form the fundamental basis for local decision-making to evacuate and rescue areas or villages that may be affected in order to prevent further losses and harm.

This new form of disaster management governance allows not only the integration of science and politics to co-produce useful information as a basis for decision-making, but also creates a collaborative interaction among agencies across different levels and scales previously not possible. Today disaster management is treated as team work, coordinated by the mayor with the participation of various governmental agencies, such as water resources, agriculture and the information office. The institutional change reveals a transformation from individualism to synergism in the ideology by which people view disaster management.

The Community-Based Disaster Management Project (CBDM)

Public participation and community perception were largely neglected in past disaster management practices. Prior to the 1999 Chi-Chi Earthquake, there were only a few community-based projects, such as the Phoenix Team, the Female Fire Safety Publicity Team and the Neighbourhood Rescue Team. These teams were generally established with the help of the fire department and emphasized emergency response. This approach often resulted in mechanistic instrumentalism as disaster management at the local level. The emphasis was on emergency response, and there was insufficient hazard-related knowledge to build coherent locally based management strategies that would be widely practised. With the higher unpredictability of multiple environmental hazards repeatedly ruining reconstructed areas in Central Taiwan and undermining reconstruction efforts, the idea of incorporating disaster mitigation within community planning and treating communities as significant links in the whole disaster management system, rather than passive actors waiting to be rescued, became mainstream.

The NCDR initiated its Community-Based Disaster Management Project (CBDM) in late 1999. The programme was originally executed by the director of the NCDR, who is also a social worker and university professor in urban planning, and has extensive experience in promoting community development with the Council for Cultural Affairs. Community participation in public affairs and planning has become more mature as a result of the reconstruction experience after the Chi-Chi Earthquake. Moreover, NCDR scientists studied foreign community-based disaster management approaches, especially the Disaster Preventive and Welfare Community Programme of Kobe, Japan, and the Project Impact of the US Federal Emergency Management Agency. They subsequently developed models and began preliminary experiments with two pilot cases.

A pilot project

In 2001, after the severe debris flows and floods induced by Typhoons Toraji and Nari caused over 200 deaths and hundreds of injuries (see Table 10.1), the 921 Post-Earthquake Reconstruction Council (921 PERC) and NDPPC formally supported the Community-Based Disaster Management Project (CBDM). This project integrated the NCDR, the SWCB, the fire department, local government agencies, local university scientists, local community associations and local residents. Its purpose was to achieve cross-boundary learning and link scientific knowledge to governmental resources to local actions. At the first stage of the programme, 921 PERC and the NDPPC selected 11 communities destroyed by Typhoons Toraji and Nari as model communities. The NCDR played an important role in the project. It implemented the project in Shan-an village, one of the 11 communities under study, and packaged practical procedures as training modules to enable other teams of experts to duplicate the experience in other communities in coming years.

Shan-an village: The local practices

This section describes CBDM's pilot study of Shan-an village, analysing in depth the collaborative working process at a local scale.

Background

Shan-an village is a rural agricultural community in Shuili Township, Nantou County, Central Taiwan. It is located on a hilly mountain slope ranging from 350m to 1100m, with Central Mountain Area to the east and Chen-Yo-Lan River to the west (see Figure 10.5). The total area comprises approximately 800ha. It has distinct seasonal variations in temperature and precipitation. The average temperature in winter is 16.74°C and in summer 26.27°C; total precipitation from November to February is 109.59mm, while it is 1350.35mm during the rainy season from May to August because of the monsoon and typhoons.

The morphology of the village developed along a New Cross-Island Highway, which was built during the 1970s, between steep hills and Chen-Yo-Lan River valleys. Several streams flow down through Shan-an village. The largest of these is Shanbu stream, which goes through the most densely populated area of the village. Shan-an village is subject to at least three severe types of environmental hazard. During the 1999 Chi-Chi Earthquake, fortunately no villagers died; but 95 houses collapsed and 77 houses were partly destroyed, amounting to one third of the total number of households. After the Chi-Chi Earthquake, some new social welfare policies were created to fund community reconstruction. The houses and the community were successfully rebuilt, and Shan-an village was selected by the Council for Cultural Affairs as a model of community reconstruction. However, the most devastating threats to the village were not earthquakes alone, but also the landslides that the Chi-Chi Earthquake set off. Figure 10.2 illustrates that on Chen-Yo-Lan River's watershed, the area prone to landslides had, by 2004, grown to eight times its size prior to the 1999 earthquake. In 2001, an unprecedented debris flow along Shanbu stream occurred as a result of the geologic crack induced by the Chi-Chi Earthquake and the heavy precipitation from Typhoon Toraji. Seventeen people died in this disaster and dozens of houses were destroyed as well. In 2004, the spill runoff from the Chen-Yo-Lan River that followed Typhoon Mindulle flooded and washed away a nearby area of 100ha.

Since the 1960s, Taiwan has experienced economic reform. Economic planning has promoted Taiwan's industrial development in the international market, on the one hand, and stimulated the need for infrastructure to deepen economic growth, on the other. The New Cross-Island Highway was built under these circumstances. As a result of this highway, Shan-an village has experienced rapid population growth since the late 1970s. New immigrants cultivate commercial agricultural products such as plums, tea, grapes and betel nuts as their main livelihood, and settle along the road to facilitate trade. The population at its peak was over 2000; however, with Taiwan's macro-urbanization and industrialization of the mid 1980s, people emigrated again from the village to nearby cities. Today the village has a total population of around 1500, 90 per cent of whom have been long-term residents living along the road.

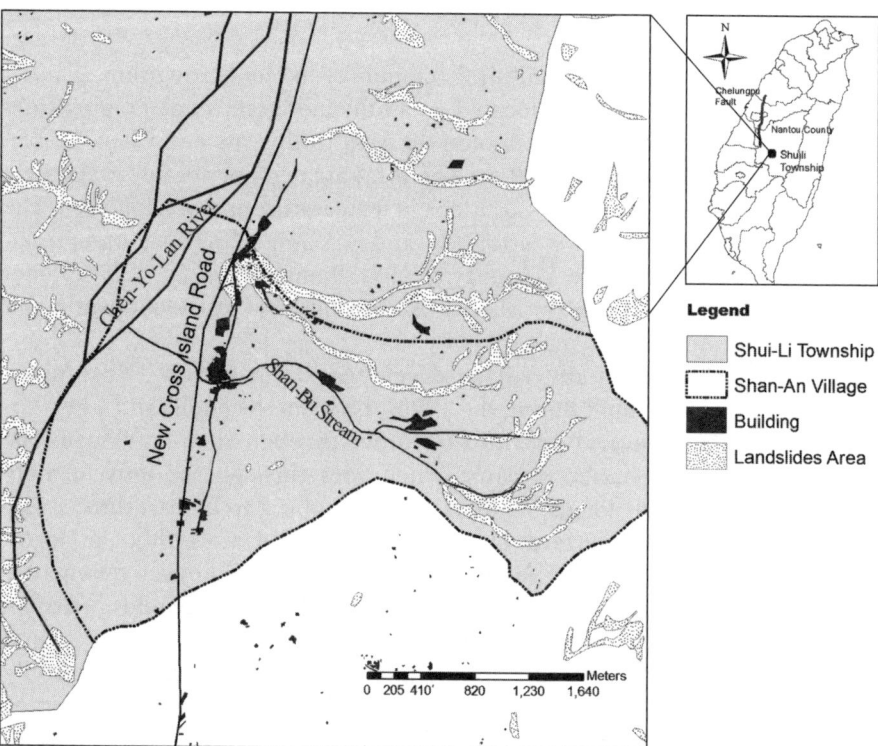

Figure 10.5 *Maps of Taiwan and Shan-an village*

Source: chapter authors

Shan-an village is subject to at least three severe types of environmental hazard. Natural stressors alone, however, did not cause its vulnerability. Human migration into the environmentally critical area imposed significant social stressors on the habitat. Soil erosion and environmental degradation may have been accelerated by the clearing of forests, while settlement along streams and rivers for irrigation and along the road for business exposed people to higher risks of hazard. Furthermore, 90 per cent of the villagers were farmers, which meant that their livelihood largely depended on the natural environment and was sensitive to environmental variations.

Involvement in the CBDM

Shan-an village started its CBDM in November 2001. In order to obtain a brief understanding about the village's background and to establish contacts, NCDR scientists first contacted the organizers of the Shan-an Community Development Association to communicate project objectives and to discuss appropriate times and frequency for meetings to conduct training activities with villagers. Since most of them are farmers and have regular working and resting hours, most of the activities were arranged for Sunday nights from 7.00 to 9.00pm, at intervals of two to three weeks. The activities in the packaged procedures comprised several steps, and were basically workshops integrating seminars, discussions, field surveys and training exercises, with wide participation from multiple stakeholders.

CBDM processes and procedures

All the procedures comprised seven steps, which included initiation, assessment, discussion, organization, planning, communication and training exercises.

Step 1: Initiation

The staff of 921 PERC, the SWCB and the fire department of Shuili Township, NCDR scientists and some tens of villagers attended the workshop. 921 PERC officials explained the project objectives to the villagers, while SWCB and fire department officials showed them how their work was connected to the project and what equipment, funding and professional skills they could provide for community disaster management. NCDR scientists gave a talk on the topic of the 'disaster-resilient community'. They explained that it was indispensable for community members to become knowledgeable about their environment and to be able to analyse important safety issues in community disaster management. There was a special activity in this workshop in which every participant was given a name tab and was taught to find their house on a topographic map and paste the name tab on it. This activity intended to help villagers learn to read maps, which was very significant for the following learning process.

Step 2: Assessment
The aim of step 2 was to assess the community's environment; its important product was a community vulnerability map. NCDR scientists and other experts educated villagers with scientific information about the causes of hazards and methods for observing and examining the environment. All of the stakeholders then conducted a half-day field survey. They were equipped with Polaroid cameras and maps to record problems found in the field. When the half-day field survey ended, all of the participants came back to the community activity centre and pasted the photos or points on a big village map. This map was loaded with locally based information showing critical environmental problems, such as the location of landslide-prone areas, weak retaining walls, the narrow Shanbu stream course, blocked gutters, etc. It also included critical social information, such as safe places for shelter, evacuation routes, the location of elders, houses closest to potential streams of debris, etc. This mapping process provided substantial opportunity for all participants to share their disaster experience and to communicate their perspectives with each other.

Step 3: Discussion
The community vulnerability map became an important communication tool for all participants in discussions about the village as a basis for developing strategies for disaster management. They categorized issues into four major groups according to the type of hazard (fire, typhoon/storm, earthquake and other), and developed adjunct issues and responding strategies under each major issue. Table 10.2 outlines some examples from the category of typhoon/storm.

These strategies were further confirmed as belonging to different management stages of mitigation, preparedness, response and recovery, and classified into household, community or government levels to outline practical actions.

Steps 4 and 5: Organization and planning
The key issues and strategies were co-produced through iterative discussions among all participants. Based on the strategies that developed, the structure of a Shan-an DPR organization was put in place and the DPR Plan was implemented. Shan-an villagers created their community EOC and organized a neighbourhood watch team, a medical and rescue team, a guard team, a mitigation team and a logistics team. Furthermore, since the village had been divided into four parts by the debris flows during Typhoon Toraji, an operation team was built for self-rescue in each part in order to manage emergency activities when isolated from other parts and to communicate with the community EOC. Meanwhile, all of the village-level information was organized into a community disaster-management database. This comprised data on the community vulnerability map; facilities and locations of elders and disabled people; evacuation routes; shelters; inhabitants with skills; and the CBDM organization. This information was remapped onto the disaster management

Table 10.2 *Examples of setting typhoons/storms as major issues and the resolution strategies following collaborative learning, discussion and communication processes in Shan-an village*

Accessory issues	Responding strategies
Serious landslides of Shanbu upstream remained untreated.	The Taiwan Soil and Water Conservation Bureau (SWCB) should conduct source treatments of the landslides.
Gutters in neighbourhoods 3 and 4 were blocked, which would put residents in danger of floods.	Ask villagers to clean and examine all gutters regularly.
Potential landslides on the slopes of Ci-Guang Temple.	Do not exploit the slope areas and try to reforest in the area.
The Chen-Yo-Lan riverbed is higher than the nearby farms, and the opposite mountain slope is collapsing, which could change the course of the river and flood nearby farms.	Avoid staying on the farms during typhoons and build higher and stronger dikes along the river.
Neighbourhoods 1, 2, 3 and 12 lacked appropriate public shelters.	Try to find a private building which is safe and big enough in which to shelter.
Many elders in neighbourhoods 1, 2, 3 and 12 were not willing to leave during typhoons.	Engage in communication with them and their neighbours; rescue teams should pay more attention.

Source: chapter authors

map, and the database and map became the basis for community disaster-management actions.

Step 6: Communication
Of the 1500 residents in Shan-an village, less than 100 participated in the workshops. As a result, there was a need to distribute the information throughout the village. Schools, community meetings and religious rituals were all effective formal activities for spreading the information, in addition to the informal vocal communication among neighbours and co-workers, which was the most common approach. Fortunately, the chief of the Shan-an Community Development Association, who was also the director of educational affairs in the village's only elementary school, had actively participated in the workshops. His position made it possible to spread the information in schools and, thus, to families. Community activities in Shan-an village also provided opportunities to distribute community disaster-management information.

Step 7: Training exercises
Knowledge and information produced collaboratively by scientists and villagers constituted the core part of community disaster-management practices. But practical exercises and technical training were also important for building the capacity of the villagers. NCDR scientists integrated governmental resources within several training seminars. The Shuili Township Health Centre and professors of medicine gave instructions on emergency medical care and instrument operation, while the Nantou County Fire Department and the

Shuili Township Fire Division together taught villagers rescuing techniques, equipment operation and ways to use communication equipment in order to obtain outside aid. A 'debris flow disaster exercise' was conducted in Shan-an village. This exercise aimed to make local villagers translate the knowledge that they had gained into successful action, and to ensure that all disaster emergency actions be based on the local disaster management map and plan. Not only were the villagers expected to act as the map and CBDM plan directed, but so were the outside support agencies. The NCDR, NDPPC, SWCB, Nantou County government, Shuili Township Office, Shuili Township Health Centre, Shuili Township Police Station, Shuili Township Fire Division, Shan-an CBDM organization, community development associations, neighbourhood rescue teams, etc. were all involved in the exercise. The conditions and sequence of a debris flow disaster were simulated; Shan-an villagers were the main actors in managing most of the emergency activities. Shan-an villagers organized the community EOC to communicate with the township EOC when the disaster warning system was activated, and observed environmental variation. Thus, timely disaster information was produced not only top down from government to local levels, but also created bottom up through local observation. Consequently, the governmental organizations that acted as outside supporters of the system learned how to work with local communities and to form an adaptive disaster management system.

Analysis of the boundary organization and effective knowledge/action links

Rapid change and ecological crises can provide windows of opportunity for creating networks and promoting new forms of governance (Folke et al, 2005). In the Taiwan case, we observed an important institutional transition for facilitating adaptive disaster-mitigation governance after the devastating earthquake and multiple environmental crises. There are at least five important conclusions.

First, flexible institutional structure can help to create successful institutional transitions in response to environmental changes and crises. Second, appropriate legislation can ensure a promising outcome from institutional transition and improve the effectiveness of the new system. Third, the process of institutional transition and legislation in the Taiwan case was primarily based on the disaster experience and collaboration among governmental agencies and scientists. Scientists recruited for government offices were part of the political culture in Taiwan, and scientific intervention in governmental systems was common. Fourth, the most relevant innovation of the new legislation referred to as the DPR Act was its special focus on public participation and scientific involvement in the disaster management system. Knowledge/action links are the heart of adaptive governance. Without a boundary organization to integrate the scientific community and expand its knowledge system into public administrative agencies, polycentric institutional design is just a simple multilevel decision-making and political organization.

Fifth, the boundary organization infused scientific information into policy-making systems and local actors, acting as a moderator to create cross-scale communication and a social learning milieu. This collaborative process of co-production of knowledge made governmental decision-making and local action more effective, credible and reliable to all the stakeholders.

Adger et al (2001) have proposed that adaptation is socially mediated and differentiated and can take place through mechanisms that can be characterized as social learning and policy learning. The Taiwan case has proven that carefully involving science in disaster management governance substantially increases the effectiveness of adaptive governance. The NCDR intentionally acted as a bridge connecting science, government and local communities. It integrated bits of scientific information scattered across different disciplines or agencies with local needs and local knowledge, and thus transformed theoretical knowledge into useful and practical knowledge widely available to all parties and practitioners.

Responding to environmental changes following the Chi-Chi Earthquake, the disaster management system in Taiwan engaged in an experimental process to create adaptive governance (see Figure 10.6). Institutional transformation and legislation were important foundations for the transition. The three-tiered polycentric management system allowed great flexibility for each municipality

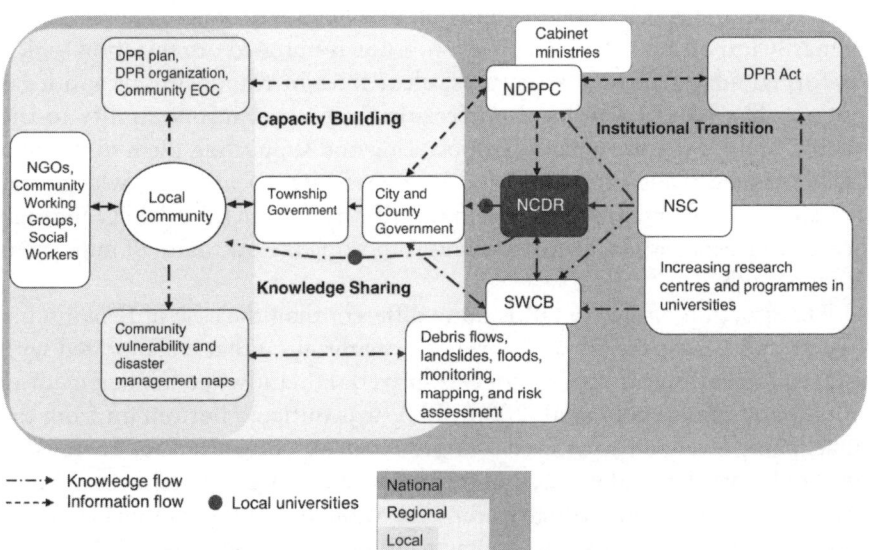

Figure 10.6 *Schematic diagram of adaptive disaster-management governance through a process of cross-scale learning, knowledge co-production and institutional transition*

Notes: DPR Act: Disaster Prevention and Response Act; DPR plan and organization: Disaster Prevention and Response plan and organization; Community EOC: Community Emergency Operation Centre; NDPPC: National Disasters Prevention and Protection Commission, Executive Yuan; NCDR: National Science and Technology Center for Disaster Reduction; NSC: National Science Council, Executive Yuan; SWCB: Soil and Water Conservation Bureau, Council of Agriculture, Executive Yuan.
Source: chapter authors

to develop policies and make decisions at its scale, but maintained the connectedness across scales at the same time. The NCDR's role in cross-boundary communication, learning and knowledge-sharing greatly improved knowledge integration and information flows in this cross-sectional and cross-scale collaborative regime. Through the Disaster Management Decision Support Project and Community-Based Disaster Management Project, a social learning mechanism was developed that empowered governments and local communities, on the one hand, and substantially contributed to more resilient disaster management governance with sound knowledge/action links, on the other.

Enabling legislation and institutional flexibility

Institutional flexibility and transformability are important for adaptation to environmental changes and crises (Olsson et al, 2006). Meanwhile, appropriate legislation can ensure the transition process of moving towards more adaptive management governance (Cash and Moser, 2000). The DPR Act regulated the polycentric and multilevel disaster management decision-making system, providing opportunities for decision-makers at different scales to develop scale-accordant disaster mitigation policies which avoided the problems of scale mismatching or institutional misfit, and forged collaboration across scales at the same time.

In multilevel governance, the function of a boundary organization is extraordinarily important. The NCDR is a top-down-initiated organization which has formal administrative structures, clear lines of responsibility and legal support. The NCDR and local universities have dual accountability to the NDPPC and local governments, empowering and supporting them to develop disaster management with useful and scale-matching scientific knowledge and technology. They obtain permission and verification from the NDPPC and sign formal contracts with local governments, reporting the outcomes of implementation to both.

The situation of the NCDR is quite different than the case of Ecomuseum Kristianstad Vattenrike (EKV), a bridging organization that collaborated with outer organizations at different levels in wetland landscape management in southern Sweden (Hahn et al, 2006). EKV was initiated bottom up from the municipality level and structured as a project-driven organization. There is no law regulating EKV and it has no enforcement status. Its authors say that political independence from ordinary administration gives EKV an unusually free mandate and flexibility to filter contacts with other municipal administrations and external collaborators. The lack of formal and legislative support to EKV, however, makes its whole management system unstable and subject to variations in human resources.

Gunderson et al (1995) have emphasized the role of such shadow networks as incubators of new approaches for governing social–ecological systems. Because members of these networks are not always under the scrutiny or obligations of their agencies, they are probably freer to develop alternative policies and think creatively about the resolution of resource problems. But Folke et al (2005) posit that even though new adaptive governance systems are performing in a resilient manner, they may be challenged and become fragile by changes in external drivers.

We argue that legal management arrangements have an indispensable importance for the successful and reliable operation of the whole management system by providing formal status or legislative support to the boundary organization. As for the problem of political lock-in or bureaucracies, we propose that social networks and social capital may alleviate these phenomena. The emergence of the NCDR and the two projects all represent creative experimentation and learning in a social network interwoven by active actors and under the adjustment of ongoing institutional change. From observations in the Taiwan case, we suggest that institutional flexibility can support experimentation and innovation in adaptive management, and that supplemental support from a source such as a boundary organization can enable institutional and financial backing for the new system, as well as monitor its progress.

Networks, leadership, partnership and socio-political culture

Adaptive governance involves devolution of management rights and power-sharing, which promotes participation; however, the devolution of management rights does not necessarily and automatically result in adaptive co-management, which requires support from social networks and social capital (Folke et al, 2005). The emergence of Taiwan's adaptive governance originates from the hazard experience and the application of a formal and informal policy orientation actors' network in which scientists and policy-makers self-organize to focus on common problems. The accumulation of hazard rescue and reconstruction experience forms a basis for social memory. These social networks and social memory represent the essence of social capital and result in informal governance systems across organizational scales that are able to stimulate collaboration, build trust, provide information and encourage the formation of common values.

Visionary and political leadership

An important mechanism that underlies the institutional transformation supporting adaptive governance may exist in political leadership (Hahn et al, 2006). In Taiwan's socio-political context, the lines between policy communities and epistemic communities were relatively blurred. This characteristic made the involvement of scientists in shaping institutional change and reorganization easier. We have illustrated many instances above. These key leaders in the political domain formed visionary political leadership in setting up the DPR Act, outlining adaptive disaster management governance and supporting the CBDM and DMDS projects.

The contribution from the director of NCDR was especially significant. As a social scientist specializing in urban planning and hazard studies, Professor

Liang-Chen devoted himself to community reconstruction fieldwork for many years, and enthusiastically promoted the integration of disaster mitigation within community reconstruction practices. He also propelled for the legislation of the DPR Act and the whole governance design, and now is a committee member on the scientific advisory board to the NDPPC.

Partnerships, key persons and social networks

Informal social networks can provide arenas for enhancing flexibility. The network structures do not replace the accountability of existing governmental institutions, but operate within them and complement them. The most important partnerships in the Taiwan case were relationships among the NCDR, local university scientists, local governments, local communities and central governmental agencies. We observe that intensive face-to-face meetings and geographic proximity aided greatly in building relationships. In addition, the NCDR and local university scientists usually had close formal and informal connections. Formally, they were colleagues dealing with the disaster management projects; informally, they were often friends or had master/apprentice-type relationships. Many local university scientists were also familiar with local government officials because their geographic proximity facilitated shared working experiences.

These key individuals operating in actor groups played leading roles in managing boundaries among different organizations involved in science and policy. They were also important educators in the contexts of learning, knowledge generation and social responses for dealing with environmental risks (Social Learning Group, 2001; Cash et al, 2003). In the domain of information and knowledge generation, they had the ability to manage existing knowledge within social networks for management and further created a learning milieu supporting a constructive learning process. In the domain of social relationships, they helped to facilitate trust-building and conflict resolution. Their networks also created more opportunities for spanning information flows across counties, cities and national organizations.

Communication, translation and mutual understanding

There were intensive regular meetings and workshops for the NCDR, local actors and government officials at different levels and scales. These activities increased mutual understanding. However, mutual understanding is often hindered by different languages, experiences and values (Cash et al, 2003). It requires personal competence in communication, as well as familiarity with the local context, social norms and languages, including folk dialects and scientific terminologies (see Chapter 4). Translation is important in such situations, not only for communicating events or phenomena, but for bridging the gap between the inner thoughts of experts, decision-makers and local actors. One NCDR scientist says that when they went into a community, they always used the local dialect because it was then easier to understand local people's identity, and people felt free to express their perspectives when using dialect.

Furthermore, in order to familiarize local actors with the mapping process, NCDR staff taught participants in community workshops to learn easy mapping languages, and confirmed that each participant had, indeed, acquired map-using knowledge.

Trust-building, norms and value formation

Trust-building and the communication of insights is crucial to the maintenance of governance in shaping interpersonal interactions and closely related to investments in social capital. The NCDR endeavoured to develop communicative skills, such as face-to-face communication, intensive meetings and workshops, and using dialect to build its relationship with each stakeholder group. These techniques fostered mutual understanding and aided in earning trust.

The formation of collective norm and value, however, takes time. As one NCDR scientist remarked: 'Norm or value discrepancy always exists in terms of different discipline training and different views of managers toward certain environmental issues'. Conflicts sometimes do occur, especially when the issues involve self-interest or personal reputation, such as was the case of the formulation of Shan-an village's DPR organization. But conflict resolution cannot depend upon improving mutual understanding alone; it requires active mediation among different views or interests. We observe that NCDR scientists earned a great deal of trust from the local people, thus enabling them to act as important negotiators in mediating different views and interests. Trust in key people and a transparent discussion process that brings all perspectives to the table are good for negotiation and conflict resolution.

Social learning, information flows and knowledge co-production

Wide participation

Adaptive co-management relies on the collaboration of multiple stakeholders, operating at different levels, to work on explicit issues or problems. In Taiwan's disaster management governance, governments at national, county (city) and township scales were all included, along with scientists and local people. This multi-stakeholders' participation created an atmosphere for constructive communication, improving cooperation and increasing the probability for success in a transparent process. However, how do we define the main stakeholders? How do we stimulate their participation? These questions remain difficult. From observations in the Taiwan case, we argue that some appropriate mechanisms may trigger a public that is willing to participate. These mechanisms include legislative support, personal interest, and attachments to social networks and social capital. Obviously, the DPR Act had, in its articles, defined the framework and levels of governance which comprised governments across scales, specialists and the public.

The unpredictable nature of disasters that bring fatal threats to people's lives and properties further hastens people's willingness to participate in disaster mitigation. The national community development policy and community reconstruction experience after disasters formed another important mechanism for social capital accumulation. Community identity was fostered through the leadership of local cultural or historical studios and community development associations, and through means of public hearings, meetings and parochial newsletters. These experiences activated local social capital and influenced public participation in the disaster mitigation projects that followed.

Social learning and the co-production of knowledge

Adaptive governance emphasizes the ability to learn. The meaning of Taiwan's disaster management governance exists not only in bridging links between people in social networks across scales, but also in the knowledge required to integrate different knowledge systems, connect them to the empirical context, and transform the understanding into useful information. Boundary objects such as hazard simulation models, county (city) DPR plans, CBDM organizations, community vulnerability maps, community disaster-management maps, etc. are all co-produced forms of knowledge and all provide important boundary functions. Through the mechanisms of key individuals managing boundaries and broad participation, this model realizes the process of wider social learning which empowers the capacity on each side of the boundary and facilitates adaptive and creative responses to environmental crises. However, we should still stress that a constructive learning environment requires leadership, social capital and changes of social norms within the management system to support its operation.

Another important point is the complementarity of local knowledge and scientific knowledge (Berkes 1999; Berkes et al, 2003). Local knowledge and scientific knowledge are structured in different ways. Local knowledge is constructed through perceiving and observing the environment, while scientific knowledge is generated through systematic measuring and testing. Currently, most scientific evaluation of the environment and disasters is focused on the larger scale, using satellite images or aerial photos to assess conditions such as debris flow or landslides. However, this approach usually cannot respond efficiently to rapid environmental changes on the community level, and cannot match environmental analysis on the smaller scale. Conversely, by means of social learning, local people have opportunities to learn systematic scientific knowledge and to convert it into part of their ecological knowledge. This process enables them to monitor the environment and translate environmental feedbacks from ecosystem dynamics into knowledge that can be used in the management system. Thus, information flows and knowledge generation cannot only be produced top down, but also created bottom up by local people observing their surrounding environment and transferring the information to scientists and managers. This condition changes policy-makers and scientists from objective managers to one of several actors in a process of learning (Folke et al, 2005).

Meaning creation

The process of knowledge co-production, in which scientists and stakeholders interact to define important questions, relevant evidence and convincing forms of argument (Kates et al, 2001), is also the process of giving the collective efforts specific meaning. We should notice that the collective production of knowledge is a negotiation process through iterative communication, translation, mediation and conflict resolution. When all stakeholders reach a coherent agreement about boundary objects, they also give special meaning to those objects. They can follow the norms, values or the knowledge that they produced. This phenomenon is quite important in the Taiwan case. Local actors reflected that through the CBDN and DMDS projects they not only created a locally based disaster-management knowledge system, but also became involved in Taiwan's polycentric disaster management system. They both respected the governance system and worked collaboratively with it.

Obstacles

Project continuity

NCDR is a formal organization that mainly serves NDPPC; but the CBDM project is a project-driven programme. In 2001, when the CBDM project started, it was an experiment. The success of the first wave of CBDM projects led to an emphasis on community-based disaster management approaches. But financial support for continuing the project was a problem. In the first wave, the NCDR obtained funding from the NDPPC and the 921 Post-Earthquake Reconstruction Council. When the project ended, however, the NCDR had no funding to keep going until the NDPPC and SWCB stepped forward in 2004 to provide stable financial support for the CBDM, thus permitting its continuation. Unless the CBDM is institutionalized into a formal administration, however, its continuity will always depend upon such financial support. The DMDS is different in this regard. Because polycentric decision-making systems are a legislative formulation, the NDPPC and county (city) governments have responsibility for maintaining continuity.

Uneven resource distribution

The serious landslides, debris flows and mud floods that occurred in Shan-an village in 2004 qualified it to become one of 11 exemplary communities where the CBDM would be implemented. The success of the Shan-an case made it an important model community, with full resources. Thus, after the CBDM ended in 2002, Shan-an village still received continuous financial support from the SWCB and other governmental agencies, such as the fire department, to maintain normal operations of community disaster management and to maintain equipment.

Meanwhile, for local Shan-an actors, the experience of working with scientists and governmental officials was a process of capacity-building in which they were not only educated by the knowledge system, but also involved in a deeper affinity with government administrative institutions. This experience contributed to local actors' capacity to know where to obtain needed resources and how to write proposals for financial support. However, not all villages exposed to potential multiple typhoon-related disasters are as fortunate as Shan-an village. There were three villages implementing the CBDM in 2004, eight in 2005 and four in 2006. Discontinuity and slow development of the CBDM were a result of time-consuming processes and intensive, lengthy involvement by experts. Thus, the problem of uneven resource distribution was widespread in disadvantaged hazard-prone areas.

Wide participation

Wide participation is fundamental to adaptive governance. It not only means involving various stakeholders; it imparts meaning to active public participation. In our interviews, nearly all the local actors who participated in the CBDM approved of its purpose and commented that they gained much valuable knowledge from the scientists. They also stated that such knowledge helped them to study their environment and design a local disaster management system.

However, in a democratic society, it is not easy to give orders to local villagers, especially in rural agricultural villages where most people do not treat events unrelated to their livelihood as a priority. Getting them to participate in certain projects can be challenging. The participation rate in Shan-an village was not high, with only tens of people contributing to the CBDM from among nearly 1500 villagers. Middle-aged people and elders constituted a majority within Shan-an village's demographic structure. We observe that middle-aged people with higher education degrees (high school degrees, Bachelor's degrees or higher) were the major force contributing to the CBDM. Elders and those who suffered less from the disaster were less willing to participate.

In Cash et al (2003), the authors discuss the problem of exclusion from knowledge co-production. Excluded stakeholders tend to reject the information produced and to question its legitimacy, regardless of its salience or credibility. We argue that a relatively low participation rate does not always indicate a low level of public concern; we find that high social capital in a local context can compensate for the insufficient direct participation of the public. Shan-an village has civic associations that existed long before the Chi-Chi Earthquake occurred. These include farmers' associations, the Community Development Association, mothering classes, parochial newsletters and annual village basketball contests that successfully integrate people's visions and circulate information. As a small-scale community, it was easier for Shan-an to concentrate its social capital for these activities. Besides, local actors were very aggressive in distributing CBDM information; their leadership in the community also played a key role in establishing shared disaster management values locally. In our interviews with the village residents who did not attend the CBDM workshops, most interviewees expressed a positive attitude towards the functions of the local CBDM organization and its plan, and felt that the project had helped to enhance resistance to hazards.

Institutional blocks

- Organizational barriers exist to the integration of various governmental agencies. Each governmental agency suffers from some narrow 'departmentalism'. Even though the DPR Act attempted to promote synergistic efforts for adaptive governance, there were still difficulties in bridging different governmental administrative agencies. For example, the SWCB and the fire department were supposed to be closely connected in the disaster management system; however, in reality, we found that direct cooperation between the two was infrequent. The NCDR or NDPPC played important roles in connecting the two agencies; but the SWCB had closer connections with the local agriculture department. Governmental agencies are still learning how to collaborate and to negotiate with each other.
- The strategies that communities developed to resist or adjust to the hazards were not reflected immediately in governmental support. For example, Shan-an villagers felt that they needed stronger and wider embankments along the Chen-Yo-Lan River to prevent serious flooding. This strategy, however, was not immediately adopted by the Water Resource Bureau because it needed to review the construction budget. Meanwhile, the views about how wide the embankment should be (and of what substance, texture and design) still differed among villagers, scientists and governmental officials. And this is another problem.

Conclusions

Multiple hazards and threats in Taiwan have not only been surprises of nature, but have created opportunities for society to reorganize its hazard management system. Through institutional transition, multilevel polycentric adaptive disaster management governance has, with legislative support, organized a science/government/local community collaborative learning environment. Taiwan's adaptive governance was formed in a top-down way, from high-level central government decision-making to local practices. This approach is quite different than that of some social–ecological management systems, where scholars assert that most of the adaptive co-management or governance is a bottom-up process linking local practice to decision-making and legislation. This point makes the Taiwan case significant among management studies. It reflected a convention of close connections between the scientific community and government, and made scientific intervention in adaptive governance more accessible. Furthermore, scholars' leadership in creating adaptive management was especially focal. NCDR scientists built positive communications and trusting relationships with government agency officials and local people, and included them in the governance circle. They were involved in producing local social norms and created more interactive information and knowledge flows across boundaries and scales. They also created a social learning milieu that gave collaboratively produced knowledge and governance practices specific meaning and value. These disaster mitigation efforts have proven effective and earned positive responses from the public. Table 10.1 shows the declining trend of harm from typhoon-related disasters. Although obstacles and conflicts did occur in the communication and governance process, most of these difficulties resulted from insufficient negotiation and institutional barriers.

If we take a longer view, we may find that despite changes in natural conditions, the development of local societies occurs in conjunction with local environmental changes and influences the occurrence of disasters. Vulnerability reduction and disaster mitigation are something that cannot be easily dealt with by looking only at surface social phenomena. We cannot ignore that disasters are always related to social development, social life, people's livelihoods and questions of political economy. However, the vulnerability reduction strategies we examined here remain limited to short-term mitigation and quick-response strategies to disasters. Long-term vulnerability reduction planning should consider longer perspectives on how to keep people away from hazard-prone areas or how to alleviate disasters that are aggravated by human activities. This applies to land-use planning and environmental policy and legislation. After Typhoons Mindulle and Aere in 2004, a series of national conferences on disaster mitigation and reconstruction were held. Instead of a traditional structural engineering approach focused on repairing roads or facilities after typhoons, scientists, decision-makers and local actors arrived at a consensus on stopping further construction in mountainous areas to allow for the recovery of nature.

A draft of a programme called the Land Recovery Act has been under discussion. Its aim is to initiate land recovery or rehabilitation, to encourage people to keep away from steep mountains and frequent flooding or potential debris-flow areas, to prohibit further inappropriate development of slope land, and to define some areas as nature reserves. This new act will empower government to apply socio-economic strategies and incentives to help people build, aided by community reorganization or resettlement to mitigate the disasters, and with a response capacity that adapts to environmental changes. We propose that boundary organizations bridging scientists, decision-makers and local actors for adaptive governance could play an essential role in cross-scale communication for a new adaptive regime. We will continue to observe future developments.

References

Adger, W. N. (2006) 'Vulnerability', *Global Environmental Change*, vol 16, no 3, pp268–281

Adger, W. N., Kelly, P. M. and Ninh, N. H. (eds) (2001) *Living With Environmental Change: Social Vulnerability, Adaptation and Resilience in Vietnam*, Routledge, London

Adger, W. N., Arnell, N. W. and Tompkins, E. (2005a) 'Adapting to climate change: Perspectives across scales', *Global Environmental Change*, vol 15, no 2, pp75–76

Adger, W. N., Hughes, T. P., Folke, C., Carpenter, S. and Rockström, J. (2005b) 'Social-ecological resilience to coastal disasters', *Science*, vol 309, no 5737, pp1036–1039

Berkes, F. (1999) *Sacred Ecology: Traditional Ecological Knowledge and Resource Management*, Taylor and Francis, Philadelphia and London

Berkes, F. and Folke, C. (eds) (1998) *Linking Social and Ecological Systems: Management Practices and Social Mechanisms for Building Resilience*, Cambridge University Press, Cambridge, UK

Berkes, F. and Turner, N. J. (2006) 'Knowledge, learning and the evolution of conservation practice for social-ecological system resilience', *Human Ecology: An Interdisciplinary Journal*, vol 34, no 4, pp479–494

Berkes, F., Colding, J. and Folke, C. (eds) (2003) *Navigating Social-Ecological Systems: Building Resilience for Complexity and Change*, Cambridge University Press, Cambridge, UK

Blaikie, P. (1985) *The Political Economy of Soil Erosion in Developing Countries*, Longman Inc, New York, NY

Blaikie, P. and Brookfield, H. (1987) *Land Degradation and Society*, Methuen, London

Blaikie, P., Cannon, T., Davis, I. and Wisner, B. (1994) *At Risk: Natural Hazards, People's Vulnerability and Disasters*, Routledge, London

Carpenter, S. R., Walker, B. H., Anderies, J. M. and Abel, N. (2001) 'From metaphors to measurement: Resilience of what to what?', *Ecosystems*, vol 4, no 8, pp765–781

Cash, D. W. (2001) 'In order to aid diffusing useful and practical information: Agricultural extension and boundary organizations', *Science, Technology & Human Values*, vol 26, no 4, pp431–453

Cash, D. W. and Moser, S. C. (2000) 'Linking global and local scales: Designing dynamic assessment and management processes', *Global Environmental Change*, vol 10, no 2, pp109–120

Cash, D. W., Clark, W. C., Alcock, F., Dickson, N. M., Eckley, N., Guston, D. H., Jäger, J. and Mitchell, R. B. (2003) 'Knowledge systems for sustainable development', *Proceedings of the National Academy of Sciences of the United States of America*, vol 100, pp8086–8091

Cash, D. W., Adger, W. N., Berkes, F., Garden, P., Lebel, L., Olsson, P., Pritchard, L. and Young, O. (2006) 'Scale and cross-scale dynamics: Governance and information in a multi-level world', *Ecology and Society*, vol 11, no 2, pp2–8

Chambers, R. (1989) 'Vulnerability, coping and policy', *IDS Bulletin*, vol 20, pp1–7

Chen, L.-C. (2006a) *Operation of the Central Emergency Operation Center Assessment Group*, Unpublished manuscript, Graduate Institute of Building and Planning, National Taiwan University, Taiwan

Chen, L.-C. (2006b) *The Interdisciplinary and Comprehensive Disaster Response Mechanism in Pre-disaster Stage of Disaster Management: The Case of Early Warning System for Typhoon in Taiwan*, The 6th Japan–Taiwan Joint Seminar on Natural Hazard Mitigation, 9–12 October 2006, Kyoto

Clark, W. C., Schröter, D., Patt, A., Gaffin, S., Martello, M. L., Neff, R., Pulsipher, A., and Selin, H. (2001). *Vulnerability and Resilience for Coupled Human–Environment Systems*, Report of the Research and Assessment Systems for Sustainability Program 2001 Summer Study, Environment and Natural Resources Program, Belfer Center for Science and International Affairs (BCSIA), John F. Kennedy School of Government, Harvard University, Cambridge, MA

Cutter, S. L. (1996) 'Vulnerability to environmental hazards', *Progress in Human Geography*, vol 20, no 4, pp529–539

Dietz, T., Ostrom, E. and Stern, P. C. (2003) 'The struggle to govern the commons', *Science, Technology, & Human Values*, vol 302, pp1907–1912

Downing, T. E. (1991) 'Vulnerability to hunger in Africa: A climate change perspective', *Global Environmental Change*, vol 1, no 5, pp365–380

Folke, C. (2006) 'Resilience: The emergence of a perspective for social–ecological systems analyses', *Global Environmental Change*, vol 16, no 3, pp253–267

Folke, C., Hahn, T., Olsson, P. and Norberg, J. (2005) 'Adaptive governance of social–ecological systems', *Annual Review of Environment and Resources*, vol 30, pp441–473

Fraser, E. D. G., Mabee, W. and Slaymaker, O. (2003) 'Mutual vulnerability, mutual dependence: The reflexive relation between human society and the environment', *Global Environmental Change*, vol 13, no 2, pp137–144

Gunderson, L. H. and Holling, C. S. (2001) *Panarchy: Understanding Transformations in Human and Natural Systems*, Island Press, Washington, DC

Gunderson, L. H., Holling C. S. and Light, S. S. (eds) (1995) *Barriers and Bridges to the Renewal of Ecosystems and Institutions*, Columbia University Press, New York, NY

Guston, D. H. (2001) 'Boundary organizations in environmental policy and science: An introduction', *Science, Technology, & Human Values*, vol 26, no 4, pp399–408

Hahn, T., Olsson, P., Folke, C. and Johansson, K. (2006) 'Trust-building, knowledge generation and organizational innovations: The role of a bridging organization for adaptive comanagement of a wetland landscape around Kristianstad, Sweden', *Human Ecology*, vol 34, no 4, pp573–592

Holling, C. S. (1986) 'The resilience of terrestrial ecosystems: Local surprise and global change', in W. C. Clark and R. E. Munn (eds) *Sustainable Development of the Biosphere*, Cambridge University Press, Cambridge, UK, pp292–317

Kasperson, J. X. and Kasperson, R. E. (2005) *The Social Contours of Risk, Volume II: Risk Analysis, Corporations and the Globalization of Risk*, Earthscan, London

Kates, R. W., Clark, W. C., Corell, R., Hall, J. M., Jaeger, C. C., Lowe, I., McCarthy, J. J., Schellnhuber, H. J., Bolin, B., Dickson, N. M., Faucheux, S., Gallopín, G. C., Grübler, A., Huntley, B., Jäger, J., Jodha, N. S., Kasperson, R. E., Mabogunje, A., Matson, P., Mooney, H., Moore III, B., O'Riordan, T. and Svedin, U. (2001) 'Environment and development: Sustainability science', *Science*, vol 292, no 5517, pp641–642

Jasanoff, S. (1990) *The Fifth Branch: Science Advisers as Policymakers*, Harvard University Press, Cambridge, MA

Lebel, L., Garden, P. and Imamura, M. (2005) 'The politics of scale, position, and place in the governance of water resources in the Mekong Region', *Ecology and Society*, vol 10, no 2, p18

Lin, C. W., Chen, H. and Lin, J. C. (2005) 'Geological hazard induced by Chi-Chi earthquake', in Jeen-Hwa Wang et al (eds) *921 Chi-Chi Earthquake*, National Science Council of Taiwan, Taipei

Lin, K. W. (2005) *The Relationships between Sediment Discharge and Landslide Induced by Typhoon and Earthquake along the Chenyoulan River*, MSc thesis, Department of Geosciences, National Taiwan University, Taipei

Lin, M. L., Chen, J. S., Lin, L. M., Chen, H. and Lin, J. C. (2004) 'A preliminary report on debris flows hazard after Mindulle typhoon along Tachia river watershed', *Magazine of the Chinese Institute of Civil and Hydraulic Engineering*, vol 31, no 4, pp19–25

McCarthy, J. J., Canziani, O. F., Leary, N. A., Dokken, D. J. and White, K. S. (eds) (2001) *Climate Change 2001: Impacts, Adaptation and Vulnerability*, Cambridge University Press, Cambridge, UK

Olsson, P., Folke, C. and Berkes, F. (2004) 'Adaptive comanagement for building resilience in social-ecological systems', *Environmental Management*, vol 34, no 1, pp75–90

Olsson, P., Gunderson, L. H., Carpenter, S. R., Ryan, P., Lebel, L., Folke, C. and Holling, C. S. (2006) 'Shooting the rapids: Navigating transitions to adaptive governance of social-ecological systems', *Ecology and Society*, vol 11, no 1, p18

Smit, B. and Wandel, J. (2006) 'Adaptation, adaptive capacity and vulnerability', *Global Environmental Change*, vol 16, no 3, pp282–292

Social Learning Group (2001) *Learning to Manage Global Environmental Risks, Volume 1: A Comparative History of Social Responses to Climate Change, Ozone Depletion, and Acid Rain*, MIT Press, Cambridge, MA

Soil and Water Conservation Bureau (2005a) *Applying Satellite Images to Monitoring Environmental Variation of Slope Land*, Nantou City, Taiwan

Soil and Water Conservation Bureau (2005b) *Maps of Potential Debris Flow Streams in Impacted Areas*, Nantou City, Taiwan

Soil and Water Conservation Bureau (2005c) *Project Report on Debris Flow Hazard Mitigation Plans in Reconstruction Areas*, Nantou City, Taiwan

Timmerman, P. (1981) *Vulnerability, Resilience and the Collapse of Society*, Environmental Monograph 1, Institute for Environmental Studies, University of Toronto, Toronto

Tompkins, E. L. and Adger, W. N. (2004) 'Does adaptive management of natural resources enhance resilience to climate change?', *Ecology and Society*, vol 9, no 2, p14

Turner, B. L. II, Kasperson, R. E., Matson, P. A., McCarthy, J. J., Corell, R. W., Christensen, L., Eckley, N., Kasperson, J. X., Luers, A., Martello, M. L., Polsky, C., Pulsipher, A. and Schiller, A. (2003) 'A framework for vulnerability analysis in sustainability science', *Proceedings of the National Academy of Sciences of the United States of America*, vol 100, no 14, pp8074–8079

Walker, B., Holling, C. S., Carpenter, S. R. and Kinzig, A. (2004) 'Resilience, adaptability and transformability in social-ecological systems', *Ecology and Society*, vol 9, no 2, p5

Watts, M. J. and Bohle, H. G. (1993) 'The space of vulnerability: The causal structure of hunger and famine', *Progress in Human Geography*, vol 17, pp43–67

Westley, F. (1995) 'Governing design: The management of social systems and ecosystems management', in L. H. Gunderson, C. S. Holling and S. S. Light (eds) *Barriers and Bridges to the Renewal of Ecosystems and Institutions*, Columbia University Press, New York, NY, pp391–427

Wisner, B., Blaikie, P., Cannon, T. and Davis, I. (2004) *At Risk: Natural Hazards, People's Vulnerability and Disasters*, 2nd edition, Routledge, London

Young, O. R. (2002) 'Institutional interplay: The environmental consequences of cross-scale interactions', in E. Ostrom, T. Dietz, N. Dolsak, P. C. Stern, S. Stonich and E. U. Weber (eds) *The Drama of the Commons*, National Academy Press, Washington, DC, pp263–291

Young, O. R. (2006) 'Vertical interplay among scale-dependent environmental and resource regimes', *Ecology and Society*, vol 11, no 1, p27

11

Participatory Evaluation of Development Interventions in a Vulnerable African Environment

Ton Dietz

Introduction

Many vulnerable areas in the world have experienced a multitude of external interventions to 'assist development', 'decrease vulnerability' or 'change people's culture and institutions'. There have been many project-specific evaluations of those interventions, but hardly any coherent multi-intervention evaluations. And despite the use of the word 'participatory' in many current development programmes, evaluations are often top-down 'professional' activities, and not at all participatory.

In the course of the last 20 years, West Pokot District in western Kenya can be regarded as a typical African example of a vulnerable area with withdrawing (and erratic) government presence, and increasing presence of foreign donor agencies – either with a government, a non-governmental organization (NGO) or a church background. Agencies from The Netherlands played a dominant role, first and foremost the Dutch-sponsored Arid- and Semi-Arid Lands Development Programme, which lasted from 1981 until 1999, and which withdrew under a cloud of anger.

Some two decades ago I was involved in a research project to reconstruct past performance of development initiatives until the early 1980s, which resulted in the PhD thesis *Pastoralists in Dire Straits* (Dietz, 1987). During those years Rachel Andiema and Albino Kotomei were my research assistants.

The same team of researchers was later involved in a participatory assessment of the impact of the development interventions, in the same administrative area, between the early 1980s and 2002. Three workshops were organized in 2001 and 2002, with about 150 participants from the area. This chapter is restricted to one of the areas, and to the results of one workshop in the most remote area: the current Alale and Kasei divisions in the upper north. It reports about approach and process, and provides some results concerning 'indigenous impact measurement'. All of it is related to the core question: did those interventions indeed diminish the vulnerability of the inhabitants according to their own judgements?

The research area and its vulnerability

The current Alale and Kasei divisions are located in the most remote, northern area of West Pokot District in western Kenya, bordering Turkana District and the Karamoja area in Uganda (see Figure 11.1 for the location of the research area). From the 1930s until 1970 it was part of the 'Karasuk' or 'Karapokot' area (currently, Kacheliba, Alale and Kasei divisions), which was administered by Uganda, under Upe County of Karamoja. For Uganda, it was a marginal zone of a marginal area. After Uganda's independence in 1962, it became a playground for the Ugandan army, under Field Marshal Idi Amin. When the area, independent since 1963, was rejoined with Kenya in 1970, very little had been done by any government agency, and hardly any non-governmental agencies or foreign churches had started any development activity either. There were virtually no schools and no health dispensaries. There were no roads, other than a few forest tracks made by a small camp of foresters. There were no shops. In 1952 the United Nations Children's Fund (UNICEF) had drilled a few boreholes; but other than a bit of famine relief (in 1965 to 1966), their yield was poor.

The population lived rather autonomous lives as pastoralists whose existence depended upon cattle, goats, sheep and camels, both economically and culturally. They lived mainly in the semi-arid lowlands. Cattle raids and counter-raids with the neighbouring Turkana and Karimojong were accepted elements of life, and a source of pride and folk culture. For boys to become men, successful participation in raids was a *rite de passage*, and economically important as a source of bride-price payments. Agriculture was practised as a fall-back strategy for the poor, as the area's hills had a sub-humid climate, allowing sorghum, millet and (later) maize cultivation during years with sufficient rainfall. However, crop cultivation was equated with poverty, and people told stories about the disastrous period around 1900, after a rinderpest epidemic and a disastrous drought killed most of the animals. Those who survived were forced to flee to the mountains with their remaining animals. This community of survivors consisted of a mixture of three ethnic groups (Pokot, Oropom and Karimojong), but was culturally dominated by the most northern section of the Pokot, one of the Kalenjin-speaking groups. After 1925

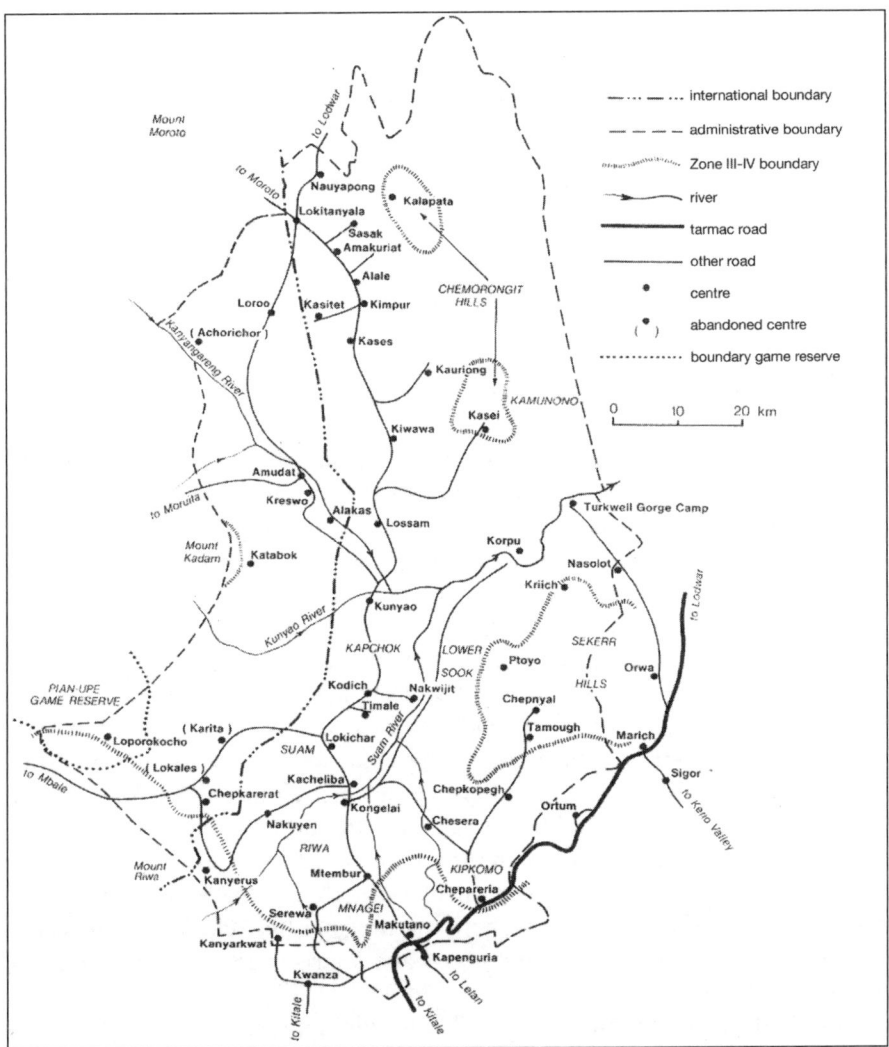

Figure 11.1 *The research area: Western Pokot in Kenya/Uganda*

Source: chapter author

they gradually recovered lost territory, and, assisted by British colonial support (see Barber, 1968), people became mobile herders again, pushing rather far into Karimojong territory in the west. They no longer lived in the mountains and foothills, but in mobile camps (*manyattas*) in the plains. For 50 years their existence was not threatened, although insecurity increased after 1950 (Dietz, 1987). During the 1950s, a group of religious refugees also settled in their midst, practising agriculture in the foothills of the mountains. These were Pokot from the southern area around Kapenguria, who were prosecuted by the British authorities in Kenya for adhering to an indigenous religious movement (*Dini ya Msambwa*) that was regarded as an anti-colonial protest cult.

During the first seven years of Kenyan administration very little changed. In 1970 chiefs had been installed in Alale and Kasei; but communications with the then divisional headquarters at Kacheliba (110km away) and district headquarters in Kapenguria (150km away) were very difficult, and during the rainy season virtually impossible because of the impassable Kanyangareng River. In 1977 some changes were in the air: a road was built, a police post was established and the Roman Catholic Church started a small primary school. By then the area had between 10,000 and 15,000 inhabitants on 2900 square kilometres, a population density of 3 to 5 inhabitants per square kilometre (Republic of Kenya, 1981, p121); but people moved freely between the neighbouring Upe County in Uganda and the 'Karapokot' area that had now become Kenyan. Then a sequence of disasters hit the area and revealed a multitude of vulnerabilities.

Throughout the period of the dry season during 1978 to 1979, grazing was very poor. In the Pokot area in Karamoja (Upe County, around Amudat), severe Karimojong raids forced hundreds of women and children to move to the east, to the Alale area. When Idi Amin's regime was toppled, part of his army fled through Karamoja and, hoping to get support, opened army stores, where new supplies of many Kalashnikovs had just arrived. The Karimojong were quick to use these arms against their Pokot enemies. To make things worse, the sorghum harvest failed and from June to August most of the goats died because of an epidemic that had killed most of the goats in Upe in May and June. The district officer urged the chief to organize a famine relief committee and to encourage parents to send their children to school. Many parents decided that their children would be better off in schools as they would have food and protection, and with the loss of animals they did not have much to do at home anymore. Next to the small Alale school, which was managed by the government, the Roman Catholics started a boarding school in Amakuriat. The number of pupils grew from 43 boys and 5 girls in 1978 to 154 boys and 11 girls in 1979.

In 1980, again the rains failed and a terrible rinderpest epidemic began to claim the lives of hundreds of cattle. Cholera reached the area and a Finnish Red Cross team started an anti-cholera campaign and provided famine-relief food. The Roman Catholic Church and the district officer also provided food, partly through Food-for-Work campaigns that consisted of building schools, water dams and roads. In June 1980 a large Karimojong force attacked the Pokot at a place just west of the growing centre of Alale. Pokot claim that 127 of them were killed and 11,000 head of cattle raided. Many people fled their houses and flocked near the famine relief centres of Alale and Amakuriat. Later, raids intensified, and Karimojong and Turkana forces even went far into the mountains to raid cattle that were hiding there. In April 1981 the Red Cross was feeding 5000 people in three famine relief centres. The total population had increased to between 20,000 and 25,000 people. School attendance had risen to 282 boys and 210 girls, many of them under the protection of the

Catholic boarding school. Probably 40 per cent of all eligible children were now in schools. The Red Cross had distributed seeds and, with better rains, a good sorghum harvest was produced. People were also making quite a lot of money by gathering miraa leaves and selling them to a few Somali traders. In October the Red Cross left the area.

When gold was discovered in 1980, a major gold rush started, attracting many people to the mountain. There was a major increase to the cash economy and an important role for Somali traders as a result. Cash opportunities were also increased by the activities of a new American missionary to the area, the African Inland Church, connected to the Reformed Church of America. When a peace treaty was arranged between Pokot and Karimojong elders, at the end of 1982, the Pokot of the Alale area had lost most of their animals and were 'pastoralists in dire straits'. They had moved from the plains to their refuge areas in the mountains and survived through a combination of sorghum cultivation and selling miraa and gold. Many children had gone to missionary schools, and many of them (and some of their mothers and a few of their fathers) either became Catholics or African Inland Church Christians (Dietz, 1987).

Summarizing the experience of vulnerability, the people in the research region had to cope with:

- occasional droughts, resulting in poor grazing, livestock deaths, crop failure and hunger;
- livestock diseases (rinderpest; contagious caprine pleuropneumonia for goats), resulting in livestock deaths and undermined livelihood security for those partly or wholly depending upon livestock for their subsistence and survival;
- human disease epidemics (e.g. cholera), resulting in health costs and human deaths;
- raids by neighbouring pastoralists, and counter-raids by the Pokot, resulting in human deaths, loss of livestock, occasional no-go areas for herding, and destruction of property; and
- army actions, resulting in human deaths, livestock confiscation and deaths, and destruction of property.

There is hardly any evidence of 'climate change' as a cause of growing vulnerability due to lower and more erratic rainfall and higher average evapotranspiration. The rainfall data also does not show a clear long-term trend; but data is scarce and unreliable. In people's perception, the most severe drought occurred during the 1979 to 1985 period; but that drought made the people vulnerable to hunger, disease and death due to its combination with all other causes of vulnerability.

Development interventions: The Dutch-funded Arid and Semi-Arid Lands programme, 1981–1999

In 1981, West Pokot District had also been adopted by the Dutch Development Agency. The Kenyan government had asked a number of donor countries to 'adopt' a district in the arid and semi-arid zones of the country. The Netherlands was eager to implement its area-based development philosophy, and chose West Pokot and another district (Elgeyo Marakwet), later to be followed by two more districts (Kajiado in the southeastern Masai area and Laikipia). In 1982 researchers from the University of Amsterdam joined the newly appointed Dutch Arid and Semi-Arid Lands (ASAL) programme adviser and the Dutch doctor who was the medical officer in charge of district healthcare. The researchers recruited a staff of local research assistants and began to make 'locational development profiles' and a district development atlas (Hendrix et al, 1985). Gradually a multi-sectoral development programme developed, first working mainly through government agencies, later trying to involve more local-level initiatives and NGO activities.

During the early years the remote parts of districts were not yet reached very well (although one of the first *rapid rural appraisals* – then called *sondeo* – in Kenya had been organized in Alale; see Dietz and van Haastrecht, 1983). Beginning in 1986, ever more ASAL projects started in the Alale and Kasei areas as well. Until about 1993 the approach thrived. The ASAL programme had become the 'oil in the district machinery', mobilizing civil servants in a large variety of sectors to develop and implement projects, and increasingly incorporating the ideas 'from the ground', as expressed in locational development committees, divisional development committees and, ultimately, the District Development Committee. These committees tried to involve indigenous leaders (chiefs, councillors, women's group leaders, school teachers, etc.) and all the external players in a particular area, mostly foreign church leaders of a multitude of churches, which had come to the district after the 1979 to 1981 disasters. It was the era of the District Focus for Rural Development, the Kenyan form of decentralization. The relative importance of the ASAL programme in the district gave large powers to the Dutch programme advisers, who, in fact, operated as leaders of a pseudo-NGO. The programme money came directly from The Netherlands Embassy, and gradually the policy changes in The Hague, and through the embassy, caused tensions between a 'bottom-up strategy', based on continuous appraisal of the ideas of the development committees, and 'requests' from 'above' to integrate every whim of the Dutch development bureaucracy. On 'environment', it created major problems.

The political problems in the district itself also caused growing tensions. Increasing political ethnicity, which the Kenyan press and foreign donors/embassies perceived to be politically manipulated by the 'King of the Pokot', Member of Parliament Lotodo, resulted in ethnic clashes around 1993. As a result, approximately 30,000 non-Pokot people fled from the southern part of the district, and there was also a major out-migration of non-Pokot teachers and other civil servants from remote areas, such as Alale. Diminishing Kenyan government finance, an increasing (Dutch) ASAL funding and growing cynicism made it too difficult for many civil servants to resist the temptation to 'eat Europeans'. From 1993 until 1999 corruption became all too visible. Added to the changing opinions in Dutch development circles about the 'lack of impact of area-based development programmes', and a preference for large-scale, nation-wide sector programmes in selected government ministries, it resulted in a decision to stop the ASAL programme in West Pokot in 1999 (see Dietz and de Leeuw, 1999). The other Dutch-supported ASAL programmes, renamed Sustainable Animal and Range Development Programme, would continue for a few more years, but all ended in 2003 as a result of the Dutch decision to get rid of Kenya as a preferential country for Dutch development aid.

From 1982 until 1991 the University of Amsterdam was involved as 'backstopper' of the ASAL programme in West Pokot (and elsewhere). However, attempts to convince the ASAL programme leadership, and the Kenyan civil servants, of the need to develop a sophisticated longitudinal (output, effect and impact) monitoring and evaluation 'infrastructure' failed ('too academic', but actually too threatening), and the 'backstopping arrangement' came to an end. In the meantime, two of the research assistants of the 1980s had been integrated within the ASAL staff, one as the programme's secretary (Rachel Andiema), and one as the programme's community liaison officer (Albino Kotomei). They would be among the few 'locals' recruited to the staff, to the growing dismay of the local population, who saw the greed of the non-local civil servants and the lack of local accountability as the main reason for ASAL's unwanted withdrawal from the district. When the programme closed, there was a lot of anger.

Follow-up research

It was decided to do an *ex-post* impact evaluation study, and to do it as a university-driven exercise (a joint venture of the University of Amsterdam, using its own funds, and Moi University's School of Environmental Studies in Eldoret, a long-term research partner). It was also decided to do it as a team of three co-researchers, with a group of local research assistants attached to them. A variety of research activities were carried out – for example:

- Make an update of parts of the district atlas, covering the 1985 to 2003 period.
- Conduct an analysis of press articles about the district.
- Study school enrolment and healthcare data.
- Make education and labour histories of all children in certain age groups who have gone to school.
- Conduct questionnaire surveys in the same villages as during 1982 to 1986.
- Construct geographical family genealogies of selected family groups.

- Carry out a study of intervening agencies and their recent history in the district.

The most important research activity, however, was the organization of three participatory impact-evaluation workshops. The researchers facilitated a local-level assessment of 20 years of 'change', of interventions and of the impact of interventions. One of those workshops took place in the Alale/Kasei area, in a village called Kiwawa, in June 2002 (Andiema et al, 2003b). This used to be the missionary station of a controversial American church group (the Associated Christian Churches of Kenya), which had to leave the country after a scandal.

The participatory impact-evaluation workshop in Kiwawa

More than 60 local leaders gathered for three days in June 2002 to discuss their ideas about the recent history of the study area. Participants came from four different sub-areas (two relatively accessible areas, Alale and Kiwawa, both on the western lowland and foothill side of the region, and two areas that are very difficult to reach: the Lokitanyala-Kalapata-Akoret-Chemorongit area in the northern and northeastern mountains and the Kasei area in the south-eastern mountains). Participants were (elected) councillors, (appointed) chiefs and assistant chiefs, local church leaders, women's group leaders, and teachers (both men and women). It became a 'local' gathering, with hardly any civil servants present, and with Pokot as the major language of discussion. Out of these more than 70 people, 52 actively participated, and this included writing a short autobiography. It appeared that, educationally, 42 per cent of them attained primary school education; 21 per cent, secondary school; 27 per cent, secondary up to college level; while 12 percent never went to school. The majority of the participants (54 per cent) were employed in one way or another, and the rest were either unemployed or still in college. Among the participants, a few were unmarried; 92 per cent were married; 46 per cent were monogamist men and 15 per cent polygamist men; while 31 per cent were married women. The participants had an average of four children per household and an average of two brothers and two sisters.

The workshop programme consisted of eight major elements:

1   introduction and a round of personal life histories, focusing on the importance of the disasters of 1979 to 1981 and of later years for their personal lives;
2   writing personal life histories (ongoing during the workshop, with some assistance from one another);
3   reconstruction of history since 1979, focusing on 'problem years';
4   reconstruction of all development projects in four sub-areas;

5   discussion about poverty and about the changes in 'capabilities' between 1980 and 2002, differentiating between natural, physical, human, economic, cultural and socio-political capabilities, following the approach of Bebbington (1999), and doing it in discussion groups for the four sub-areas, separately for men and women;
6   assessment of the impact of projects and activities upon each of these six groups of 'capabilities', and upon their importance for poverty alleviation;
7   grading of all projects per sub-area, per sub-group of men and women, and selecting the ten 'best' and the ten 'worst' projects;
8   final discussion about the development prospects of the area and about the virtues and vices of donor dependence.

Reconstruction of the recent past

The participants of the Kiwawa workshop were able to recall the events (good and bad) that the community encountered between the years of 1979 and 2002 (see Table 11.1). Stories of raids and other aspects of insecurity dominated the accounts of many. Most of the recalled raids were when the Karimojong and Turkana seized large numbers of their livestock between 1979 and 1982 (a period now known as the 'dark age'). Life without livestock was no life and had no meaning for the pastoral Pokot, and so went the saying: 'A Pochon who has no livestock/cow is as good as a dead one'. Their life rotated around their livestock and therefore what threatened cattle (livestock) threatened the Pokot pastoralists as human beings and people with cultural characteristics that are unique to them.

As raiding is a traditional exercise of the pastoralists and has been there since time immemorial, it has become part and parcel of their lives. The pastoral Pokot participate in raiding their neighbours, who, in turn, raid the Pokot. In both cases these operations are carried out as actions planned by the elders and executed by their warrior sons. During the past these raids were carried out throughout the dry season because the herds were usually far from the villages at that time. During those dry periods there was, and still is, often severe competition over water and pasture. In the traditional 'scale of tribal values', the highest is the ability to increase one's herd through intelligence, force and even cunning. Therefore, whoever remains without livestock for a certain period gives a sign of having lost those skills and is put aside, ignored and sometimes even despised. The pastoralists feel that whoever endangers the safety or existence of livestock automatically becomes an enemy, to be neutralized or eliminated. However, during the period from 1979 to 1981, the community experienced raiding at a much larger scale, and with much more sophisticated weapons. They lost.

In addition, there were other calamities as well. Many human lives were lost because of the outbreak of diseases such as cholera, meningitis, dysentery and malaria. Workshop attendants also mentioned many cases of death because of premature births and caesarean operations for women. During the

Table 11.1 *A chronology of events, 1979–2002*

1979–1980	Insecurity/raids; rinderpest; drought/famine; army operations; cholera; Red Cross expands activities
1981	Same; Red Cross services; no dowry payments; gold mining (Korpu); Associated Christian Churches in Kenya (ACCK) and Reformed Church in America/Africa Inland Church (RCA/AIC) start activities
1982–1983	Raids; gold mining (many places); military coup; 'home guns' provided by government for self-defence; Pokot-Karimojong peace treaty
1984–1985	Raids (Turkana); major army operations; drought/famine; exodus to the South; Pokot-Karimojong peace
1986	Major army operations; famine; start Turkwel Dam construction; start Kasei dispensary
1988	Election problems; leaders rejected; political instability; famine (Anglican church intervened)
1989	Pokot massacre during Karimojong raid; big raid in Alale; army assisted the Pokot defenders
1990	Big raid on Masai by Karimojong; meningitis; bush clearing in Turkwel catchment; people chased away; first Pokot MP in Upe/Pokot County, Uganda
1991	Meningitis outbreak; to Turkana for relief food; big raid in Uganda
1992	Meningitis; big raid in Kiwawa; policemen died; people fled; multi-party elections; insecurity; famine; assistance from World Vision and Anglican Church of Kenya (ACK)
1993–1994	Army worms; Turkana raid; people moved to East Kasei; earth tremor; children drowned in new Turkwel Lake; registering of guns on Uganda site by Museveni
1995–1996	Ruby found in Alale
1997	President Moi visited Alale; 50 Pokot children killed in Uganda raid by Karimojong; successful counter-raid by Pokot; elections; harvest failure; torture
1998	Construction of new divisional headquarters in Alale, El Niño floods; landslides; water sources destroyed; roads damaged; children drowned
1999	Drought/famine; POKATUSA formed as an NGO
2000	Pokot leader and MP Lotodo died; famine; peace activities by POKATUSA and justice and peace groups
2001	Elections in Uganda; cost-sharing started in dispensaries; relief food; late rains and then floods; new Kanyangareng bridge threatened; insecurity problem in Turkwel area

Source: chapter author

above-mentioned years, there were very few health facilities and a shortage of drugs and health personnel. The traditional herbalists were not able to treat some of these diseases because they were new to them (e.g. cholera and meningitis). The community also lost most of its non-raided livestock from various diseases during the years under review because of insufficient veterinary services in the area. The worst diseases were rinderpest and East Coast fever. Prolonged droughts were also mentioned among the most disturbing problems, as there had been no harvest at all for several years; this caused livestock loss as well. Due to the severity and length of the droughts, even fruits and roots were no longer available. This forced the community to look for other ways of survival. Because of these problems, some community members decided to migrate to other places, especially to the south (the highlands of southern West Pokot and the large farm area of Trans Nzoia). This was not an easy decision to make, but there was no choice. Their problems were not solved, however,

Table 11.2 *Perceived positive and negative changes in six capability domains*

Capability domain	Perceived positive change	Perceived negative change
Natural	Permanent settlement is found in more fertile areas where more land is used for agriculture; through the use of fertilizer and manure, the land has improved. There is also enough pasture, improved afforestation and sufficient water supply (boreholes and gravity). Land is still communally owned. Because of the improved availability of drugs for livestock, their numbers have increased.	Water catchments have dried due to deforestation in some areas; soil erosion has increased because of population pressure; soil infertility is a result of overgrazing; documented loss of lives and displacement of people (e.g. at the man-made Lake Turkwel Gorge and in mining areas). Spread of diseases has increased in mining areas because of the interaction with outsiders. The topography of the land was destroyed due to mining. The climate has rapidly changed due to prolonged drought. Wildlife is increasingly vulnerable to poaching; scarcity of wild fruits due to persistent drought and no more shifting cultivation.
Physical	Improved infrastructure. More roads have been constructed. Communication devices have been introduced, alongside improved road networks, houses, farming technology and cattle dips. More guns were bought between 1981 and 2001 for defence purposes.	The roads are poor and are frequently a cause of accidents. There is no electricity produced from Turkwel Gorge. Shortages of drugs exist in Government of Kenya dispensaries; building materials and management of boreholes are expensive. People in Upe were shocked that they were forced to surrender guns to the Ugandan government.
Economic/ financial	Many more businesses. Some income through miraa and mining of gold and rubies, and this has initiated interaction with other communities from Kenya and even beyond. Increased possibility of transacting business because of employment of teachers, nurses, chiefs, etc. More organizations and donors have come to assist the people. Money is an accepted medium of exchange by everybody and it is durable: people feel superior when they have it, it improves one's living standard and as such one becomes a role model to the community. The availability and exchange of commodities have improved the development of the area.	Low employment and lack of job opportunities; poor production of both livestock and crops and inflation of commodity prices; no financial bodies to give sufficient loans to local businessmen/women; unlike stolen cattle, money is not traceable and can easily be stolen, creating poverty and envy. Civil servants who are employed far from home can easily divorce; there is increasing spread of diseases and use of drugs by youth. Loans without proper planning leads to stress.
Human	The population has increased, alongside improved health facilities, more schools and higher school enrolment. Pokot are courageous by nature and have improved their skills to defend themselves against attacks from their neighbours. Population increase is a result of reduced mortality rates.	There are new diseases (e.g. HIV/AIDS and cancer).

Table 11.2 *continued*

Capability domain	Perceived positive change	Perceived negative change
Social/ political	'Since independence the government and their elected leaders have done very little to help them as a community.' This community hopes that there might be positive changes in this multi-party era. More Pokot became national leaders. More local people in local leadership positions, as well as more organizations (such as women's groups, youth groups).	Little has been done by the elected leaders and the government. The community feels that they have been neglected for many years by their elected leaders as a result of greed and corruption. The government has also imposed leaders upon the people. Elected leaders frequently live far away from the people. Nepotism and tribalism occur.
Cultural	Increased adherence to Christianity and Islam; many more churches; increased conservative dress codes; improved language abilities; better food diets; fewer 'evil practices'; increased Pokot pride.	Threatened ethnicity and erosion of cultural traditions occur, although traditional religion has kept people together. Cultural styles of dress have been lost; new 'modern' clothes are expensive. Vernacular language skills have disappeared; there is a lack of differentiation between married and unmarried people. There is a heightened sense of immorality and an increase in crime; dowry payments rarely occur due to diminishing numbers of livestock among the poor.

Source: chapter author

because the migrants faced many adversities – finding decent housing, discrimination and exploitation, as many were casual labourers. After numerous experiences perceived as 'dehumanizing', many of them went back to their original homes, with a grudge.

Perception of change

Table 11.2 highlights the perceived positive and negative changes in the area's living conditions during the last 20 years. We have organized it according to the six capability domains discussed above, although it is obvious that some changes in one domain also cause changes in another, and it is possible (and sometimes perceived as such) that some positively evaluated changes in one domain do impact negatively upon another.

The perception about 'the government'

During the participatory evaluation workshop people discussed the roles of various external agencies in contributing to change. It became very clear that many had a 'grudge' against 'the government'. Due to the continuous raids between the Pokot and their neighbouring communities, the government decided to disarm them a few times and this was not an easy task. The Pokot resisted and the government decided to use power. In the process, many lives, both human and livestock, were lost. Everyone remembers the military operations of 1984 and 1986, and for many years 'the government' was equated with the army, who killed their people and animals. The government was also negatively connected with the way in which a large-scale hydro-electricity project (Turkwel Gorge, under the pseudo-government Kerio Valley Development Authority) was taking shape without any compensation for the Pokot, whose land had been expropriated. The government was active in the area through the provincial administration (each 'division' was headed by a district officer; from 1970 to 1985 the Alale area was under the Kacheliba Division; in 1985 a new Alale division was formed, and in 1996 this was split into the Alale and Kasei divisions). The district officer was responsible for the (appointed) chiefs of locations and sub- or assistant chiefs for sub-locations; but (mostly coming from among the local people) these chiefs were often caught between two fires. The local people also elected local councillors for the West Pokot District Council; but their powers (and money) were very limited. The council was responsible, though, for granting trade licences (although most of the trade in the area went beyond these licences to trade in livestock, gold, rubies, miraa and arms). Somali traders played an important role, but Pokot traders gradually increased their importance.

The district officer was supposed to coordinate the various representatives of line ministries in the area; but the Kenyan government did not provide those civil servants with needed project money or facilities, and non-donor money and purchasing power of their salaries dwindled to very little over the course of time. However, teachers (both trained and untrained) were increasing rapidly in numbers, and their salaries were paid by the government's Ministry of Education. The district officer was also responsible for coordinating famine relief operations in the area and for supporting and coordinating '*harambee*' fundraising activities for 'development projects'.

The perception about non-governmental agencies

People were much more positive about the many non-governmental agencies that had come to their area. They easily mention the churches, with their acronyms: the Catholic Mission, the Reformed Church in America/Africa Inland Church (RCA/AIC); the Anglican Church of Kenya (ACK); the Pentecostal Full Gospel Church of Kenya (FGCK); the Evangelical Lutheran Church in Kenya (ELCK); the Associated Christian Churches in Kenya (ACCK); the National Council of Churches of Kenya (NCCK); the Baptist Kenya Assemblies of God (KAG); and other Baptist missions. Connected to the Christian donors were NGOs such as the Christian Children's Fund (CCF) and World Vision. Non-Christian foreign-donor agencies also became visible entities in the area: the Red Cross, the World Food Programme, UNICEF, The Netherlands Development Organization SNV, the Dutch-funded Netherlands Harambee Foundation for Health, and another foundation for water. All NGOs active in the area were involved in 'development

Table 11.3 *Development projects by NGOs, including churches*

Sector	Projects
Agriculture	Tractors for ploughing; provision of seeds and pesticides; horticulture in field demonstration plots; provision of farm tools.
Livestock/veterinary	Provision of drugs; training of para-vets; mobile treatment services; disease identification and vaccination; supply of hand-spray pumps; mobilization of peace-keeping; introduction of community-based animal health workers who later sold animal drugs to the community.
Forestry	Provision of tree seedlings to farmers; planting of trees in water catchment areas and near schools and homes; introduction of tree nurseries and conservation of natural resources (forests).
Education	Construction of primary and pre-schools; provision of boarding facilities; sponsoring poor children from primary to higher education levels; employment of Parent–Teacher Association (PTA) teachers; provision of food, clothing and books; training and employment of pre-school teachers; payment of subordinate staff.
Health	Construction and renovation of dispensaries; primary healthcare; provision of drugs; employment of nurses; mobile clinics (flying clinics); sponsoring nurses in training colleges.
Water	Drilling and renovation of boreholes; construction of sub-surface dams and ponds; piped gravity water; purchase of solar panels.
Religion	Building churches; employing evangelists and patrons in schools.
Famine relief	Supplying food to people during famine; coming together to work on a communal project (e.g. mudding a classroom, doing some road work or putting up a church) when there is food for work.
Social services	Assisting women's groups in income-generating activities; registration certificates for women and youth groups; fundraising for women and youth groups; establishment of youth workshops and hardware; employment of social workers.
Public works/roads	Churches and NGOs also play a role in maintaining some roads and constructing air strips.
Energy	Provision of solar panels in schools and health facilities.

Source: chapter author

projects' as well (see Table 11.3), some in only one or two sectors (often education), others playing a role as 'pseudo government' in particular areas, with projects in many sectors (even as far as 'peace-keeping' and, in the case of the Anglican Church in Kenya before they were ousted, arms maintenance).

Development activities of the Arid and Semi-Arid Lands programme

The local people also regarded the Dutch-funded Arid and Semi-Arid Lands (ASAL) programme as an NGO, although most of its work was carried out as part of the district government apparatus, the so-called 'line ministries' (agriculture, livestock/veterinary, forestry, education, social services, etc.). With a bit of exaggeration we may say that the perception of the Pokot was that anything 'bad' was connected to the government, and anything 'good' to NGOs, so even projects that were regarded as 'good', but came from the

Table 11.4 *Arid and Semi-Arid Lands (ASAL) projects*

Sector	Projects
Agriculture	Staff houses; demonstration plots; supply of seeds; introduction of animal traction; tours; seminars/workshops for farmers.
Livestock/veterinary	Provision of drugs; purchase of solar panels and fridges; vaccination and branding; construction of crushes and dips; growing Napier grass.
Forestry	Planting of trees in various areas (e.g. schools); provisions of tree seedlings to the community; installation of water tanks for tree nurseries; tree demonstration plots; provision of water cans; community training on forest conservation and the environment; provision of soil-conservation tools.
Education	Construction of classrooms, dormitories, water tanks, toilets, kitchens; provision of desks and text books; in-service, workshops and seminars for primary school teachers; provision of material for mother tongue booklets, school atlases, Pokot/English dictionaries; sponsorship for needy secondary and college female students and for both male and female university students; training of Parent–Teacher Associations (PTAs).
Water	Sub-surface dams; rehabilitated boreholes; drilling boreholes; training of water committees and borehole attendants; water committee tours to other districts.
Social services	Construction of roof catchments, water jars and rental houses for women's groups; supporting income-generating activities for women; training women's groups on management; support for youth groups (e.g. buying tools).
Public works/roads	Construction of the Kanyangareng Bridge; repair and maintenance of the road between Konyao and Alale.
Energy	Introduction of energy-saving cooking stoves (jikos) through women's groups.

Source: chapter author

government, could not be seen as 'government' and were perceived as being related to foreign donors and their 'NGO-like' approach. The ASAL programme was a typical 'area development programme', with lots of small-scale projects in a variety of fields (see Table 11.4). Due to the donor's mandate (activities in the Ministry of Health were already supported through another Dutch development programme), health projects were excluded.

Assessing status and impact upon capability domains

Four geographical sub-groups made an inventory of all 'development projects' in their area since 1979. They classified all relevant sectors, the period during which the project lasted, the 'sponsor' (government, ASAL, churches, other NGOs), a first assessment of the project's status, and the type of 'capability' they thought the project would enhance (see Tables 11.5 and 11.6). In total, these four groups listed 294 different projects. Men and women conducted a separate assessment (hence, a minimum of 584 project scores). Here we present a summary of the assessments about the status and capability domains of the collective projects. Projects could receive more than one score (in total: 839 scores on status and 1265 on capability). More than one score on 'status' meant that members in a group had different opinions and could not agree. More than one score on 'capabilities' meant that a project was perceived to have an impact upon more than one capability.

Table 11.5 *Status assessment of development projects in northwest Pokot, according to type of donor*

Donor	Number of projects	No. of Project scores (total)	No. of on-going Project scores (category 3 only)	No. of Finished projects	No. of finished Projects (percentage per status category)*			
					1	2	4	5
Government	72	203	76	127	47	19	17	17
ASAL programme	43	121	15	106	42	10	40	8
Churches	123	339	161	178	48	8	35	9
NGOs	56	176	58	118	47	27	22	3
Total	294	839	310	529	47	15	29	9

Notes: * Status: 1 = project never really started or was negligible; 2 = project existed, but had no lasting impact, 'nothing to be seen on the ground' and was unsustainable; 3 = project is still ongoing; no impact yet; 4 = project was finished and had an impact that is perceived to be positive; 5 = project was finished and had an impact that is perceived to be negative.
Source: chapter author

Many projects were still ongoing in the area (310 scores out of 839), and the workshop members decided that they could not give an impact assessment of these projects at this time. Of the finished projects, many were regarded as so small that their impact was seen as negligible (47 per cent), and there were also a number of past projects that were 'a waste of time and effort', as nothing substantial remained (15 per cent). For an impact assessment exercise, those projects that were ready and that were perceived to have had an impact are most interesting: 29 per cent of the status assessment scores were positive and 9 per cent negative. It is interesting to note that men judged differently than women, and in the 'most developed' areas (Alale and Kiwawa), there were major differences of opinion: in Alale, particularly among the men, and in Kiwawa, both among the men and among the women. It is remarkable that the men in Alale and in Kiwawa had outspoken negative opinions about a considerable number of projects, while the women in those areas did not give any negative impact score at all.

We differentiated among four types of donors. In terms of numbers of projects, the churches were most active (123 projects, with 339 project scores), followed by the government (72, with 203 scores), non-church NGOs (56, with 176 scores) and finally by the ASAL programme (43 projects, with 121 scores). If we look at the status assessment data among the four types of donors, there are interesting differences. Projects that had been organized by the government (including the government administration, the county council, the Kenya African National Union, the Kerio Valley Development Authority and the Rural Development Fund) had a higher than average score on projects that were not sustainable, a much lower than average score on positive impact, and a remarkably high score on negative impact. Projects that were a result of the ASAL programme had a remarkably high score on positive impact, and much lower than average scores on negative and unsustainable impact. The same was true for projects organized by churches. Finally, non-church NGOs

Table 11.6 *Capability assessment of development projects in northwest Pokot*

Donor	Capability scores number	Natural (%)	Physical (%)	Economic and financial (%)	Human (%)	Cultural (%)	Social and political (%)
Government	281	7	16	23	28	9	16
ASAL programme	217	15	31	16	21	6	11
Churches	515	7	21	17	30	10	14
NGOs	252	5	23	20	26	8	18
Total	1265	9	23	19	26	8	15

Source: chapter author

had a remarkably high score on unsustainable projects, but a remarkably low score on negative impact.

The workshop members regarded the impact of all projects combined upon 'human capability' (skills, knowledge level, health) most pronounced. But their impact upon physical, economic/financial and social/political capability was also perceived as considerable. Less impact was noted upon natural and cultural capabilities. In all groups, women were much more inclusive than men: many projects were regarded as having an impact upon more than one capability. Women showed a much more 'holistic' approach in discussing the impact of projects. If we compare the impact assessment scores for the four different types of donors there is a striking overall resemblance in which all four types of donor agencies, including the churches, were, in fact, active in all domains and had a perceived impact upon all capabilities. However, there are a few interesting differences. The government had a higher than average score on economic capabilities and a lower than average score on physical capabilities. The ASAL programme had a higher than average impact upon natural capabilities ('the environment') and physical capabilities, and a lower than average impact upon the other four capabilities. The churches had a slightly higher than average impact upon human and cultural capabilities, and a slightly lower than average impact upon economic capabilities. Finally, the non-church NGOs had a lower than average impact upon natural capabilities and a higher than average impact upon social and political capabilities.

Assessment of the most positive and negative impacts

Finally, in each of the area groups, the workshop members, mostly men and women separately, were asked to choose ten projects that they regarded as the best ones (with most positive impact) for their area and ten projects that they regarded as the worst (with most negative impact, or the largest difference between expectations and outcome).

There are major differences among the groups and also between men and women from the same area. In some cases, projects that were regarded as a very positive contribution to capability development, by the women, were

Box 11.1 Overview of development projects with the most positively perceived impact

N = northern area; A = Alale; K = Kasei; W = Kiwawa; m = men; w = women.

Provision of tree seedlings and water cans (Km); training farmers to make terraces (Kw); provision of veterinary drugs (Kw); livestock vaccination (Wm, Ww); construction of roads (Km); drilling of boreholes (N, Am, Aw, Wm, Ww); construction of piped gravity water (Aw); construction of sub-surface dams (Kw); construction of primary and pre-primary schools (Am, Aw, Km, Ww); construction of dispensary (Am, Aw, Km, Kw, Wm, Ww); new road building (Ww); building of churches (Wm); vaccination of children (Am, Aw, Km); medical treatment (Am); provision of medicines (N, Aw, Wm, Ww); cost-sharing of drugs (Wm); mobile clinics (N, Am, Aw, Km); 'flying' mobile clinics ('Helimission mobile') (N, Am, Aw); building of schools (N, Wm); provision of teachers (Kw); feeding and paying nursery school teachers (Kw); lessons about dress making (N); extension about growing crops (N); sponsoring students (Wm, Ww); registration of women and youth groups (Wm); evangelization (N, Kw); relief food (Km, Ww); school feeding programme (Wm); providing security (Kw); peace-keeping mobilization (Wm).

Box 11.2 Overview of development projects with the most negatively perceived impact

N = northern area; A = Alale; K = Kasei; W = Kiwawa; m = men; w = women.

Provision of forestry personnel (N); tree planting (Am); provision of seedlings and water cans (Kw); provision of tree seeds (Wm, Ww); training to make terraces (Km); provision of soil conservation tools (Wm); soil conservation (Ww); provision of seeds (Am, Km, Kw, Ww); provision of fertilizers and pesticides (Ww); extension to grow Napier grass (Wm, Ww); provision of livestock pasture and hay (Wm); provision of crop seeds and new varieties (Wm); building agricultural extension office (Km); supplying oxen ploughs (Km); animal vaccination (Am); training on forest conservation (Km); training on the 'timing of rains' (Wm); provision of engines for grinding of maize (N); provision of energy-saving *jikos* (Wm); maintenance/gravelling of the main road (N, Am, Aw, Wm, Ww); construction of Turkwel Gorge Dam (Km, Kw); renovation of boreholes (Kw); construction of water dams (Ww); provision of school building materials (Aw, Kw); provision of school desks (Kw); construction of a dispensary (N); improving the buying and selling of livestock and goods (Km); provision of loans (N, Am, Aw); sponsoring of nursing students (N); provision of school milk (N, Aw); provision of school books (N, Ww); sponsoring poor children's education (N); training of pre-school teachers (N); providing extra-curricular activities at schools (Ww); women's awareness training (Aw); harambee for women's groups (Aw, Ww); harambee for youth groups (Aw); employment of nursery school teachers (Am); employment of party youth wingers (Am, Aw); peace initiative (Am); relief food supply (Am, Aw); school feeding programme (Kw); enforcement of law and order (Aw, Km, Kw); registration of party membership (Wm).

regarded as a very bad contribution to capability development and as having a major negative impact in other areas, by the men.

We compared the 'best' and 'worst' project scores for the four different types of donors (see Table 11.7).

Table 11.7 *'Best' and 'worst' projects for four types of project donors; separate assessments by men and women (all research areas combined)*

Donor	Men		Women	
	Best	Worst	Best	Worst
Government	7	21	6	24
ASAL programme	2	6	2	2
Churches	24	6	21	5
NGOs	5	4	4	8
Total	38	37	33	39

Source: chapter author

Conclusions

Impact assessments

- Both the men and women regarded the churches as the best 'development agency', and 'the government' as the worst.
- 'Impact assessment' does not depend only upon reaching the targeted result of a project, but upon the way in which a project was started and implemented.
- Projects that raised major expectations and could only fulfil a minor part of those were often evaluated negatively, even if they accomplished something.
- Projects that did not treat the local population with respect were also valued negatively.
- Projects that created (or increased) tensions in the local community were often seen as very negative, especially if 'outsiders' created these tensions and were no longer there to assist in restoring peace;
- The activities of some of the missionaries, who had stayed in the area for a long time and who had shared the area's problems, were generally evaluated very positively, even by those who did not belong to the particular church affiliation of the missionary.

The long-term commitment to providing water, healthcare, veterinary care and education was valued most positively; hence, the overall positive judgement of church-based NGOs, who provided these services in a bottom-up way and with a long-term commitment.

Development agencies were particularly valued positively if they were flexible enough to change timing, spacing and content to the major fluctuations in the area's environment, and if they provided some form of counselling to discuss the challenges, which the population faced (including harsh government/army behaviour).

Where ASAL and some other government projects had the same 'style' of flexibility and counselling, they were also valued positively. Where projects were perceived as 'hit-and-run', top-down implemented hobbies of some external donors, the overall assessment was often very negative.

Mitigating vulnerability

In an area such as northern West Pokot, mitigating vulnerability means preparing for drought- and epidemic-related crises, preventing war and violence, and assisting the people in defending themselves. In the first domain, various interventions were regarded as useful. The most important one was the provision of a sustainable water infrastructure, preferably one which did not have high maintenance costs, and which did not make people dependent upon an untrustworthy public water agency. Down-to-earth provision of veterinary care and accessible and dependable healthcare were important as well, as were support for drought-resistant crops and animals and fast recovery support after a crisis. The provision of education was favoured, as it supplied a long-term escape route which could also function as a means to geographically and sectorally widen the support structure. This was true both for remittance support to livelihoods and for political support, as educated people could become advocates for the plight of their 'home area', not only in government circles but also in NGO and church circles, in human rights agencies, and in their communications with potential donors.

Gradually, the focus among donors and among the local population shifted to the second domain during the period under review: providing basic security against violence, including that of government agencies. Human rights groups and churches provided important support to form a potential *cordon sanitaire* against outbreaks of violence (e.g. peace-building conferences; confronting army and police atrocities); but the people's own defence forces were also important in cases where the 'state monopoly of violence' did not work (when armies did not provide security), or worked counterproductively (where army and police agencies were part of a predatory and rent-seeking force). In a situation where state violence did create havoc once in a while, other state agencies, 'bringing development', were often treated with caution or downright disrespect. Non-governmental agencies that had become rooted in local institutional life and had shown long-term commitment were seen as much more useful. In our research area, it was mainly churches that were regarded as allies. But churches, and the many local and international non-governmental agencies that became 'swarms of support', particularly in the aftermath of crises, were often treated with caution, as it was never clear from the outset if they could be trusted or if they formed part of a rent-seeking and distorting external threat to long-term survival. The people's perception was not a simple matter of 'bad government' versus 'good NGOs'. It was a matter of building and maintaining mutual trust and providing long-term commitment. These ingredients can be available in both government and non-government agencies, but they often were not.

Bridging science and practice in vulnerability research

In our long-term research project, we started close to a government agency, as was usual during the 1970s and early 1980s, among many 'development researchers'. We received research money from a donor agency, allied ourselves with a local development programme and local research institutions, and soon found out that a lot of the 'development initiatives' were rather donor-driven – part of perceived wisdom in the government-donor nexus, with its rapidly changing donor speak and ever-changing prevailing approaches. Despite the fact that the ASAL programme for which we conducted research started with a 'process approach', and that we soon initiated various participatory research approaches, the initial orientation was very much on strengthening the government machinery. Only gradually did the approach shift to include more NGOs and to take local initiatives more seriously. And only later did the importance of combining development initiatives with security and peace initiatives become more evident, as well as the importance of finding and working with local peace and development brokers. During the early years of our research work we spent considerable time and energy seeking to understand the 'institutional logic' of the intervening agencies, and particularly of the government machinery on the ground. We should have spent more energy on understanding the 'institutional logic' of people's behaviour and its roots in culture. Bridging the evident gap between researchers/practitioners focusing on culture (and often standing with their back to the development industry) and researchers/practitioners focusing on development (and often standing with their back to culture) is one of the major challenges in vulnerability research and practice.

Acknowledgements

With thanks to the participants of the Kiwawa workshop in June 2002, in West Pokot, Kenya; to the participants of the livelihood sessions of the Ceres Summer School, June 2003, at the Royal Tropical Institute, Amsterdam; to participants at the International Geographical Union (IGU) conferences in Durban 2002 and Glasgow 2004; and to research partners Rachel Andiema and Albino Kotomei. An earlier version of this chapter appeared in the *Proceedings of the 2003 Ceres Summer School* (see Andiema et al, 2003a). Comparable publications appeared in a Canadian publication (Andiema et al, 2008), and in *The Netherlands Yearbook on International Collaboration* (Dietz and Zaanen, 2009). Experiments with this method of participatory assessment of development continued in Ghana and Burkina Faso in a project together with Tamale University of Development Studies and Expertise pour le Développement du Sahel, from 2007 onwards (for first results, see Dietz et al, 2009).

References

Andiema, R., Dietz, T. and Kotomei, A. (2003a) 'Participatory evaluation of development interventions for poverty alleviation among (former) pastoralists in West Pokot, Kenya', in Ceres (ed) *Faces of Poverty: Capabilities, Mobilization and Institutional Transformation, Proceedings of the International Ceres Summer School 2003*, Ceres/UvA/VU/KIT, Amsterdam, pp193–209

Andiema, R., Dietz, T. and Kotomei, A. (2003b) *Workshop Proceedings of the Participatory Impact Evaluation of Development Interventions on People's Capabilities in Alale and Kasei Divisions, West Pokot District, Kenya*, West Pokot Research Team, Kapenguria, Amsterdam/Kapenguria

Andiema, R., Dietz T. and Kotomei A. (2008) 'Participatory evaluation of development interventions for poverty alleviation among (former) pastoralists in West Pokot, Kenya', in C. R. Bryant, E. Makhanya and T. M. Herrmann (eds) *The Sustainability of Rural Systems in Developing Countries*, Laboratoire de Développement Durable et Dynamique Territoriale, Géographie, Université de Montreal, Montreal, pp117–131

Barber, J. P. (1968) *Imperial Frontier: A Study of Relations between the British and the Pastoral Tribes of North East Uganda*, East African Publishing House, Nairobi

Bebbington, A. (1999) 'Capitals and capabilities: A framework for analysing peasant viability, rural livelihoods and poverty', *World Development*, vol 27, no 12, pp2021–2024

Dietz, T. (1987) *Pastoralists in Dire Straits: Survival Strategies and External Interventions in a Semi-Arid Region at the Kenya/Uganda Border*, Western Pokot, 1900–1986, PhD thesis, Netherlands Geographical Studies, Amsterdam

Dietz, T. and de Leeuw, W. (1999) 'The Arid and Semi-arid Lands Programme in Kenya', in J. Sterkenburg and A. van der Wiel (eds) *Experiences with Netherlands Aid in Africa*, Ministry of Foreign Affairs, The Hague, pp37–57

Dietz, T. and van Haastrecht, A. (1983) *Rapid Rural Appraisal in Kenya's Wild West: Economic Change and Market Integration in Alale Location, West Pokot District*, Working Paper 396, Institute for Development Studies, University of Nairobi, Nairobi

Dietz, T. and Zaanen, S. (2009) 'Assessing interventions and change among presumed beneficiaries of "development": A toppled perspective on impact evaluation', in P. Hoebink (ed) *The Netherlands Yearbook on International Collaboration*, Van Gorcum, Assen, pp145–164

Dietz, T., Obeng F., Obure J. and Zaal F. (2009) 'Subjective truths: Participatory development assessment', *The Broker*, issue 15, August 2009, pp19–21, 31; longer version at www.thebrokeronline.eu

Hendrix, H., Mwangi, M. S. and de Vos, N. (1985) *District Atlas West Pokot*, Ministry of Planning and National Development, ASAL Programme, Kapenguria, Kenya

Republic of Kenya (1981) *Kenya Population Census, 1979, Volume 1*, Central Bureau of Statistics, Ministry of Economic Planning and Development, Nairobi

12

Science and Indigenous Knowledge in Resource Management in the Canadian Arctic

Johanna Wandel, Barry Smit, Tristan Pearce and James Ford

Introduction

Like many parts of the world, the Arctic is sensitive to changing environmental conditions and changing political economies and cultures. Resource management in Canada's north occurs in an environment of multiple stresses and contexts, including land claims, aboriginal self-government cultural change, mineral exploration, ecological changes and, increasingly, climate change. Significant climate-related changes have already been recorded in the Arctic during recent years, including warming, decrease in summer sea ice, increased frequency and magnitude of storms, sea-level rise, accelerated coastal erosion, permafrost thaw, and changes in the health and distribution of arctic flora and fauna (Krupnik and Jolly, 2002; Einarsson et al, 2004; ACIA, 2004). Climate scientists predict that these changes will continue and it is projected that the effects of future climate change will be experienced earlier and acutely at high latitudes (ACIA, 2004; Arzel et al, 2006). Climate change has implications for biological systems and for the natural resources upon which most Arctic communities depend, and residents have expressed growing concern about these trends (Nunavut Tunngavik Incorporated, 2001; Ayles et al, 2002; Simon, 2004). Communities are dealing with these environmental changes at the same time as they are grappling with the growing influence of the global economy, the encroachment of southern culture, threats to traditional identity, and evolving resource management regimes. These broader forces are affecting how communities interact with environmental conditions and their capacity to manage change (Duerden, 2004; Kofinas, 2004; Ford et al, 2006a, 2006b).

There are many kinds of policy and decision-making that relate to the ability of Arctic communities to maintain livelihoods in the face of these multiple forces of change. Decisions relating to natural resource management, economic development, community and social programming, training and development initiatives and school curricula, for example, have the potential to reduce or exacerbate vulnerability to changing conditions. These involve formal and informal institutions from the local level through regional and national to international levels, and involve to varying degrees the participation and input of local peoples. Additionally, in the Canadian Arctic, land claims agreements, particularly with the Inuit population, have created new fora within which indigenous peoples become active partners in decision-making in the north. Increasingly, it is recognized that decisions that affect the management of northern resources and, consequently, the livelihoods of indigenous peoples are more likely to succeed if they incorporate both scientific and indigenous knowledge (Huntington, 2000; Moller et al, 2004).

The land claims of Canadian Inuit have been settled during the last 30 years, resulting in the creation of four Inuit regions: Nunavik in 1975, Inuvialuit Settlement Region in 1984, Nunavut in 1999 and Nunatsiavut in 2006. Each land claim agreement has provided various degrees of self-government. Local institutions (including hunters and trappers committees/ organizations and local councils), co-management institutions and, in Nunavut, a new territorial government have considerable authority for resource management, including wildlife and resource extraction, and other jurisdictions such as social, cultural and economic development. The federal government, however, retains joint responsibility in many areas, particularly fish and wildlife management and natural resource development.

The type of information employed in these decision-making processes varies greatly. Most decisions relating to federal resource management in Canada employ various types of Western science, for example, to assess changing climatic conditions, location and stocks of species, and stability of coastal areas or building sites. In addition, in the context of the Canadian north, indigenous knowledge is often employed in assessments to complement scientific perspectives and to reduce distrust of resource management mechanisms (Moller et al, 2004). Indigenous knowledge is certainly important for the design and acceptance of practices aimed at increasing sustainability of both human communities and the biophysical environments upon which they depend, and indigenous knowledge is explicitly recognized in the joint management agreements, James Bay and Northern Quebec Agreement, Inuvialuit Final Agreement, Nunavut Land Claim Agreement and Labrador Final Agreement.

In this chapter we illustrate several of the ways in which science and indigenous knowledge have contributed to resource management in the Canadian Arctic, particularly as it relates to climate and environmental changes in the context of socio-political-economic transformations of Inuit communities. First we review the concept of indigenous knowledge and outline its place in the institutional arrangements that have been developed to deal with resource management issues in the Canadian Arctic. We then provide insights from research that address the complementarity of indigenous knowledge and Western science in resource management in the Canadian Arctic.

Indigenous knowledge and science in research and practice

This analysis uses the term 'indigenous knowledge' to capture the knowledge, experiences, practices and beliefs of indigenous communities as they apply to resource management and adaptation to climate stress. Considerations of indigenous knowledge in science and policy are commonly captured in discussions on traditional ecological knowledge (e.g. Ferguson and Messier, 1997; Pierotti and Wildcat, 2000; Usher, 2000; Watson et al, 2003; Moller et al, 2004). Traditional ecological knowledge is frequently understood as a knowledge/practice/belief complex which is focused on the relationship of living beings with one another and with the environment (Berkes et al, 2000). Indigenous knowledge includes traditional ecological knowledge, other (non-environment-related) traditional knowledge, and non-traditional local knowledge held by indigenous peoples in the north (Stevenson, 1996). In Nunavut, the term 'Inuit knowledge' (*Inuit Qaujimajatuqangit*) is used to capture an epistemology which includes knowledge of the environment and its interrelationships, as well as social, kinship and ethical obligations (Arnakak, 2001). The term 'indigenous knowledge', which is broader than traditional ecological knowledge and not specific to a particular region of the Arctic (Wenzel, 1999), is used in this discussion.

Academic researchers have included indigenous knowledge in analysis for decades; but the integration of indigenous and scientific knowledge in research is relatively new. Research that included indigenous knowledge was often conducted with indigenous knowledge as the research subject through the lens of anthropology and ethnography and focused on indigenous/environment interactions (Wenzel, 1999; Cruikshank, 2001). Research about indigenous knowledge is not the same as its inclusion in scientific analysis on northern environments. Traditionally, the bulk of research on northern environments has dealt with natural science, notably glaciology, oceanography, climatology and biology, only occasionally addressing human dimensions, and rarely including indigenous knowledge as a substantive part of the analysis. More recently, however, research that is focused on humans and their environments and uses indigenous knowledge as both a source of information and a subject of investigation has begun to emerge to document changes in climate

(Riedlinger and Berkes, 2001), caribou (Ferguson et al, 1998; Kendrick et al, 2005), gulls (Chardine et al, 2004), hunting (Condon et al, 1995), social networks (Collings et al, 1998), and to characterize community vulnerability (Ford et al, 2006a, 2006b).

Although indigenous knowledge was first incorporated within institutional processes, with its inclusion in guidelines on environmental impact assessment only during the 1980s, it has been recognized as an input into wildlife monitoring programmes for many decades. The crisis over the Beverly and Qamanirjuaq herds of barren-ground caribou during the late 1970s illustrated the need for the inclusion of both indigenous and scientific knowledge in both monitoring and management. The two herds had reached perilously low numbers. However, indigenous groups who rely on barren-ground caribou as a key food source refused to accept Western scientific data, and the slaughter of thousands of animals in northern Saskatchewan during the winter of 1979 to 1980 was blamed on indigenous overharvesting (BQCMB, 2006). It became clear that sustainable resource management would have to respect both scientific and indigenous perspectives, supporting the argument that harvest prohibitions are doomed to failure without local support (Moller et al, 2004). The Beverly and Qamanirjuaq Caribou Management Board (BQCMB) was created in 1982 to monitor and manage the herds across provincial/territorial boundaries and indigenous groups (the herds travel in traditional Inuit, Métis and Cree lands), with the aim of full participation of communities and governments and an explicit emphasis on both scientific and indigenous knowledge. The board is credited with successful management while respecting traditional use of the animals. Unlike conventional policies focused solely on harvest prohibitions, the current board's goals include not only the management of caribou habitat, but encouraging the wise use of caribou (BQCMB, 2006).

The role of indigenous knowledge has evolved from an information input to (in theory) full and equal consideration in resource management during recent land claims agreements (Usher, 2000). Effective strategies for adapting to change require accurate assessment of human resource use (a key component of vulnerability), and relevance to communities to ensure acceptance of, and compliance with, institutional strategies. Policy has more relevance if it appropriately considers the cultural as well as scientific context within which it operates (Huntington, 2000; Moller et al, 2004). The meaningful integration of both indigenous and Western scientific knowledge can contribute to accuracy, relevance and appropriateness of policy initiatives in the Canadian Arctic.

Indigenous knowledge provides insights into the current state of affairs in the north (via species abundance estimations and baseline monitoring activities), and also a holistic understanding of the dynamic relationships in Arctic environments. Furthermore, the process of including indigenous peoples and their knowledge in research and policy development establishes a common understanding of the context and mutual respect for expertise among southern scientists, northern knowledge holders and resource management practitioners (Huntington, 2000; Moller et al, 2004). Effective resource management and adaptation to the challenges posed by environmental and social change requires cross-scale institutional cooperation from the community to the national and international level.

Indigenous knowledge contributes to resource management and policy-making not only as input into southern-led scientific assessment, but also as an important element in the process of negotiation and consensus-building. Public participation is generally recognized as a key aspect of resource management and policy formulation as it increases trust, mutual cooperation and willingness to accept less than ideal outcomes (Lind and Tyler, 1988). This is particularly important in cross-cultural contexts such as the Canadian Arctic, where research and policy formulation have historically been primarily in the hands of non-indigenous southerners. The nature of participation and the consideration of alternate voices to provide fair representation and meaningful integration are not always met in public participation processes if these do not go beyond public hearings and token participation (Smith and McDonough, 2001). This is particularly important in contexts involving indigenous peoples, as they may be predisposed to distrust Western science (Moller et al, 2004). Furthermore, traditional resource use is intimately linked with indigenous cultural identity, and externally determined resource management plans do not necessarily consider negative impacts which, through the lens of Western science only, may not be seen as related to resource-use decisions. Meaningful integration of indigenous and scientific knowledge in policy generation in the Arctic requires participation beyond southern-led research and policy-formulation processes to include holistic Inuit perspectives in order to achieve relevance, acceptance and appropriateness.

Institutionalization of indigenous knowledge in practice: Inuvialuit and Nunavut

Canada's Inuit land claims regions formally aim for full and equal integration of indigenous knowledge and scientific research in the policy formulation process through co-management and community-based resource management. Inuit land claim agreements establish systems of resource management and regional policy formulation that include multiple actors at a range of scales, from indigenous community organizations to federal departments, in order to generate resource management policies. This is illustrated with examples from two of the four Inuit land claim regions: Inuvialuit Settlement Region and Nunavut (see Figure 12.1).

The integration of indigenous and scientific knowledge in Inuvialuit and Nunavut has evolved from the inclusion of indigenous knowledge as data inputs to impact assessments to an emphasis on co-management and community-based resource management as part of institutionalized land claims settlements. Both the Nunavut Land Claims Agreement and the Inuvialuit Final

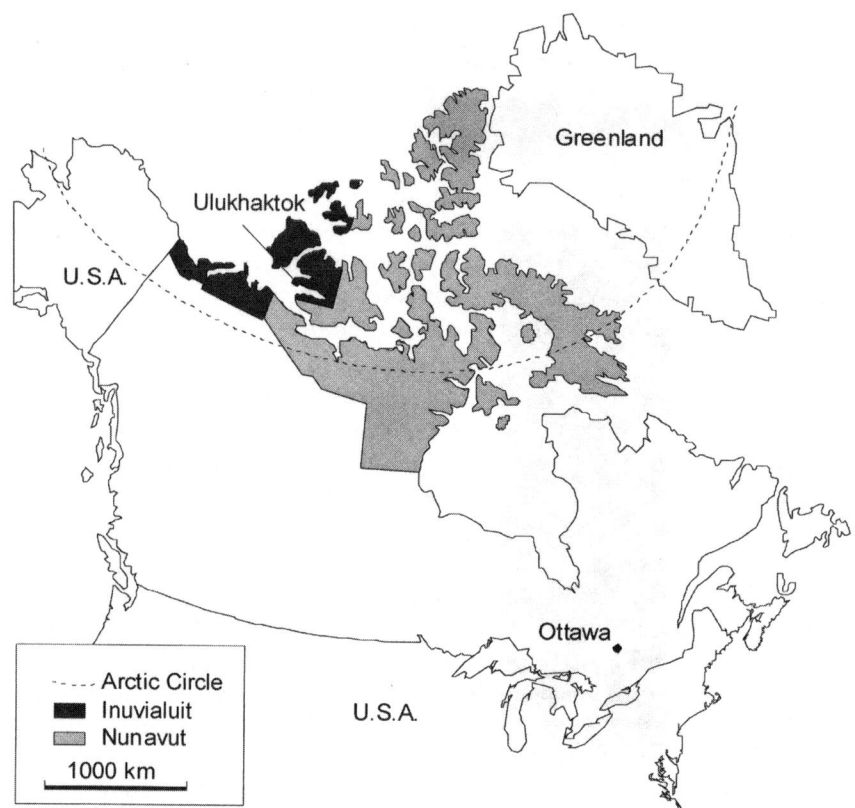

Figure 12.1 *Location of Nunavut, Inuvialuit and Ulukhaktok*

Source: chapter authors

Agreement outline processes which explicitly incorporate indigenous knowledge within institutionalized resource management.

In Nunavut, wildlife management principles are set out by the Nunavut Land Claims Agreement, which explicitly recognizes that 'there is a need for an effective system of wildlife management that complements Inuit harvesting rights and priorities, and recognizes Inuit systems of wildlife management' and that Inuit play a role in all aspects of wildlife management, including research (Article 5.1.2). Indigenous knowledge is treated in a holistic manner, with the recognition in the agreement that harvesting must be sustainable within the ecosystem, but also support cultural objectives and customary practices (Armitage, 2005b). The agreement established the Nunavut Wildlife Management Board, which works in collaboration with community-based hunters and trappers organizations, the territorial land claims administration Nunavut Tunngavik Incorporated and federal departments to set allowable harvest limits (see Figure 12.2). Certain species (polar bear, narwhal, beluga, musk ox) were previously subject to community quotas, which set a cap on the annual allowable harvest. The establishment of Nunavut and the ratification of

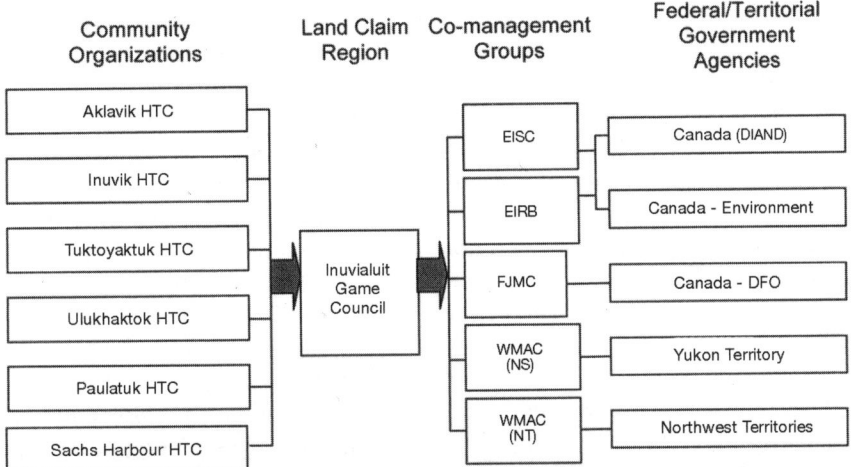

Figure 12.2 *Actors in resource management in Inuvialuit*

Source: adapted from Fast et al (2001) and Inuvialuit Joint Secretariat (2006)

the Nunavut Final Agreement represented a renegotiation of the relationship between the Inuit and the federal government, and envisioned indigenous and southern scientific knowledge as equal contributors to governance, including resource management (Armitage, 2005a). This has led to a number of community-based natural resource management efforts, in which local indigenous organizations (hunters and trappers) participate in the planning and management of wildlife harvesting, contribute to monitoring of populations, and are expected to mediate local harvesting activities (Armitage, 2005b).

Community-based natural resource management in Nunavut was initiated in 1999, and has resulted in changes to resource management rules and structures. Armitage (2005a) outlines changes in narwhal management in Baffin Island communities. The narwhal harvest is significant not only for traditional livelihood strategies as an important food source, but also the commercialization of natural resources through the sale of the male narwhal's ivory tusk. Previously, annual allowable harvesting limits for narwhal (the quota) were set by federal guidelines. Under community-based natural resource management, hunters and trappers organizations contribute to the rules and guidelines in reference to considerations which include, but are not limited to, narwhal population size and stability to provide a link to actual requirements for food, guidelines for avoiding wastage of the animals, respect for animals, and greater flexibility of intra- and inter-annual timing of harvest (Armitage, 2005a). In addition, the participation of community members (particularly elders) and locally led reporting and monitoring activities have the potential to result in greater compliance with regulations (Armitage, 2005a).

An integrated approach to wildlife management is formalized in the Inuvialuit Final Agreement, a land claim agreement between the Canadian government and Inuvialuit people, signed in 1984. Sections 14.2 and 14.4 of the agreement recognize the importance of integrated wildlife and land management regimes with the implicit integration of indigenous knowledge within all bodies, functions and decisions pertaining to wildlife and land management in the Inuvialuit Settlement Region. To meet this need, a series of five government and Inuvialuit joint wildlife and land co-management bodies was created. Figure 12.2 illustrates the relationships between communities and provincial and territorial authorities via the co-management bodies. The advice and recommendations of the co-management bodies assist and guide government policy decisions and the direction of renewable and non-renewable resource management in the region (Inuvialuit Joint Secretariat, 2006). Indigenous knowledge is represented in wildlife management by the Inuvialuit Game Council, which speaks for the collective Inuvialuit interests in wildlife, and the community hunters and trappers committees, which represent community interests (Fast et al, 2001). Additional co-management bodies have jurisdiction over specific sectors of wildlife and land management and include both government and Inuvialuit representatives. An example of co-management in practice in Inuvialuit is presented in the following section.

Data inputs in co-management processes in Inuvialuit and Nunavut, however, are not limited to indigenous knowledge; rather, indigenous and scientific knowledge complement one another. For example, it is difficult for Western science alone to fully monitor ongoing species health and detailed locally specific animal behaviours. Similarly, however, indigenous hunters have intimate knowledge of a finite harvesting territory, and do not necessarily have the ability to observe the animals when their migratory ranges take them outside of this. Furthermore, Moller et al (2004) note that indigenous knowledge is frequently relational in nature and consequently focuses on events that are outside the normal range. Scientific knowledge, on the other hand, focuses on long-term variation in resource monitoring and emphasizes long-term changes in abundance, but may miss isolated extreme events due to its macroscale orientation (Moller et al, 2004). Linking indigenous knowledge and science in practice results in a richer, more comprehensive picture. In addition, the ability to revise rules and regulations based on community feedback on an ongoing basis has the potential to lead to higher adaptive capacity in light of both environmental and socio-economic changes in the Canadian north. This is illustrated via an example from Ulukhaktok, Northwest Territories, a community in the Inuvialuit Settlement Region.

Indigenous and scientific knowledge in resource management in Ulukhaktok

The community of Ulukhaktok (formerly Holman) is located in the western Canadian Arctic, on Victoria Island, approximately 523km (325 miles) north of the Arctic Circle (see Figure 12.1). Ulukhaktok is one of the smaller Inuvialuit communities with a population of about 420 people, 95 per cent whom are Inuvialuit (the designation used by Inuit of the western Canadian Arctic).

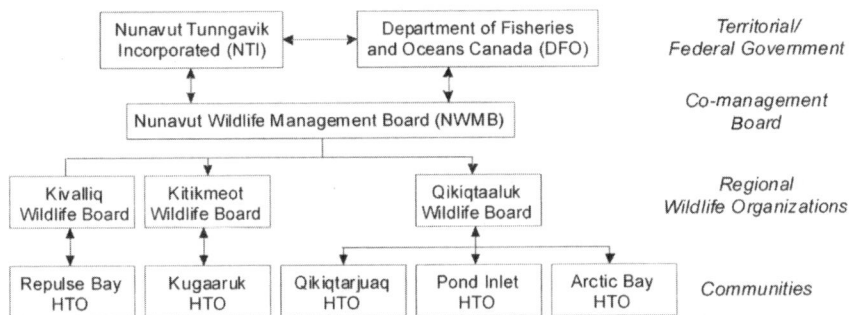

Figure 12.3 *Actors in Narwhal Resource Management in Nunavut*
Source: adapted from Armitage (2005a)

Victoria Island and the surrounding region are characterized by a wealth of natural resources, including terrestrial and marine wildlife species, and subsistence harvesting of these resources continues to be an important activity in the lives of Inuvialuit living in Ulukhaktok today. Subsistence harvesting ('country foods') contributes significantly to the community's food source and local economy, and harvesting and consuming country foods are also clearly essential to cultural identity and community well-being. The importance lies in the activity of harvesting, spending time with family members, the fulfilment and pride associated with the harvest, and also in the sharing of country foods throughout the community (Collings et al, 1998).

Inuvialuit in Ulukhaktok are experiencing significant changes in the local environment and these, together with changing social, political and economic conditions, are affecting people's ability to perform harvesting activities. Inuit in the Canadian Arctic were settled in permanent villages during the early to mid-20th century and many of their conditions changed (e.g. the introduction of mandatory schooling, the loss of the ability to follow the migratory patterns of game over decades, the introduction of southern ('store') food, and greater pressure on some resources due to higher local population densities). The implications of these changing conditions for harvesting activities include the relative availability and abundance of game near communities and less time spent in hunting/gathering activities among the younger generations. Subsistence hunting remains a viable economic and food security strategy in Ulukhaktok (Condon et al, 1995). Furthermore, involvement in hunting is seen as a means of cultural preservation to help younger generations cope with rapid social change brought about by the introduction of the wage economy, compulsory schooling and southern mass communication (Condon et al, 1995).

Social and economic changes have been coupled with unprecedented environmental changes, including climate warming during recent years. Inuvialuit in Ulukhaktok have observed significant changes in the health and distribution of some wildlife species in light of social and environmental changes, including climate variability (Harwood et al, 2000; Joss and Inuktalik, 2000; Stirling and Smith, 2004; Nickels et al, 2005). These changes include a drastic decline in the number of young ringed seals harvested near the community, which scientists report is associated with an increasingly earlier timing of sea ice break-up (older ringed seals are very low in body weight and often not suitable for human consumption); Perry caribou, which were once plentiful near the community, have shifted their migration route to a location far from the community at the end of Prince Albert Sound, a location which is accessible only during summer months by boat; there are fewer musk ox near the community and harvesters must travel longer distances to find wildlife; more polar bears are coming onto the land in search of food; and there has been concern about fish stocks including Arctic char, whitefish and lake trout.

The *Olokhaktomiut Community Conservation Plan* of 2000 (Holman formally changed its name in 2006 and settled on the current spelling of Ulukhaktok; previously, a variety of spellings was in use) outlines 22 species which are managed for conservation. This example illustrates how indigenous and scientific knowledge have combined, in practice, to solve the issue of resource sustainability of one population of these species, Arctic char (*Salvelinus alpinus*). Inuvialuit in Ulukhaktok have traditionally fished for landlocked Arctic char (commonly known in the south as blueback trout, silver trout or white trout) at Fish Lake, located 75km north of the settlement, in both spring and autumn. In spring, families camp at Fish Lake and engage in ice fishing; during autumn, the focus is on net fishing (Condon et al, 1995).

Over 3000 fish were harvested from Fish Lake annually during the late 1980s. Community members began reporting declining catch rates and

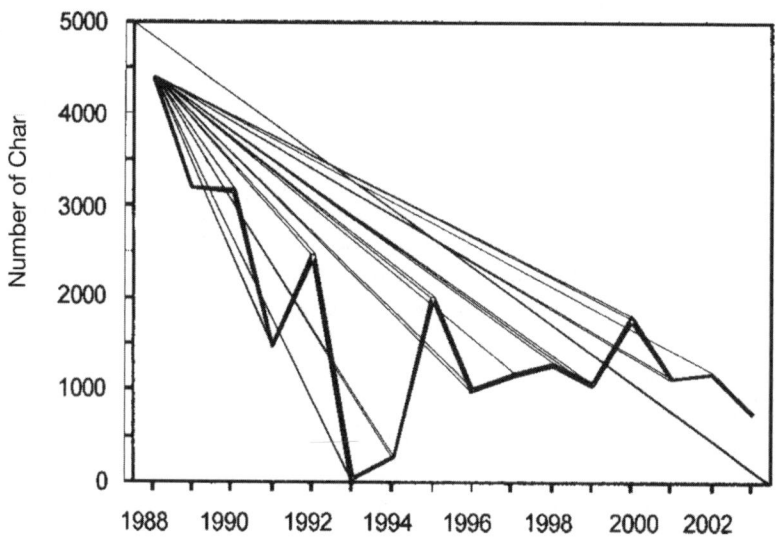

Figure 12.4 *Arctic char harvests from Lakes Fish and Red Belly, 1988–2003*
Source: HCWG (2004)

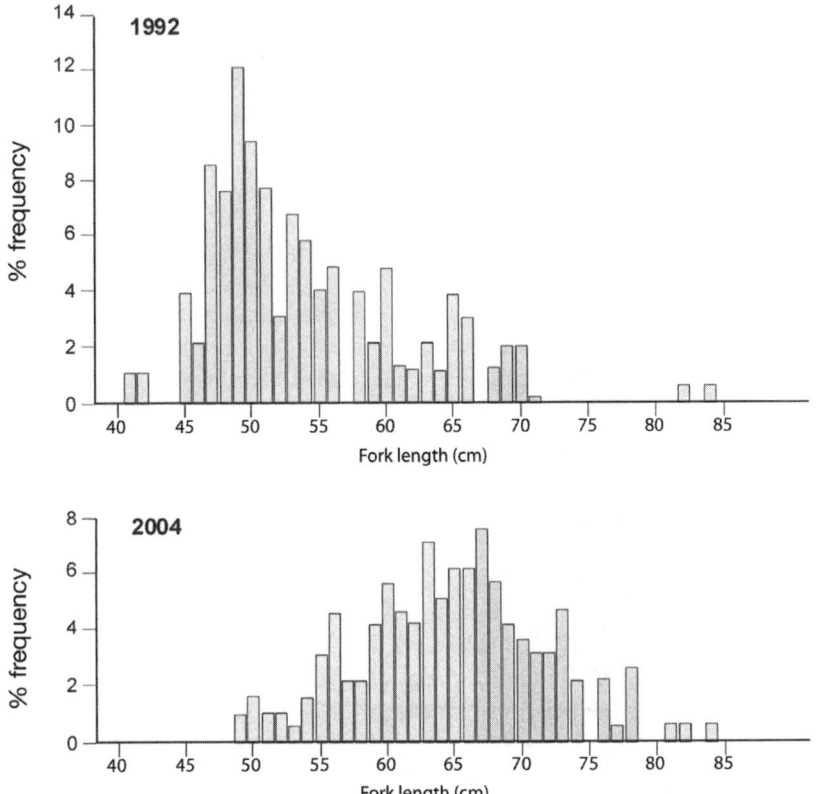

Figure 12.5 *Length of Arctic char harvested at Fish Lake, 1992, 2004*

Source: adapted from Harwood (2010)

younger, smaller char at this time (HCWG, 2004) (see Figure 12.4). Monitoring of char in Fish Lake began in 1992, at which time scientific research conducted by Department of Fisheries and Oceans scientists through the Fisheries Joint Management Committee confirmed that the bulk of the take was relatively small (less than 55cm in length) (see Figure 12.5). Consequently, the Fisheries Joint Management Committee closed Fish Lake to all fishing activity from 1993 to 1995, and set a limit of 50 fish per household for 1996.

Figure 12.4 illustrates the decline in char harvested from Fish Lake (including the connected Red Belly Lake) from 1988 to 2003. It should be noted that the closure did not result in full community compliance, as is evidenced by the harvest data for 1994 and 1995. Since Fish Lake and its char population represent a very important traditional food source that is readily accessible to the community and the closure did not result in full compliance, the Holman Char Working Group (HCWG) was founded in 1996. This management body consists of members of the Holman hunters and trappers committees, both Inuvialuit and scientist members of the Fisheries Joint Management Committee, and a federal Department of Fisheries and Oceans representative.

The Holman Char Working Group's goals are to review available indigenous knowledge and scientific data on the local char fisheries and to discuss options for their future management (including the implementation of the *Char Fishing Plan*), with the goal of incorporating local knowledge and concerns within the plan.

The Holman Char Working Group represents a locally based activity with scientific input which allowed community members to be updated on and respond to the working group's recommendations. Approximately 25 per cent of the adult population of the community attended the working group's open house in July 1996. Fisheries biologists had determined that the safe harvest level of Arctic char in Fish Lake was approximately 1000 fish per annum. The working group recommended that char fishing at Fish Lake should be resumed, but with caution, and a (voluntary) limit of 25 fish per household from 1 October to 1 November to allow fish stocks to recover. In order to meet community food needs for char, the hunters and trappers committees decided to bring 25 fish per household from nearby Prince Albert Sound to the community to supplement 1996 winter food supplies. In addition, the Holman Char Working Group initiated the Char Household Needs Survey to more accurately determine how much char each household needed for subsequent years.

The community, via the Holman Char Working Group, continues to monitor and re-evaluate allocations from Fish Lake. Recently, the limit for Fish Lake was set at 30 char per household per annum. The working group's data collection and monitoring activities combine both indigenous and scientific knowledge. For example, the recorded harvest for Fish Lake in 2003 appears very low (see Figure 12.4). Scientific research alone might interpret this as a lower population in the lake; however, indigenous knowledge from the community notes that there were several known harvesters missing from the recorded list and that the take was actually similar to the preceding five years. The date of fish harvesting remains somewhat flexible, as autumn fishing should not commence until after spawning. Monitors are in place on Fish Lake to check on the progress of spawning through the ice in early October.

While there have been instances of non-compliance with respect to both date of harvesting and timing of activity (some locals express concern that 30 per household is not enough, and that they like the fish when they are full of eggs), the general consensus of the Holman Char Working Group in 2004 was to remain with the existing limit to allow recovery to continue and that the overall plan has been working well. This consensus is supported by the harvest data, which shows harvests at well below the rates evidenced before the closure during the 1990s (see Figure 12.4) even though the catch per unit effort has remained consistent (HCWG, 2004). Furthermore, data on both average age and length of Arctic char sampled at Fish Lake from 1992 to 2004 shows an increase, indicating recovery of the population (see Figures 12.5 and 12.6).

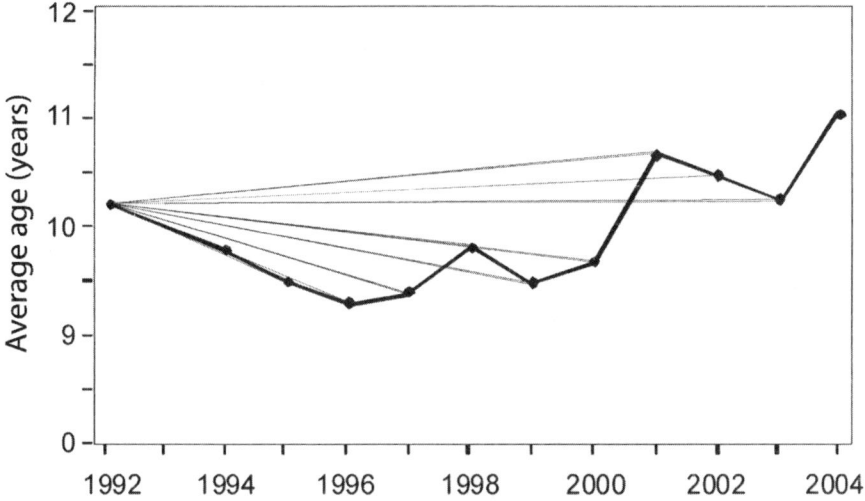

Figure 12.6 *Average age of Arctic char sampled at Fish Lake, 1992, 1994–2004*

Source: adapted from Harwood (2010)

Conclusions

This chapter used several examples from Nunavut and Inuvialuit to illustrate the value of indigenous and scientific co-management in practice. The benefits of integrating scientific research and indigenous knowledge within wildlife management will become more evident as climate change and other stresses continue in the Arctic. Existing co-management bodies provide a functional setting for the meeting of scientific and indigenous knowledge to occur.

The Ulukhaktok example highlights the ways in which the integration of scientific and indigenous knowledge in resource management practice leads to effective strategies. Management of the landlocked char fishery was prompted by observed declines in the availability, size and age of fish (indigenous knowledge) and confirmed by scientific surveys. The initial closure of Fish Lake by the Fisheries Joint Management Commission incorporated both science and indigenous knowledge. Scientific research was a key component of determining the safe harvest level. The community's need for the resource for food and cultural reasons, as identified by indigenous knowledge, prompted the managed reopening of the fishery. Ongoing monitoring draws upon both science and indigenous knowledge.

The char example supports the point that the combination of science and indigenous knowledge leads to greater accuracy in resource management. Although the safe harvest level draws on the science of fisheries biologists, the dates at which the fishery opens after spawning is determined locally through observation and indigenous knowledge. The *Char Fishing Plan* has a high degree of relevance, as it goes beyond safeguarding the population of char to include households' needs for fish in the determination of harvesting limits. The inclusion of both indigenous knowledge and science leads to greater appropriateness, as it draws on indigenous knowledge's holistic understanding of the resource: the resource must be managed not only to safeguard its future stability, but also to allow indigenous peoples to engage in traditional activity and to ensure food security. The process of co-management via the Holman Char Working Group, which operates locally and holds open houses, leads to a high degree of community participation, and decisions are made on the basis of consensus. This, in turn, contributes to community acceptance of harvesting limits and, consequently, increased compliance. Finally, ongoing monitoring and revision of harvesting limits as appropriate means a high degree of flexibility in resource management.

References

ACIA (Arctic Climate Impact Assessment) (2004) *Impacts of a Warming Arctic*, Cambridge University Press, Cambridge, UK

Armitage, D. R. (2005a) 'Adaptive capacity and community–based Narwhal natural resource management', *Environmental Management*, vol 35, no 6, pp703–715

Armitage, D. R. (2005b) 'Community-based Narwhal management in Nunavut, Canada: Change, uncertainty, and adaptation', *Society and Natural Resources*, vol 18, pp715–731

Arnakak, J. (2001) 'What is Inuit Qaujimajatuqangit?', *Canku Ota: An Online Newsletter Celebrating Native America*, www.turtletrack.org/Issues01

Arzel, O., Fichefet, T. and Goosse, H. (2006) 'Sea ice evolution over the 20th and 21st centuries as simulated by current AOGCMs', *Ocean Modelling*, vol 12, nos 3–4, pp401–415

Ayles, G., Bell, R. and Fast, H. (2002) 'The Beaufort Sea Conference 2000 on the renewable marine resources of the Canadian Beaufort Sea', *Arctic*, vol 55, supplement 1, ppiii–v

Berkes, F., Colding, J. and Folke, C. (2000) 'Rediscovery of traditional ecological knowledge as adaptive management', *Ecological Applications*, vol 10, no 5, pp1251–1262

BQCMB (Beverly and Qamanirjuaq Caribou Management Board) (2006) *24th Annual Report, 2005–2006*, www.arctic-caribou.com

Chardine, J. W., Fontaine, A. J., Blokpoel, H., Mallory, M. and Hofmann, T. (2004) 'At-sea observations of ivory gulls (*Pagophila eburnea*) in the eastern Canadian high Arctic in 1993 and 2002 indicate a population decline', *Polar Record*, vol 49, no 215, pp355–359

Collings, P., Wenzel, G. and Condon, R. (1998) 'Modern food sharing networks and community integration in the central Canadian arctic', *Arctic*, vol 51, no 4, pp301–314

Condon, R., Collings, P. and Wenzel, G. (1995) 'The best part of life: Subsistence hunting, ethnicity, and economic development among young adult Inuit males', *Arctic*, vol 48, no 1, pp31–46

Cruickshank, J. (2001) 'Glaciers and climate change: Perspectives from oral tradition (of Athapaskan and Tlingit elders)', *Arctic*, vol 54, no 4, pp377–393

Duerden, F. (2004) 'Translating climate change impacts at the community level', *Arctic*, vol 57, no 2, pp204–212

Einarsson, N., Larsen, J. N., Nilsson, A. and Young, O. R. (2004) *Arctic Human Development Report*, Stefansson Arctic Institute, Akureyri

Fast, H., Mathias, J. and Banias, O. (2001) 'Directions towards marine conservation in Canada's Western Arctic', *Ocean and Coastal Management*, vol 44, pp183–205

Ferguson, M. A. D. and Messier, F. (1997) 'Collection and analysis of traditional ecological knowledge about a population of Arctic tundra caribou', *Arctic*, vol 50, no 1, pp17–28

Ferguson, M. A. D., Messier, F. and Williamson, R. G. (1998) 'Inuit knowledge of long-term changes in a population of Arctic tundra caribou', *Arctic*, vol 51, no 3, pp201–219

Ford, J., MacDonald, J., Smit, B. and Wandel, J. (2006a) 'Vulnerability to climate change in Igloolik, Nunavut: What we can learn from the past and present', *Polar Record*, vol 42, no 221, pp127–138

Ford, J., Smit, B. and Wandel, J. (2006b) 'Vulnerability to climate change in the Arctic: A case study from Arctic Bay, Nunavut', *Global Environmental Change*, vol 16, no 2, pp145–160

Harwood, L. (2010) *Monitoring Char Harvests at Ulukhaktok (Fish Lake) 2009*, unpublished report, Department of Fisheries and Oceans, Canada

Harwood, L., Smith, T. and Melling, H. (2000) 'Variation in reproduction and body condition of the ringed seal (*Phoca hispida*) in western Prince Albert Sound, NT, Canada, as assessed through a harvest-based sampling program', *Arctic*, vol 53, no 4, pp422–539

HCWG (Holman Char Working Group) (2004) *Minutes of the Char Fishing Plan*, Meeting, 9 June 2004

Huntingon, H. P. (2000) 'Using traditional ecological knowledge in science: Methods and applications', *Ecological Applications*, vol 10, no 5, pp1270–1274

Inuvialuit Joint Secretariat (2006) Inuvialuit Joint Secretariat homepage, www.jointsecretariat.ca

Joss, S. and Inuktalik, B. (2000) 'Effects of climate change on wildlife at Holman Island', in *Climate Change Impacts and Adaptation Strategies for Canada's Northern Territories: Final Workshop Report*, 27–29 February 2000, Prepared for Natural Resource Canada and Environment Canada, GeoNorth Ltd Explorer Hotel, Yellowknife, Northwest Territories

Kendrick, A., Lyver, P.O'B. and Lutsël K'É Dene First Nation (2005) 'Denésoliné (Chipewyan) knowledge of barren-ground caribou (*Rangifer tarandus groenlandicus*) movements', *Arctic*, vol 58, no 2, pp175–191

Kofinas, G. (2004) 'A research plan for the study of rapid change, resilience, and vulnerability in social-ecological systems of the Arctic', *The Common Property Resource Digest*, vol 73, pp1–10

Krupnik, I. and Jolly, D. (2002) *The Earth Is Faster Now: Indigenous Observations of Climate Change*, Arctic Research Consortium of the United States, Fairbanks, Alaska

Lind, E. A. and Tyler, T. R. (1988) *The Social Psychology of Procedural Justice*, Plenum Press, New York, NY

Moller, H., Berkes, F., Lyver, P.O. and Kislalioglu, M. (2004) 'Combining science and traditional ecological knowledge: Monitoring populations for co-management', *Ecology and Society*, vol 9, no 3, p2, www.ecologyandsociety.org/vol9/iss3/art2

Nickels, S., Furgal, C., Beull, M. and Moquin, H. (2005) *Unikkaaqatigiit – Putting the Human Face on Climate Change: Perspectives from Inuit in Canada*, Joint publication of Inuit Tapiriit Kanatami, Nasivvik Centre for Inuit Health and Changing Environments at Université Laval and the Ajunnginiq Centre at the National Aboriginal Health Organization, Ottawa

Nunavut Tunngavik Incorporated (2001) Elders Conference on Climate Change, NTI, Iqaluit

Pierotti, R. and Wildcat, D. (2000) 'Traditional ecological knowledge: The third alternative', *Ecological Applications*, vol 10, no 5, pp1333–1340

Riedlinger, D. and Berkes, F. (2001) 'Contributions of traditional knowledge to understanding climate change in the Canadian Arctic', *Polar Record*, vol 37, no 203, pp315–328

Simon, M. (2004) *The Arctic: A Barometer of Global Change and a Catalyst for Global Action*, Speaking notes for Mary Simon for the Inuit Circumpolar Conference, April 2004, New York, NY

Smith, P. D. and McDonough, M. H. (2001) 'Beyond public participation: Fairness in natural resource decision making', *Society and Natural Resources*, vol 14, pp239–249

Stevenson, M. G. (1996) 'Indigenous knowledge in environmental assessment', *Arctic*, vol 49, no 3, pp278–292

Stirling, I. and Smith, T. (2004) 'Implications of warm temperatures and an unusual rain event for the survival of ringed seals on the coast of southeastern Baffin Island', *Arctic*, vol 57, no 1, pp59–67

Usher, P. J. (2000) 'Traditional ecological knowledge in environmental assessment and management', *Arctic*, vol 53, pp183–193

Watson, A., Alessa, L. and Glaspell, B. (2003) 'The relationship between traditional ecological knowledge, evolving cultures, and wilderness protection in the circumpolar north', *Conservation Ecology*, vol 8, no 1, article 2, www.consecol.org/vol8/iss1/art2

Wenzel, G. W. (1999) 'Traditional ecological knowledge and Inuit: Reflections on TEK research and ethics', *Arctic*, vol 52, no 2, pp113–124

13

Reducing Vulnerability of Rural Communities in the Philippines: Modelling Social Links between Science and Policy

Lilibeth Acosta-Michlik and Victoria Espaldon

Introduction

Global processes such as climate change and trade liberalization present great challenges to both science and policy because of the unequal distribution of benefits and costs among countries, among sectors, among communities and among people. Governments develop and implement regional and local policies that can help to balance the negative impacts of these processes and reduce the vulnerability of affected areas and communities. However, emerging patterns of vulnerability are not the consequences of policies alone, but are also a manifestation of human adaptive behaviour to the impacts of these policies. Human beings possess cognitive abilities to exhaust or economize social, economic and natural resources to adapt to any changes in the environment. Understanding vulnerability requires knowledge of adaptation processes, and the reduction of vulnerability demands appropriate adaptation measures. While policy should be able to provide measures to help local communities adapt in a sustainable manner, science has the challenging task of informing policy about the sustainability of these measures. Building an effective science/policy interface is thus a prerequisite for reducing the vulnerability of communities to the impacts of global processes. In this discussion, vulnerability and adaptation are understood as an outcome of both institutional and social links and decision processes between science and policy, on the one hand, and among people in affected communities, on the other.

In order to trace manifestations of how adaptation measures reduce the vulnerability of rural communities to the impacts of global processes, a process-based tool that can model visible links and adaptive decisions of actors in science, policy and affected communities is very useful. This case study in the Philippines aims to apply such a tool by adopting an agent-based 'intervulnerability' framework (Acosta-Michlik and Rounsevell, 2005) that emphasizes the importance of considering the interaction of the impacts of global processes and the interconnection of global to local changes in assessing vulnerability. Agent-based models can combine geographic information system (GIS)-based biophysical maps with agent-based socio-economic surveys; thus, the social characteristics of a geographic location are made explicit in assessing vulnerability. Combining survey- and GIS-based information is important because both social and physical environments influence human adaptive behaviour (Acosta-Michlik and Rounsevell, 2008).

The Philippine case study is an extension of a SysTem for Analysis, Research and Training (START)-supported pilot project on intervulnerability assessment to include agents from science and policy, in addition to agents from rural communities. The case study areas were villages located in a municipality where scientists and politicians are actively implementing local adaptation measures, either jointly or individually, formally or informally. The degree of complementarity or incompatibility of their motivations and the instruments implemented to address the needs of the vulnerable groups were examined. The research procedures involved the identification of relevant policies and concerned authorities, outreach activities of selected scientists to policy and communities, interviews with representatives from science and policy, social surveys of farmers, and extension of the agent-based model.

The case study area

The case study area was Tanauan City, one of the municipalities of Batangas Province in the Philippines (see Figure 13.1). Batangas Province was identified as one of the provinces on Luzon Island with the highest level of vulnerability to the impacts of globalization and climate change (Acosta-Michlik, 2005). The total land area of Tanauan City is 10,716 square kilometres (or 20,716ha), representing 3.38 per cent of the total Batangas land area. The population density is 1096 persons per square kilometre and more than 80 per cent of the residents live in rural areas (National Statistics Office, 2000). In terms of land use, agriculture covers 63 per cent of the total land area; built-up areas are 23 per cent; industrial areas, 6 per cent; and forest, 5 per cent. Tanauan is strategically located near major growth centres, such as Metro Manila and Batangas City, which are driving forces for its urbanization. Access to markets, development inputs, and skilled and semi-skilled labour within and outside the

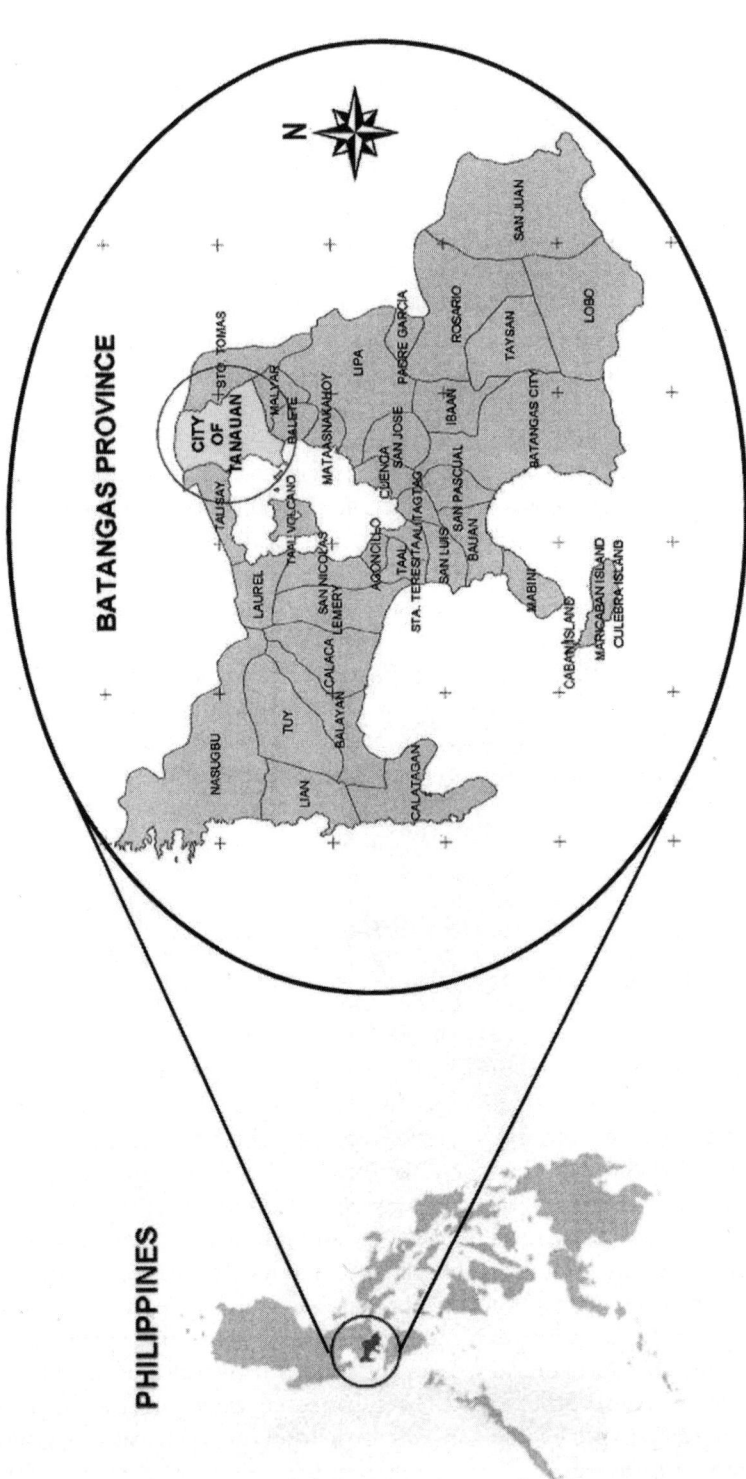

Figure 13.1 *Geographical location of the case study area in the Philippines*

Source: chapter authors

province is good. Moreover, the presence of Taal Lake at the southwestern part of the city and Mount Makiling at the eastern part provide additional opportunities for pursuing eco-tourism development. About 12 per cent of the total labour force is engaged in agriculture (National Statistics Office, 2000). Although the entire land area of Tanauan City is suitable for agriculture, the area devoted to agriculture shrank by 15 per cent from 1993 to 2001 for several reasons. Out of the 48 villages (known as *barangays*) in Tanauan City, only two villages have irrigation systems. Rain-fed agriculture thus characterizes the municipality, contributing to very low productivity despite its good soil quality. Another important reason for the low return on investment in farming is the intense market competition that reduces the prices of agricultural commodities. Because Tanauan City is a trading centre in Batangas Province, producers from other municipalities also go there to sell their agricultural produce.

Three villages were selected for case studies in Tanauan City. Each village represented a particular ecosystem: Cale (agricultural), Natatas (urban) and Gonzales (coastal). Natatas, the closest to the city centre, has experienced fast urbanization during the last decade and only 33 per cent of its total land area is devoted to agriculture. Cale and Gonzales are far away from the centre, with a distance of 5.4km and 9km, respectively. Farming, fishing and eco-tourism are the main sources of income of the residents of Gonzales. With its plain terrain, agriculture is the primary source of income in Cale. Agriculture production and marketing in Tanauan City remain predominantly conventional and arduous, with little use of mechanical equipment and modern transport systems (see Figure 13.2).

Methods

Surveys and cluster analysis

Cale, Natatas and Gonzales, where the social surveys were conducted, have a total land area of 685ha, accounting for 3.3 per cent of the total land area of Tanauan City. Ninety-nine farmers, representing 24 to 31 per cent of all the farmers in each village, were interviewed in these three villages. Information gathered included the farmers' social and economic attributes. The social attributes are, among others, farmer's age, education, location of residence, household size, dependency ratio, availability and number of farm successors, sources of information, willingness to sell land, and how they value their land. Meanwhile, the farmer's economic attributes are income level, other jobs, other assets, farm size, crops cultivated, type of landownership, time spent on the farm, employment of agricultural workers, markets for their produce, and land conversion from rice to other crops. Their personal views on the effects of globalization and climate change, as well as their adaptive measures to these effects, were also gathered.

Figure 13.2 *Conventional production and marketing systems in Tanauan City*

Source: chapter authors

A global positioning system (GPS) was used for ground validation of farm ownership and land use in the villages. This was necessary because cadastre surveys are not available, nor are the land uses clearly captured by the satellite images. The information collected from the GPS survey was used to develop maps of farm ownership and land use in GIS formats with a 10m x 10m resolution. This resolution corresponds to other GIS maps, which contain information on elevation, slope, soil properties, erosion potential, fertility, rivers and road networks. Selected socio-economic and biophysical attributes of the 99 farmer respondents were subjected to cluster analysis to identify the appropriate number and description of farmer typologies. The clustering technique employed here followed a two-step approach (Heir et al, 1995): first, the use of a hierarchical procedure through a dendrogram, and, second, the use of initial seed points in the k-means (non-hierarchical) analysis. The cluster analysis provided empirical support for building farmer typologies, which are assumed to have different adaptive behaviour. Detailed discussions of the techniques and results of the cluster analysis in the case study area are given in Acosta-Michlik (2005).

Interviews and network analysis

In addition to the social surveys, interviews with representatives from local authorities and research institutions were carried out to identify knowledge and information that support adaptation and to see how they are transferred to the farmers. Municipal planners and village administrators, as well as representatives from farmers' associations, were consulted to acquire information about the sources of vulnerabilities and the available adaptation measures for the farmers. Researchers from academic and research institutions (e.g. University of the Philippines at Los Baños, Southeast Asian Regional Centre for Graduate Study and Research in Agriculture and Mandala Agricultural Development Corporation), as well as extension officers from private companies (e.g. East-West Seed Company, Office of Environmental Protection and Office of Agriculture, Environment and Natural Resources) were interviewed about the technologies and farming practices that they transfer to the farmers and the methods for transferring the know-how. Cross et al (2002) argue that an important, yet often overlooked, component of people's information environments is the relationships that they use to acquire information and knowledge. The importance of personal connections in the construction and acquisition of information is emphasized in social network research (Granovetter, 1973; Burt, 1992; Rogers, 1995; Shah, 1998; Hansen, 1999).

Social network analysis is a useful tool for visualizing and understanding the myriad relationships that can either facilitate or impede knowledge creation and transfer. It suggests that 'who you know has a significant impact on what you come to know' (Cross et al, 2002, p2). Networks are a set of actors (or agents) connected by a set of ties. The actors, which are also often called 'nodes', can be people, teams, organizations, concepts, etc. 'Ties connect pairs of actors and can be directed (i.e. potentially one-directional, as in giving advice to someone) or undirected (as in being physically proximate) and can be dichotomous (present or absent, as in whether two people are friends or not) or valued (measured on a scale, as in strength of friendship)' (Borgatti and Foster, 2003, p992). Based on social network analysis, diagrams can be designed to map the actors (or agents) and the social ties that link them together. The network diagram can help to identify to whom the agents turn for advice and how information flows among individuals within and beyond their social network (e.g. a personal relationship, institutional affiliation and the like). Such a diagram was developed in this study to identify the links between science and policy, on the one hand, and the outreach of their activities to the local communities, on the other.

Modelling agents' decisions and interactions

Using the information generated from the cluster and network analyses, an agent-based model was applied to estimate vulnerability futures based on two social network scenarios (i.e. current and idealized networks). Agent-based models are a promising tool not only for analysing the complexity of human links to their environments, but also for evaluating the consequences of the pattern and structure of social networks on human adaptation to changes in the environment. Because agent-based models allow representation of human cognition, they recognize that knowledge utilization is fundamentally a social process (Wenger, 1998). 'To utilize this knowledge in the solution of problems and the creation of new knowledge, organizational members must know who knows what, and interact with each other in order to utilize and combine knowledge' (Cross et al, 2001, p216). The use and flow of knowledge in a cognitive process is represented in decision rules. The cognitive process in this chapter consists of observation, income maximization, interaction and social comparison (see Acosta-Michlik, 2005, for details). For example, the decision rules for income maximization are the knowledge of expected changes in income and the economic capacity to take immediate action. In this case, the relevant farmer attributes are education, profession, farm size and income. The decision rule that is important for interaction is the membership in the social network, while for social comparison the number and location of members in the network are important. Moreover, the attributes of the members in the social network, such as education, profession, farm size and income, are important for social comparison. The decision rules thus determine the types of farmers who can engage in a particular type of cognitive process. The cognitive process is represented in a behavioural model drawn from the 'consumat' approach (Jager et al, 2000). Different groups of farmers, which are represented in the farmer typologies generated from the cluster analysis, are assumed to follow different cognitive strategies. Details on the construction of cognitive process and decision rules are available elsewhere (Acosta-Michlik, 2005; Acosta-Michlik et al, 2005).

In the agent-based model, the farmers are equipped with mental and memory maps, which enable them to perform cognitive processes. The mental map allows the farmers to compute income Y_t^k from farming and non-farming activities and to select the combination of activities that yields the highest income. Meanwhile, the memory map allows them to remember their income from the previous years (i.e. Y_{t-1}^k, Y_{t-2}^k ... Y_{t-n}^k):

Model 1 $\sum(Y_t^k) = \tau Y_{f,t}^k + (1 \times \tau) Y_{nf,t}^k$ where : $\tau = 1$

$Y_{f,t}^k = [(P_{i,t} Q_{i,t}) * (P_{j,t} Q_{j,t})] * A_{i,t}^k$ st. $F_{i,t}^k \geqslant \delta_F$

$Y_{nf,t}^k = W_{x,t}$ st. $E_{w,t}^k \geqslant \delta_E$

The level of farming and non-farming income depends upon the time allocated to these activities (i.e. $\tau = 1$ full-time and $\tau = 0.5$ part-time farmers). $Y_{f,t}^k$ is the farm income of farmer k at time t from crop i. $P_{i,t}$ and $P_{j,t}$ are the prices per unit of crop i and input j, while $Q_{i,t}$ and $Q_{j,t}$ are the quantities of crop i (i.e. expressed in terms of yield per hectare) and input j at time t. The total farmer's income from crop i depends upon the size of the area planted $A_{i,t}^k$ to this crop. The choice of alternative crops is restricted by the financial capacity $F_{i,t}^k$ of farmer k to invest in alternative crop i, as measured by parameter δ_F. The non-farming income $Y_{nf,t}^k$ is a function of the wage per month $W_{x,t}$ from employment x at time t, and is subject to meeting the education requirement of the non-farming job.

The farmers' mental map also allows them to interact with other farmers, as well as science and policy agents who are part of their social networks. Their memory map helps them to remember the attributes of the other members and the ability of these members to advise or to help. For example, a farmer who needs financial help to invest in new crops will interact with farmers in his social network who have high income and, thus, the capacity to provide credit:

Model 2 $\sum(SC_t^k) = Z_t^k + P_f^k + Q_f^k{}_t$

The size of the social network Z_t^k refers to the number of farmers who have common attributes (e.g. family relationship, profession and neighbourhood) at time t. The spread of the social network P_t^k refers to the distance between the residences of the members, while the quality of social network Q_t^k refers to the intellectual and economic capacity of the members. Farmers with a higher value of social connectivity SC_t^k will be more likely to engage in social comparison.

Results and discussion

Current vulnerabilities and adaptations

The cluster analysis identified four farmer typologies:

1 traditional;
2 subsistence;

3   diversified; and
4   commercial (Acosta-Michlik, 2005).

Most of the traditional farmers in the case study area had small- to medium-size farms and cultivated traditional crops such as rice and corn. Moreover, they had low levels of education, income, assets, diversification and information. Subsistence farmers, who provide food for their own consumption or feed for livestock production, had the lowest average farm size (1.37ha). Compared to traditional farmers, they had higher incomes as they also had non-farming jobs, mostly as non-skilled labourers. The diversified farmers also had jobs other than farming. Many of them (30 per cent) were skilled workers with large heterogeneous farms, high levels of information and high income levels. Of them all, the commercial farmers owned the largest farms.

The highest proportion (40 per cent) of the interviewed farmers fell under the subsistence typology (Acosta-Michlik, 2005). Most subsistence farmers are located in Cale, whereas a large number of diversified and commercial farmers are in Natatas. The latter can be attributed to the proximity of the municipal public market in Natatas, making large-scale farming more profitable. Traditional farmers, who accounted for the smallest number of farmers, are found mostly in Gonzales.

Analysing the profiles of the farmers (see Table 13.1) leads to the assumption that traditional and subsistence farmers are those most vulnerable to the impacts of globalization and climate change. Their income is low, assets are minimal and they have little farm and job diversification. These farmers also have low education and are poorly informed. Such attributes limit the application of adaptation measures. Vulnerability, however, is dependent not only upon a farmer's personal attributes, but also upon the economic and physical characteristics of his location. Traditional and subsistence farmers are assumed to be very vulnerable in Cale, but not necessarily in Natatas and Gonzales. Natatas is close to the Tanauan city centre and public market, while Gonzales offers alternative sources of income such as fishing and ecotourism. Cale has

Table 13.1 *Profiles of the farmer typologies in the case study area*

Profiles	Farmer Typologies			
	Traditional	Subsistence	Diversified	Commercial
Level of education	L	L	M	H
Level of household income	L	M–H	M–H	H
Non-farming source of income	L	H	H	M
Size of household assets	L	L	M–H	H
Size of agricultural land	L-M	L	M	H
Level of farm diversification	L	L	H	L–M
Land planted to traditional crops	H	L–M	L	L
Information on global change issues	L	M	H	M–H
Application of adaptation measures	L	L	M	M–H

Note: L = low or small; M = moderate; H = high or large.
Source: chapter authors

very good biophysical attributes; but farms are very far from the market and the main highway, making production and marketing more costly. Moreover, lack of irrigation in this village deprives the farmers of the opportunity to increase agricultural productivity. Relative to traditional farmers, diversified and subsistence farmers have better capacities to adapt to the negative impacts of global processes upon agriculture because farming is not the only source of their income. Subsistence farmers in Cale are least affected by high transport costs because their crops are used mainly for home consumption.

Table 13.2 shows the different actions carried out by farmers in Tanauan City to adapt to the negative impacts of climate change (mainly droughts) and globalization (i.e. price fluctuations, high production costs and low farm prices). The top three adaptation measures adopted by all types of farmers in order of importance are undertaking spiritual practices; cultivating crops other than rice; and changing food consumption habits (i.e. eating cheaper food). Commercial farmers showed the highest rate of adoption of all these measures, traditional farmers the lowest. In terms of changing production activities and consumption habits, the traditional farmers show that they are the least flexible. Turning to relatives and neighbours for help is an important adaptation measure for all farmers, emphasizing the role of social relationships in decreasing vulnerability. Among the traditional farmers, seeking assistance from government officials appeared to be unpopular, with only 25 per cent of them consulting village and municipal officials. None sought assistance from government officials such as agriculture officers.

None of the traditional and subsistence farmers are willing to migrate to other places in search of better opportunities, which may be attributed either to the social values that they attach to their lands or the low expectation of finding alternative jobs, considering their low level of education. Both types of farmers are also less likely to sell their farms and to start a new business. Implementing technical solutions (e.g. cultivating new crop varieties and applying new farming techniques) to solve farming problems ranked only 8, particularly among the traditional farmers, with only a 33 per cent rate of adoption (see Table 13.2).

The main reasons for not cultivating new crop varieties are lack of money and knowledge (see Figure 13.3). Among traditional farmers, a lack of opportunity was also given as a reason for not adopting other crops, which further emphasizes that they are not well informed – they lack information about the availability of new crop varieties. Many commercial farmers are also not adopting new crop varieties due to a lack of knowledge. As for the application of new farming techniques, lack of money and knowledge were the main reasons mentioned by subsistence and traditional farmers, respectively. Some diversified farmers also mentioned that applying new farming techniques was against their principles. Overall, lack of knowledge and money was the most important reason for not adopting drought-related technical measures. Both the government and scientists have been providing support to farmers to provide them with some human and/or capital assets for adaptation.

Table 13.2 *Percentage of interviewed farmers in Tanauan City carrying out the adaptation measures ranked in terms of adoption rate, by farmer typology*

Adaptation Measures	Rank	Farmer Typologies			
		Traditional	Subsistence	Diversified	Commercial
Undertake spiritual practices (e.g. pray)	1	75	79	85	90
Plant crops other than rice	2	67	75	82	85
Change usual food consumption	3	58	82	82	85
Look for additional sources of income	4	67	64	69	90
Get help from family or relatives	5	58	75	77	75
Borrow money from relatives	6	58	68	69	75
Ask for help from neighbours	7	50	71	72	75
Implement new farming techniques	8	33	50	62	50
Consult village/municipal officials	9	25	46	62	60
Sell valuables (e.g. jewellery, animals)	10	42	50	54	35
Join farmers' co-operative	11	33	32	46	30
Send a family member to the city to work	12	25	25	33	50
Seek government assistance	13	0	21	44	35
Send a family member abroad to work	14	8	18	33	30
Temporarily stop farming to work in the city	15	0	21	15	35
Temporarily work as agricultural labourer	16	8	11	15	30
Migrate to another region or city	17	0	0	5	20
Borrow money from bank	18	8	4	3	10
Sell the farm and start other business	19	0	0	5	5

Source: chapter authors

Policy and science support

The results of the interviews are summarized and analysed in this section. The views from policy are first presented, followed by those from science, on how vulnerable farmers have been supported to adapt to global changes. Finally, this section provides an analysis of the structure and characteristics of the policy/science interface that builds a bottleneck to a sustainable reduction in vulnerability in the case study area.

Views from administrators and planners

Farm productivity
The very low level of agricultural productivity due to the absence of irrigation systems contributed to the low competitiveness of farming in many villages in Tanauan. In 2004, the local government consulted with the engineering depart-

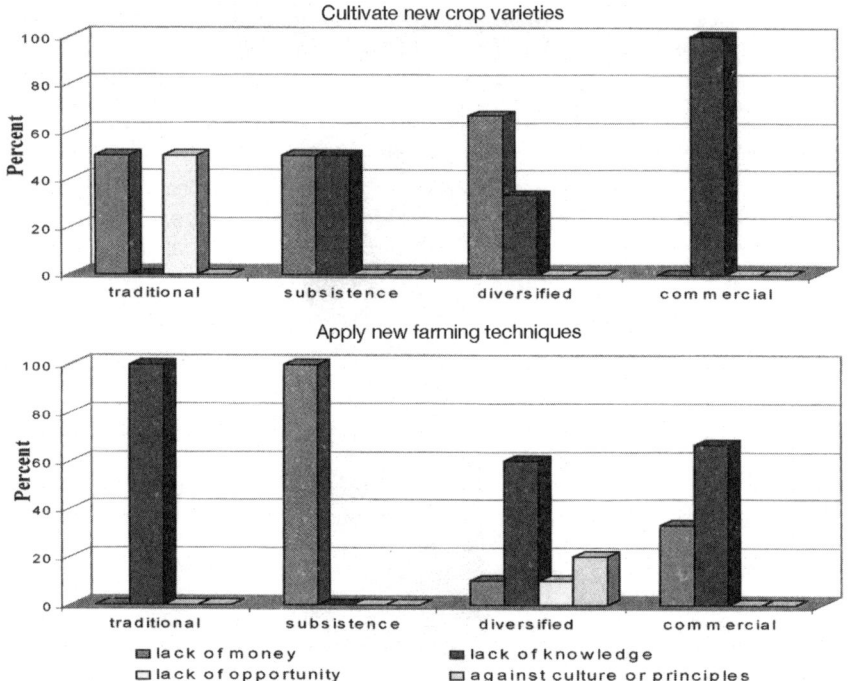

Figure 13.3 *Percentage of farmers not adopting adaptation measures, by typology and reasons*

Source: chapter authors

ments of the University of the Philippines and the Bureau of Soils and Water Management about the potential for developing an irrigation system in Tanauan. However, no feasibility studies were undertaken because of monetary constraints. The development of an irrigation system would have been very costly due to the deep groundwater level in the municipality. The absence of irrigation is particularly problematic for Cale because of its distance from the San Juan River and Taal Lake, which are sources of irrigation water for other villages in the municipality.

Furthermore, there are no calamity funds to support the farmers in cases of drought. Drought-resistant crops, however, are being introduced by the City Agriculture Office to help the farmers increase productivity, mainly through extension services. A somewhat informal link exists between the seed company and the local office of the Department of Agriculture to promote the use of drought-resistant crop varieties in the municipality. The Department of Agriculture has contacted the seed company to inform and train the farmers on the use of these crops.

Market facilities
With political support from the local government, market facilities in Tanauan City have been expanded and improved. The Tanauan public market, which

was constructed through a loan package from the World Bank, accommodates more than 1000 stalls and allows market operations for 24 hours a day for seven days (City of Tanauan, 2004). Because Tanauan City has become an important trading centre, not only for Batangas Province but also for adjacent areas, including Metro Manila, producers from other municipalities go there to sell their agricultural produce. Although an excellent market for vegetables has developed in Tanauan City, the small-scale diversified farming system is not able to respond to this development. Traditional sustainable farming is not oriented towards the very competitive and commercialized Tanauan public market.

Extension services
The local government sends extension services to the villages at least once a month, depending upon the availability of budget and new technologies. Moreover, the number of extension workers depends upon the political support of the mayor for agricultural development *vis-à-vis* other sectors, such as industry and urban development. The extension workers often have difficulty in encouraging farmers to attend meetings and to try new farming practices. Most farmers prefer to continue their traditional practices or emulate other farmers who have experienced success in adopting new practices. There is a general lack of trust of government extension services among the farmers because they are often regarded as outsiders without an understanding of local farming conditions. Moreover, political interests and personal rivalry among heads of the villages sometimes hinder the extension of successful adaptation strategies from one village to another.

Farmer associations
The president of the Tanauan Federation of Farmers on Conservation Farming is planning to adopt a research-oriented approach for disseminating knowledge and technologies on conservation farming in the municipality. The federation will pattern this approach from previous studies conducted by research institutions and fertilizer companies, which have planted different varieties of rice and use organic fertilizers. The federation will also promote the use of organic farming in the village through education about new farming technologies. They will seek financial assistance from the government for demonstrations and lectures on new technologies from agro-chemical companies. The president of the Tanauan Federation of Farmers on Conservation Farming is aware of scientific studies conducted in the village, but no records are available. For example, there were studies on crop susceptibility to pests and diseases and methods to control them; however, the results of the studies were not disseminated.

Land-use plans
The law mandates municipalities to have a comprehensive land-use programme, which guides local government not only in the efficient use of

land, but also in the effective delivery of development programmes. The first programme for Tanauan City was prepared by the technical services of the Housing and Land-Use Regulatory Board of the Philippine government. The preparation of the second comprehensive land-use programme was offered for bidding, which allowed the participation of scientists through project contracts. The Mandala Agricultural Development Corporation, a private research consultancy company, prepared this second comprehensive land-use programme. The draft programme had to be approved by the mayor, who then endorsed the approved plan to the legislative body for further endorsement to the Housing and Land-Use Regulatory Committee. Before the endorsement of the plan to the legislative body, the city government presented the results of the study to local officials through a stakeholder dialogue. The dissatisfaction with the study among some heads of villages was attributed partly to poor understanding of the local environment by the scientists.

Views from academics and companies

Farm productivity

Research projects that promoted the use of drought-resistant varieties and increased crop diversification were carried out in selected villages in Tanauan City. The adoption rate of high-yielding rice varieties was low because, with their shorter straws, chickens can reach and easily eat the panicles. Moisture-resistant varieties of tomatoes were introduced to allow farmers to cultivate them during the rainy season, when prices were higher. Seedlings and crop managements of improved crop varieties were introduced through action research. To diversify crop production, the scientists also introduced a drought-resistant variety of sorghum, which was used as feed for livestock. However, the adoption of drought-resistant crops was hindered by the lack of mechanical threshers and/or markets.

The scientists sought approval from local and village officers for carrying out projects in particular areas; projects so endorsed were easily accepted by the farmers, who then participated actively. However, as the environment changed over time, the technologies evolved as well. When a specific project ended, the farmers were unable to catch up with the new technologies as they had no additional external support from local governments, either by funding or the provision of extension services.

Market facilities

Most farmers did not own vehicles and relied on renting tricycles or jeeps (usually used for human transport) for transporting their produce to the market (see Figure 13.2). Due to the high costs of transportation, many farmers preferred to sell their products to pick-up buyers. In order to encourage the adoption of improved varieties, scientists included market links in their projects. They talked with feed millers who were willing to pick up and buy the sorghum produce of the farmers, and linked the farm association to the feed millers to ensure that the former got regular buyers and reasonable prices for their produce and the latter had enough sorghum for milling. The scientists also linked the vegetable (e.g. tomato and cucumber) producers to food-processing companies. However, the links between farmers and buyers were not sustained after the projects ended. The scientists emphasized the need for bridging this missing link; technology alone did not help the farmers if the market was not available or the price was not reasonable.

Extension services
Action research, which applies farm techniques or technologies that have been proven relevant for farmers, enabled scientists to provide extension services to the farmers. During the project periods, scientists stayed in the villages and tried new farming practices at selected farms. Thus, the scientists developed a deeper understanding of the needs of the local communities and their reasons for not adopting new technologies, while the farmers could consult the scientists at any time and directly saw from the experimental fields how successful were the new practices. This was in contrast to the government services, which sent extension officers to the farms depending upon the available budget or extension programmes.

Farmer associations were also effective tools for promoting extension activities of the scientists to the farmers, and they also helped to connect the farmers to suppliers of inputs and buyers of crops. Associations were particularly useful when the project concluded, as the government extension services were not able to take over the responsibilities of the scientists.

Farmer associations
The farmers, through the research projects, organized into an association to improve the dissemination of new farming techniques and technologies (many of them did not voluntarily join and had to be encouraged by the scientists). The association helped them to apply for crop loan programmes from the government. This facilitated the process, which was cumbersome if sought by individual farmers. At the initial stage of a project, the scientist in charge approved the association's farm planning budget and endorsed it to the extension officer of the rural bank. The head of the association eventually took over the evaluation of loan applications and the scientists remained as endorsers to the bank. As mentioned above, the association could assume the responsibilities of the scientists, such as linking farmers to buyers, after the termination of a project. However, when the farmers in the association who had been trained by the scientists retired or died, they did not have successors to continue the work. The children of many farmers were not interested in farming and the know-how was not transferred over to the next generation.

Land-use plan
Researchers from the Mandala Agricultural Development Corporation had to consider the issues that were raised during the dialogue in the final report of the

comprehensive land-use programme. They generally did not agree with the comments given by the local officers, particularly with respect to categorizing areas according to different ecosystems. For example, some areas that were not completely covered with trees and were highly inhabited were designated by the local government as forest because of the development policy for such a type of ecosystem. The views of the scientists thus were not completely represented in government development plans, although they had actively participated in the design of such plans. Moreover, after the study had been completed and the contract terminated, the link between the Mandala Agricultural Development Corporation scientists and Tanauan planners also ended. The former were thus not involved in any post-evaluation of the land-use plan.

Bottlenecks in the policy/science interface

Diverse understanding of vulnerability and adaptation

There was a different understanding between officials and scientists on the appropriate strategies to reduce the vulnerability of the farmers. For example, while the former viewed the Tanauan public market as a structure that offered an opportunity to sell crops, the latter viewed it as a system that could harm those who did not have the capacity to compete. The scientists viewed the market as more than a structure where farmers could bring and sell crops; it was a system that allowed the profitable satisfaction of demand and supply, subject to the constraints confronting the buyers and sellers.

Take the case of sorghum production and marketing: the farmers could not afford to invest either in vehicles to transport or in millers to process the sorghum. The feed millers needed large amounts of sorghum to buy, but there were not enough domestic suppliers, as many millers import sorghum. In this case, the sorghum market did not exist in the Tanauan public market. It existed only at a point where a number of farmers agreed to satisfy the quantity demanded by the millers, and the millers were able to pick up and buy the sorghum. This was a form of contract-growing system with the farmers and was widely applied by many food-processing companies in the Philippines. A deeper understanding of the reasons and sources of vulnerability of local communities was very important; but these differed from one community to another because of differences in social and geographical characteristics, as exemplified by the three villages in Tanauan City. Scientists who lived with the community during a project developed a relationship that allowed them not only to acquire knowledge on the vulnerability of the people, but also to transfer this knowledge back to them in the form of adaptation measures. The government extension officers lacked this social relationship.

Lack of common objectives and interests

The diverse conceptions in addressing the adaptation needs of the communities may be attributed to a lack of common objectives and interests between policy and science. In performing their tasks, officials aimed to address the well-being of the general public during their incumbency, whereas the scientists aimed to address the specific needs of a group of people within the project life. Political interest in agriculture development could thus be low in a municipality such as Tanauan, where the share of the agricultural labour force was low and the rate of urbanization and industrialization was high. Addressing agricultural vulnerability could thus be linked more to achieving other development goals, such as the expansion of public markets, which serve not only the farm producers, but also the growing urban consumers.

Although academic interest in agriculture development was high, the impacts of research were often limited to a small group of people within a short duration of time. Development policy thus aimed to find long-term solutions that maximized the number of beneficiaries, while action research aimed to find practical and immediate solutions to the problems of specific communities. The interviews revealed that sustainable farm management was often difficult to achieve after the research ended because farmers were faced by new challenges. However, the different objectives of policy and science may not necessarily hinder a sustainable reduction of vulnerability. With an effective coordination of policy and actions, the long-term adaptation policy could complement the short-term adaptation measures.

Weak institutional and structural links
Long- and short-term adaptation measures can be coordinated effectively when policy and science institutions are strongly linked. At the national and regional levels, science is often an inherent part of the structure of government agencies. Moreover, these agencies have built important institutional links to universities and research institutions, which serve as their consultants or collaborators. A feedback system exists in most collaborative work between scientists and policy-makers; but such a link is very weak at the local government level, if it exists at all. With the exception of the comprehensive land-use programme, there were no other plans or programmes of the municipal office of Tanauan through which scientists were actively involved. Similarly, action research in the municipality lacked the strong commitment and involvement of local officials. Within a devolved system of government, local planners and administrators play an even more important role in providing the necessary social and capital support to communities. However, creating a contiguous policy/science interface at lower administrative levels was perhaps more difficult here because, in contrast to the national and regional authorities, local officials had closer contact and more direct responsibilities to their constituents. Science may thus have little space in an already complex structure of local governance with links to a large number and different groups of people in the community, industry, business and politics. As long as scientists take a passive role in this structure, they will not be able to transfer the knowledge that they acquire from the community to local planners and administrators. In such a case, scientists may need a link to a 'pressure group' who will facilitate the flow of knowledge from science to policy, or even vice versa, in order to provide opportunity for a feedback process.

Lack of recognized and influential intermediaries
As Chapter 1 of this book makes clear in its discussion of 'spider webs', science needs connecting links to local governments. Such intermediaries should be recognized as capable of influencing the decisions of the latter. Farmers' associations or community groups with specific interests can play intermediary roles between scientists and local authorities. They are important to the scientists as partners in transferring knowledge to the community and to the authorities as constituents with the power to vote and influence political decisions. Many Philippine agricultural co-operatives, which were organized through government development programmes, could not be effective intermediaries because they had been conditioned as beneficiaries and passive actors in science and policy. A new generation of farmer associations such as the Tanauan Federation of Farmers on Conservation Farming, which has strong interests not only in exploring and disseminating modern farm technologies, but also in seeking financial support from the government to realize its objectives, are indispensable as an effective intermediary between science and policy. The new generation of farmer associations should develop skills in managing farm business, assessing economic and environmental risks, negotiating interests and operating as independent entities. They should have programmes for recruiting and training new members to transfer not only skills, but also contacts, in order to ensure the sustainability of the associations as recognized and influential intermediaries.

Lack of close partnership and constant dialogue
Weak institutional and structural links between science and policy hinder the development of close partnerships and opportunities for constant dialogue. Partnership and dialogue are indispensable in the transfer of knowledge between science and policy actors. The intermediaries should thus facilitate the knowledge transfer through their close partnership with both. Stakeholder dialogue or community consultation is an important process to build consensus on issues, especially where science and policy have different views.

Social network analysis

To identify the most important social network for each typology, farmers were asked about the sources of information of their farming practices, such as the use of organic fertilizers, intercropping and improved crop varieties. Table 13.3 shows that family and neighbours are the most important social network that the farmers turn to for advice. The family is the most important source of information for commercial farmers and neighbours for subsistence farmers. Among the four farmer typologies, traditional farmers turned out to consult the least with other people as far as farming techniques were concerned. The role of extension officers and local officials in transferring information to farmers was not very significant. None of the traditional farmers had been involved with or contacted members of co-operatives or their farm communi-

Table 13.3 *Sources of information on farming practices, by farmer typology and in percentage*

Sources of Information	Farmer Typologies			
	Traditional	Subsistence	Diversified	Commercial
Family	33	61	51	75
Neighbours	17	54	33	30
Extension services*	8	18	8	10
Local officials	8	14	5	5
Co-operatives	0	7	5	5
Farm community	0	4	3	5

Note: * Provided through services of action research, private companies and government.
Source: chapter authors

ties about ways to improve farming practices. The other types of farmers had links to these social networks, albeit very insignificant. The role of co-operatives in transferring information to farmers was not very important because they lacked the capacity to convince and recruit new members. As discussed above, farmer associations should be improved to be able to respond to the needs of their members and to bridge the missing link between science and policy. The latter would help co-operatives to become sources and propagators of new farming practices. Unlike those in big cities, people in small villages in the Philippines are usually characterized not only by close family ties, but also close community relationships. For example, community is an important source of credit and useful daily information. However, as Table 13.3 reveals, the community network is not currently being tapped for the transfer of information on improved farming practices.

The existing social network in the case study area does not support a widespread transfer of knowledge from either research or government institutions to the farmers. The foregoing discussion reveals that, in many cases, the links between the farmer, policy and science agents were unidirectional and lacked feedbacks, as depicted in Figure 13.4. Extension services through action research have been more targeted towards traditional and subsistence farmers. However, most research that aims to provide extension services on the use of new farming techniques is not action oriented and research that is purely scientific has very little outreach to farmers. Researchers contacted the local government and village heads only to inform them of the objective of the work and to obtain permission to conduct studies. In most cases, this was the only link that existed between the researchers and local officials.

The village heads do not have records on the studies carried out by research institutions in their areas, which shows the lack of cooperation among them. The farmers also mentioned that they did not get any feedback from the researchers after the studies had been completed, which resulted in a lack of interest from some to collaborate with them. Extension services provided by private companies, such as seed companies, have been more useful for diversified and commercial farmers. Since the main objective of a seed company is to

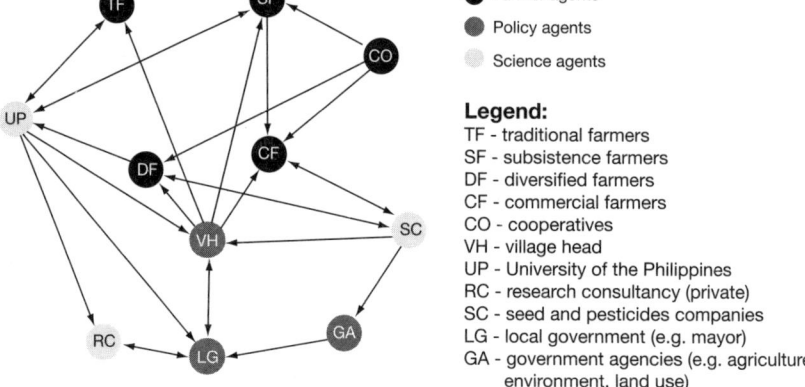

Figure 13.4 *Network for transferring knowledge among farmers, policy agents and science agents*

Note: The arrows show the direction of knowledge transfer
Source: chapter authors

increase sales, they usually target the farmers who not only show an interest in using the seeds, which are free, but are also able to cover the costs of fertilizer. According to the sales representative of a seed company, other farmers are encouraged to buy new seeds only when they see the proven success of fellow farmers in adopting them. The transfer of knowledge is thus through comparison and emulation within a farmer's social network. As the traditional farmers are the least integrated in the network, they are less likely to adopt the new seeds.

Because a lack of knowledge is one of the most important reasons given by all types of farmers for not adopting new crops or farming techniques, improving the current social network would help many of them to increase their adaptive capacity. As discussed above, a better link between science and policy at the local level through intermediaries could improve the overall social network in the case study area. Other recommendations to improve the flow of information among the different agents are as follows:

- The head of the village could assign one of his councillors to observe how the research is conducted and to understand the benefits of the results. The research results should be presented to the farmers, particularly those who have served as informants to the researchers, by the village heads and scientists. Each village should employ a qualified councillor who coordinates the activities of research institutions and keeps records of the studies conducted in the areas. This would ensure that the knowledge generated from various projects conducted in the case study areas would be continuously available for use not only by the local communities, but also by other scientists who would conduct similar studies in the future.

- The local government could work together with private companies to support the needs of the farmers while achieving the latter's corporate goals (as shown by the example of the seed company). The participation of some farmers in the technical training on seed trials was hindered by their lack of budget for fertilizer and pesticides. Although the local government did not have the budget to subsidize the production inputs of the farmers, which are necessary for the farmers to try out new seed varieties, it could contact the companies producing organic fertilizer to work together with the seed companies. Free samples of organic fertilizer and new seeds could be given to the farmers so that they could experiment on the viability of the new technologies on their farms. It would be easier to convince the farmers to invest and adopt new crops and farming techniques if they could see the success on their own farms, rather than on demonstration farms. The local government should take advantage of the companies' extension services, which aim to promote the sale of their products.
- The preparation of the comprehensive land-use programme in Tanauan City is a good example of combining the practical expertise of scientists and the policy objectives of the local government. Similar policy exercises that need some scientific inputs could be promoted in the future. However, the collaboration between science and policy should not stop after the delivery of the endorsed comprehensive land-use programme. The scientists should be given the opportunity to carry out regular monitoring during the implementation phase of the programme and conduct a post-evaluation at its conclusion. The former would allow appropriate adjustments to the programme while considering the fast-changing physical and economic environment, while the latter would support the preparation of a better update of the programme in the future. Such a system of monitoring and evaluation exists for development projects funded by international organizations and coordinated by national or regional government agencies.

Figure 13.5 depicts an idealized network that considers an intermediary of farmer associations and the above-mentioned recommendations to improve the transfer of knowledge to farmers. The suggested social network would allow for feedback effects between science and policy.

Vulnerability and network scenarios

Using the results of the social network analysis (i.e. Figures 13.4 and 13.5), the agent-based model was applied to identify the impacts of improved links between science and policy agents on reducing the vulnerability of farmers in the village of Cale in Tanauan City. To simulate trends in vulnerability to global changes, trade and climate scenarios were used in the agent-based model. Changes in prices and yield represented the impacts of trade liberalization and climate change. These changes were based on the results of the trade

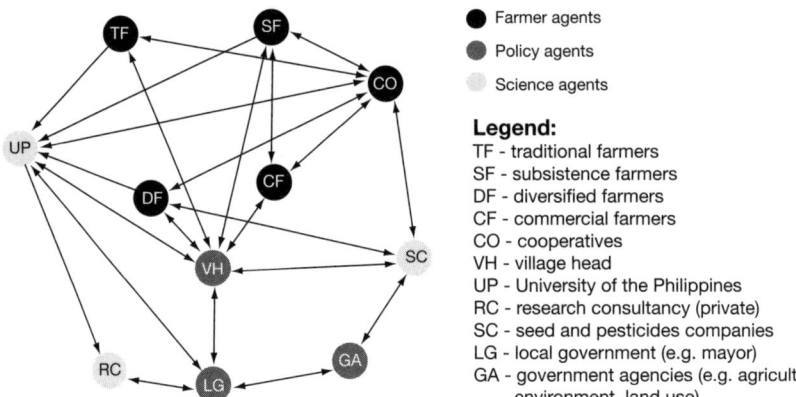

Figure 13.5 *Idealized network for transferring knowledge among the agents*

Note: The arrows show the direction of knowledge transfer
Source: chapter authors

model of Conforti and Salvatici (2004) and the crop model of Centeno et al (1995). The increase in paddy prices in the Philippines in 2013 was 1.7 per cent; cereals, 0.7 per cent; vegetables and fruits, 4.4 per cent; sugarcane, 2.2 per cent; and other agricultural crops, 0.5 per cent. The yield of paddy declined by about 6.27 per cent for a 1 per cent increase in temperature. The other crops were assumed to experience the same yield effects in the model. The global change scenarios suggest that while economic risks resulting from decreased yield would be very high, economic opportunities from increased prices would be very low.

The interviews suggest that there are opportunities to reduce the vulnerability of the farmers through the adoption of new seed varieties that are adapted to the current climate and economic condition in Tanauan City. However, these opportunities are limited to a small number of farmers due to poor transfer of knowledge. Figure 13.6 reveals that the level of vulnerability of the farmers in Cale village with respect to the different components (i.e. economic gain, community integration, crop diversification and farm management) is higher in the current social network than in the idealized social network scenario. In the agent-based model, economic gain is expressed in terms of the average income of the farmers as a result of adopting new crops that are resistant to drought and/or have higher prices. Community integration represents the number of interactions among farmers, science agents and policy agents, which results in an increase in the rate of transfer of knowledge. Crop diversification measures the ratio between the areas planted to the different crops and the total agricultural area. Farm management captures the changes in the overall yield of the farmers resulting from their choice of crop and use of production inputs. With the idealized social network scenario, the level of vulnerability with respect to economic gain would decline from high to low during the first half of 2010 due to an increase in the income of many farmers.

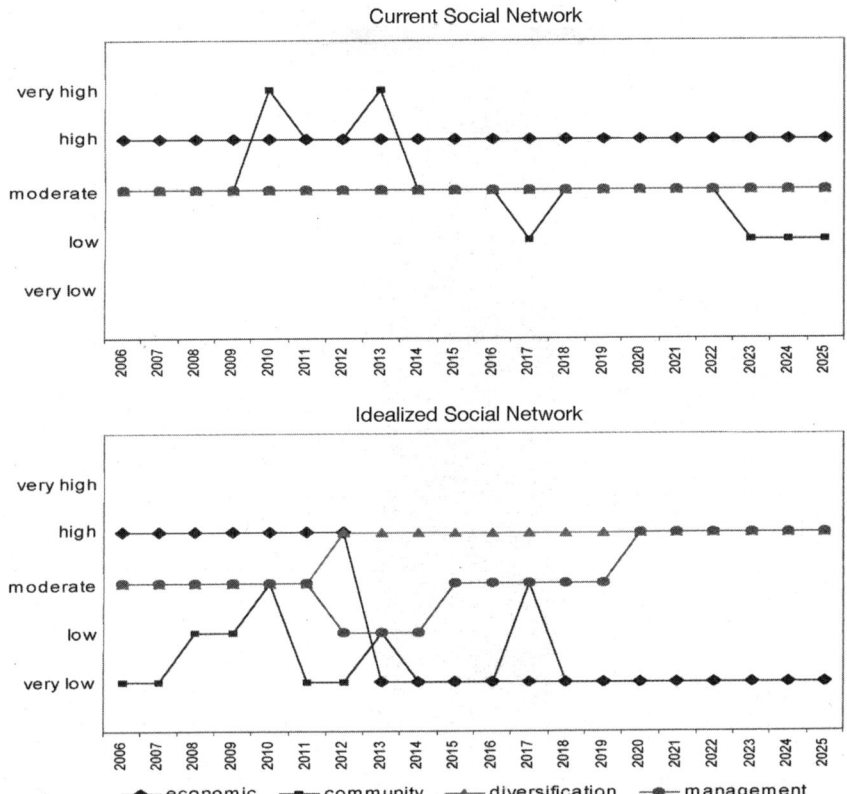

Figure 13.6 *Trends in vulnerability components with the current and idealized social network scenarios, 2006–2025*

Source: chapter authors

However, vulnerability with respect to crop diversification would increase from moderate to high. The poor level of crop diversification in Cale village is attributed to the effects of a contract-growing system.

Figure 13.7 compares the trends in vulnerability in Cale village from 2006 to 2025 under the current social network and idealized social network scenarios. The level of vulnerability of the farmers would decrease with the latter scenario because the transfer of knowledge about new crops, which are suitable to the changing climate and economic conditions, would significantly increase by improving the links between science and policy. By supporting the transfer of new knowledge about crops developed in research institutions and private companies, the government would increase the number of farmers who would adopt these technologies. Verification of the model results was made using sensitivity and uncertainty analyses. Incremental changes on the income parameters were mapped against the spatial land-use patterns that were generated from the model for the year 2025 to analyse the sensitivity of the model to the baseline values of the parameters. A similar mapping procedure was

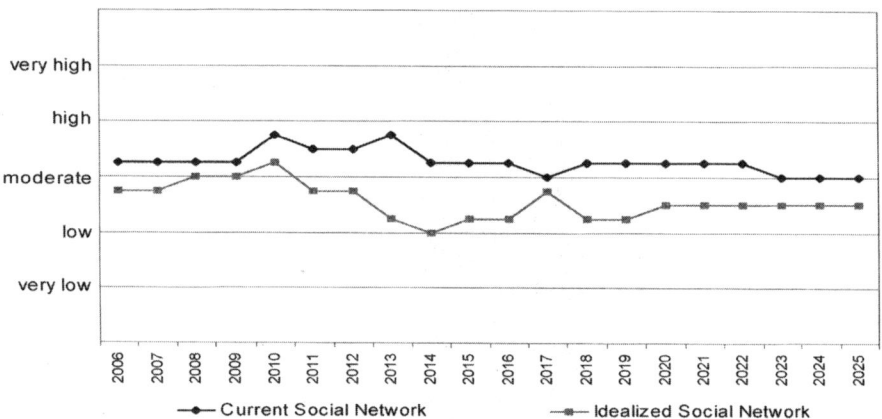

Figure 13.7 *Trends in vulnerability with the current and idealized social network scenarios, 2006–2025*

Source: chapter authors

applied to the decision rules (e.g. farm size and education level) to verify the sensitivity of the model results to the changes in these rules. Uncertainty analysis is particularly important for agent-based models as a means of accounting for errors and uncertainties in GIS maps. Such errors may have been generated from classifying land use through remote sensing and identifying farm boundaries through GPS. The degree of error propagation and uncertainty was investigated through random distortion of land use and systematic changes in farm size and boundaries.

The application of an agent-based model in this chapter has been useful for evaluating scenarios not only of environmental changes and adaptation options, but also of social networks through an improved science/policy interface. The model has combined a large amount and different types of qualitative and quantitative information, including spatial digital maps, time-series data, agent profiles, cognitive strategies, decision rules and scenarios. It has provided a means of formalizing information that emerged through social surveys and interviews. Implementing an agent-based model is thus time consuming and data hungry. This characteristic limits its empirical application to well-documented case study areas.

Conclusions

Science has the challenging task of informing policy about the sustainability of measures that will help local communities to adapt to the impacts of global changes. Building an effective science/policy interface, as argued throughout this book, is thus a prerequisite for reducing the vulnerability of communities. A process-based tool that models visible links and adaptive decisions of actors in science and policy was applied in this analysis to trace evidences on how adaptation measures reduce the vulnerability of rural communities in the municipality of Tanauan City in the Philippines. The case study area is characterized by the good quality of agricultural lands and the availability of large public markets. However, the productivity of farms is affected by problems of soil erosion and frequent droughts, while the marketing of produce is affected by the intensive competition with farmers from neighbouring provinces and cities.

The application of cluster analysis from the social survey data revealed that among the four farmer typologies (traditional, subsistence, diversified and commercial), the traditional farmers in the village of Cale are the most vulnerable. Two of the most important reasons for their high level of vulnerability are lack of money and knowledge. While science, through action research that involves social learning, was able to gain a deeper understanding of the adaptation needs of the vulnerable, the knowledge has not been effectively propagated and sustained because of the existing bottlenecks in the science/policy interface. The bottlenecks in the case study area include diverse understandings of vulnerability and adaptation; lack of common objectives and interests; weak institutional and structural links; lack of recognized and influential intermediaries; and lack of close partnerships and constant dialogue.

The social network analysis revealed that the existing network in the case study area does not support a widespread transfer of knowledge to farmers from either research or government institutions. Suggestions have been offered to improve the social network and, thus, the flow of information. These could be techniques such as improving the link between science and policy at the local level through intermediaries (e.g. farm associations); employing a qualified village councillor who would liaise the activities of research institutions and keep records of the studies conducted in his area of jurisdiction; collaborating with private companies to support the needs of farmers; and allowing the participation of scientists not only in the development of local government plans, but also in monitoring and evaluation of those plans.

There is an adequate knowledge base, both from scientific and action research, to support short-term efforts for reducing vulnerability and increasing adaptation. However, the results of the agent-based model suggest that a sustainable reduction of the vulnerability of farmers in the case study area could be achieved in the long run only with an improved social network among science, policy and farmer agents. Both the generation and distribution of knowledge require a close relationship with the community to build trust, which is necessary to encourage the use of appropriate adaptation measures among the farmers. The most important sources of information on technical adaptation options among the farmers are currently the family and neighbours. The structure of adaptation should thus be able to extend this level of trust outside family and neighbour relationships to include extension services (i.e. from science, government and companies), local officials, co-operatives and the farming community. Social learning thus requires trust. Family and neighbours are sources not only of knowledge about adaptation, but also a source of credit that enables farmers to implement the measures. For many farmers, technical assistance from new varieties of seed and modern farming techniques should be complemented with input and/or market support. Thus, in addition to improving the social network, the appropriate bundle of support should be given to those who need it to increase their capacity to adapt to the combined impacts of climate change and globalization. Inability not only to identify and provide, but also to update this bundle to meet the challenges of future global changes, would cause any successful science/policy interface to fail.

Acknowledgements

The science/policy component of this project was supported by a grant provided to Harvard University by the David and Lucile Packard Foundation. The Intervulnerability Project, which provided the groundwork for the Philippine case study, was supported by the global change SysTem for Analysis, Research and Training (START), in partnership with the International Institute for Applied Systems Analysis (IIASA) and the International Human Dimensions Programme (IHDP), with the financial support of the David and Lucille Packard Foundation.

References

Acosta-Michlik, L. (2005) *Intervulnerability Assessment: Shifting Foci from Generic Indices to Adaptive Agents in Assessing Vulnerability to Global Environmental Change (a Pilot Project in the Philippines)*, Report on a Project of the Advanced Institute on Vulnerability to Global Environmental Change, www.start.org/Program/advanced_institutes_3.html

Acosta-Michlik, L. and Rounsevell, M. (2005) *From Generic Indices to Adaptive Agents: Shifting Foci in Assessing Vulnerability to the Combined Impacts of Globalisation and Climate Change*, IHDP Update 2005, Bonn, Switzerland

Acosta-Michlik, L. and Rounsevell, M. (2008) 'An agent-based framework for assessing vulnerability futures', in A. Patt, D. Schröter, R. J. T. Klein and A. C. de la Vega-Leinert (eds) *Assessing Vulnerability to Global Environmental Change: Making Information Useful for Adaptation Policy and Decision-Making*, Earthscan, London

Acosta-Michlik, L., Rounsevell, M., Van Doorn, A. and Bakker, M. (2005) 'Modeling agent's socio-economic and ecological environment: An agent-based approach for developing land use change scenarios', in *Crop Science and Technology 2005 Proceedings*, Glasgow, 31 October–2 November 2005, www.bcpc.org/Congress2005/Congress2005_home.asp

Borgatti, S. P. and Foster, P. C. (2003) 'The network paradigm in organizational research: A review and typology', *Journal of Management*, vol 29, no 6, pp991–1013

Burt, R. (1992) *Structural Holes*, Harvard University Press, Cambridge, MA

Centeno, H. G. S., Balbarez, A. D., Fabellar, F. G., Kropff, M. J. and Matthews, R. B. (1995) 'Rice production in the Philippines under current and future climates', in R. B. Matthews, M. J. Kropff, D. Bachelet and H. H. van Laar (eds) *Modeling the*

Impact of Climate Change on Rice Production in Asia, CAB International, Wallingford and the International Rice Research Institute, Manila, pp237–250

City of Tanauan (2004) *Socio-Economic Profile: City Of Tanauan*, Municipal Office of Tanauan City, Batangas, the Philippines

Conforti, P. and Salvatici, L. (2004) *Agricultural Trade Liberalization in the Doha Round: Alternative Scenarios and Strategic Interactions between Developed and Developing Countries*, FAO Commodity and Trade Policy Research Working Paper No 10, Rome

Cross, R., Borgatti, S. P. and Parker, A. (2001) 'Beyond answers: Dimensions of the advice network', *Social Networks*, vol 23, pp215–235

Cross, R., Parker, A. and Borgatti, S. P. (2002) *A Bird's-eye View: Using Social Network Analysis to Improve Knowledge Creation and Sharing*, IBM Institute for Business Value, New York, NY

Granovetter, M. (1973) 'The strength of weak ties', *American Journal of Sociology*, vol 78, pp1360–1380

Hansen, M. (1999) 'The search-transfer problem: The role of weak ties in sharing knowledge across organizational subunits', *Administrative Science Quarterly*, vol 44, pp82–111

Heir, J. F., Anderson, R. E., Tatham, R. L. and Black, W. L. (1995) *Multivariate Data Analysis*, Prentice-Hall, Upper Saddle River, NJ

Jager W., Janssen, M. A., De Vries, H. J. M., De Greef, J. and Vlek, C. A. J. (2000) 'Behaviour in commons dilemmas: *Homo economicus* and *Homo psychologicus* in an ecological-economic model', *Ecological Economics*, vol 35, no 3, pp357–379

MADECOR (Mandala Agricultural Development Corporation) (2005) *GIS Maps for the Comprehensive Land Use Plan and Tax Mapping Project for the City of Tanauan*, Laguna, the Philippines, MADECOR Environmental Management Systems Inc., College, Los Baños

National Statistics Office (2000) *NSO Census*, Republic of the Philippines

Rogers, E. (1995) *Diffusion of Innovations*, 4th edition, Free Press, New York, NY

Shah, P. (1998) 'Who are employee's social referents? Using a network perspective to determine referent others', *Academy of Management Journal*, vol 41, no 3, pp249–268

Wenger, E. (1998) *Communities of Practice*, Oxford University Press, Oxford, UK

14

Addressing Vulnerability in the European Programme for Food Aid and Food Security: Knowledge Gaps, Obstacles and Opportunities Across the Science/Practice Interface

Elia Machado

Introduction

This chapter presents a case study from the Vulnerability and Resilience in Practice (VARIP) research project. As an integral part of the open-ended network Initiative on Sustainability Science and Technology, VARIP focused on the linkage of science and practice in sustainability science endeavours within the area of vulnerability, adaptation and resilience to global environmental change. To this end, a number of core questions were drafted in this domain, which were the focus of several case studies.

This study approaches the issue of the science/practice gap by focusing on the process of knowledge transfer about vulnerability and food security among the different actors of the European Programme for Food Aid and Food Security in Nicaragua (EP-FAFS). Here, knowledge transfer is used in its broader sense, including both tacit and explicit knowledge (as per Davenport and Prusak, 2000) and refers to exchanges between individuals and groups of researchers and decision-makers. Effective knowledge transfer has implications for the success of an organization by promoting the identification and dissemination of best practices and the co-production of new knowledge, among other potential benefits (Argote and Ingram, 2000; Davenport and Prusak, 2000).

In particular, our interest is the identification of the gaps, obstacles and opportunities for improvement of the knowledge-transfer process. Insights on these issues are investigated by tracing the articulation of vulnerability, adaptation and resilience concepts in the EP-FAFS, and focusing on how these are conceptualized, communicated, transformed and operationalized through its web of actors.

First, we provide an overview of the context in which the programme operates. Subsequently, we describe the web of actors involved in the programme. A characterization of the science/practice interface follows and, next, we discuss the VARIP core questions. Finally, we suggest how the science/practice interface could be improved in the context of the EP-FAFS.

The findings suggest that the major obstacle to food security in Nicaragua does not lie in deficiencies in the process of knowledge transfer or in the external food security programmes such as the EP-FAFS *per se*, but rather in this country's institutional and political capacity to address the structural causes of food insecurity.

Food security is an inherently complicated issue that relates to many other issues; among the most important are environmental resources, but also political governance and the market economy. In Nicaragua, political climate and governance tensions have permeated important aspects of food security policies (e.g. access), thereby constraining their performance.

Research approach and objective

This research seeks to inform the following core questions defined by the VARIP group (adapted to this case study) as follows:

- How adequate is the knowledge base within the EP-FAFS about vulnerability? Who has generated that knowledge? How is that knowledge distributed among the actors involved in this programme?
- How may the EP-FAFS's science/practice interface be characterized? Who are the principal actors? What are their roles and interests?
- To what extent do the actors involved in the EP-FAFS make use of the knowledge available to them? How relevant and pertinent is the knowledge to the needs of decision-makers and other actors?
- What barriers and failures limit the transfer of knowledge among the actors across the science/practice interface? How important are the intermediaries?
- How does the nature of institutions hosting the EP-FAFS shape its science/practice interface? To what extent is institutional fragmentation a problem? How permeable are the boundaries of institutions to new information?

- How can the science/practice interface be improved?

An extensive documentary and policy analysis were conducted, and insights on these questions were drawn by examining the articulation and operationalization of vulnerability, adaptation and resilience concepts throughout the programme. To this end, a study of the actors (i.e. the persons and institutions involved in the EP-FAFS) and their relationships within the programme were critical. Some examples of the documents examined include legal frameworks of the European Commission (EC) food security policy, EC assessments of the EP-FAFS, field reports from EC delegations, the EC's food security monitoring documents from the EC and its partners – for example, the United Nations Food and Agriculture Organization, and background material on Nicaragua. Those interviewed included officials, staff from EC headquarters and delegations, and experts contracted by the EP-FAFS (whose identity has been protected). The insights presented in this chapter pertain to the programme as per the European Council Regulation (EC) No 1292/96 of 27 June 1996 on Food Aid Policy and Food Aid Management, and special operations in support of food security. As of 1 January 2007, the regulation was replaced by Regulation (EC) No 1905/2006 of the European Parliament and of the Council December 2006 Establishing a Financing Instrument for Development Cooperation (European Parliament and of the Council, 2006).

All of these questions are addressed below.

Background

Socio-ecological context

Nicaragua is a country historically marked by political and socio-economic instability as well as by high exposure to natural hazards. Following the peace agreements of the 1990s, an aggressive reform, focused on strengthening the political system and the market economy, started. This period was characterized by some recovery of the agricultural sector, but also an increase in migration (and remittances), as well as an expansion of the *maquiladora* industry.

Democracy brought up the resumption of massive international aid, upon which the country is highly dependent, and to which the European Union (EU) is a major contributor. In 2005, the EU was funding 35 food security-related projects, including rural development initiatives such as the Food Security and Local Development Programme (PRODELSA), adding up to more than 42 million Euros in 16 northern Nicaraguan municipalities.

During the last 20 years, the country has undergone radical changes, involving the hereditary dictatorship of the Somozas, the Sandinista revolution and four democratically elected governments. The latter have focused on performing market economy reforms, regional economic integration with Central American countries, and reducing the national debt after achieving highly indebted poor country (HIPC) completion point (IMF, 2000). These policies have had significant implications for the food security situation of the country.

A critical fact that shapes Nicaragua's history is its high exposure to natural hazards and severe droughts. Two events have been particularly dramatic: the 1972 earthquake (5000 deaths) and Hurricane Mitch in 1998, which caused 3800 deaths and more than 800,000 victims (NOAA, 2005).

The EC focuses its policies on the socio-economic situation of the country. Some relevant facts are as follows:

- Nicaragua is the poorest country in Latin America and the Caribbean after Haiti. Although its relative poverty indicators fell between 1993 and 1998, the absolute number of poor people increased.
- Poverty is mostly rural. About 65 per cent of the population is poor in rural areas versus 33 per cent in urban areas; but poverty is increasing in urban and decreasing in rural areas (except for the Pacific rural). This is due, in part, to an increase of agricultural exports in addition to the rapid increase of immigrants' remittances (pers comm). The low poverty incidence in Managua (0.25 per cent) is attributed to economic expansion and the increase of the service sector. The richest 10 per cent of the population receive 45 per cent of the country's total income, whereas the poorest 40 per cent receive only 10 per cent (the income Gini coefficient was 53 in 2001) (World Bank, 2000).
- The HIPC decision point was reached in 2000; but Nicaragua's trade deficit remains chronic. The country is also highly sensitive to market price fluctuations, particularly coffee. The prevalence of undernourishment in Nicaragua is higher than in Central America, Latin America and the Caribbean. Although the proportion of undernourished has decreased, the percentage is still high, and the number of undernourished people has increased compared to the 1990 to 1992 period (FAO, 2006; see also Figure 14.1, opposite).

The causes of food insecurity in Nicaragua have been related to persistent structural problems such as high levels of poverty and unemployment, among others, including:

- competition difficulties in the external market and high technology dependency;
- low educational level of the rural population;
- unresolved conflicts over land property resulting in landless families, as well as a lack of support to subsistence agriculture and organizational deficiencies among the producers;
- environmental degradation and high exposure to natural hazards;
- spatial and temporal concentration of agricultural production in the Pacific and Central regions (about five to six months per year);
- low exports volume – the main exports are concentrated in a few products such as coffee, meat, sugar, fishery products, prawns and lobsters.

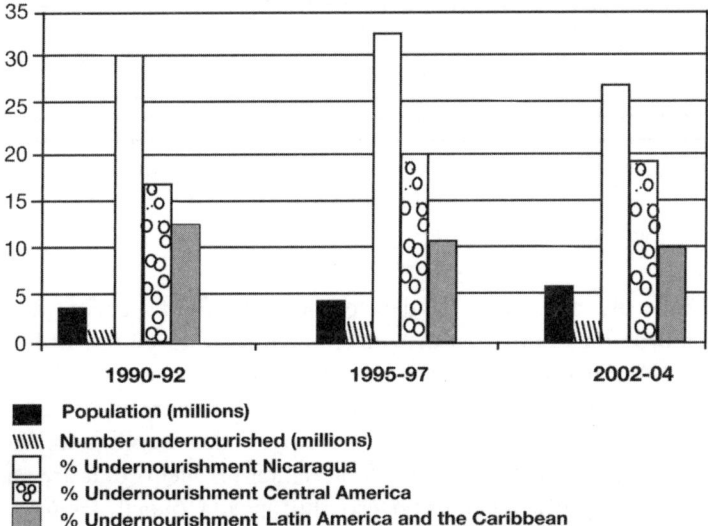

Selected statistics	1990-92	1995-97	2002-04
Population (millions)	3.9	4.6	5.5
Number undernourished (millions)	1.2	1.5	1.5
% Undernourishment Nicaragua	30	33	27
% Undernourishment Central America	17	20	19
% Undernourishment Latin America and the Caribbean	13	11	10
GDP per capita	694	-	793
Population below 1 US dollar per day (%)	48	-	45

Figure 14.1 *Selected population food security trends in Nicaragua*
Source: FAO (2006)

From an institutional perspective, the EP-FAFS is heavily influenced by three major policy domains (i.e. cooperation, development and trade from regional to international scales) (see Figure 14.2).

The European Programme for Food Aid and Food security (EP-FAFS)

Historical overview

European food aid started with the 1967 International Convention on Wheat, serving the main purpose of managing the community's stocks. As such, the aid was strongly conditioned by agricultural surpluses. Since then, the EC's food aid has evolved from being mostly part of poverty alleviation initiatives and

> **INTERNATIONAL**
> Agreements in environment, trading and foreign aid: Nicaragua is a member of the World Trade Organization and Sistema de Integración de Centroamericana (SICA) and signatory of a number of bilateral trade agreements, including the Central America Free Trade Agreement (CAFTA) with the United States. This country received about USD 600 million/year in aid during the 1990s. The main sources of development financing are the World Bank, Inter-American Development Bank and the Central Economic Integration Bank. The EC and its member states are Nicaragua's main net donor. Japan and the United States are significant contributors, but the latter has reduced its assistance.
>
> **NATIONAL**
> Nicaragua's Strengthened Growth and Poverty Reduction Strategy Paper, 2001, which aimed at:
> - Broad-based economic growth and structural reform
> - Investment in education and health services
> - Better protection for vulnerable groups, access to public services, and good governance.
>
> The second generation SGPRS (2004) focused more on reducing poverty.
>
> Trade relations: Nicaragua is a beneficiary of the EU's system of generalized preferences.
>
> **REGIONAL**
> EC *Country Strategy Paper* (1998–2000) outlines EU's priorities for Nicaragua:
> - Support for democratic governance
> - Developing agricultural production and livestock
> - Promote social cooperation
>
> EC *Country Strategy Paper* (2002–2006) emphasizes support of national programs for:
> - Rural development
> - Education governance and civilian security, as well as economic cooperation
>
> EU - Central America Political Dialogue and Cooperation Agreement (2002–2003)

Figure 14.2 *Institutional context of the European Programme of Food Aid and Food Security*

Source: chapter author

oriented towards food aid to the development of a food security strategy that finances a broadened range of instruments within a development framework (European Commission, DG Development, 2001).

The EC's legal framework of the EP-FAFS describes it as an instrument to address the poverty dimension of food security whose goal is to bring assistance to developing countries facing food deficit problems that are temporary and mostly structurally linked to poverty (European Council, 1996).

The programme is regularly submitted to the European Parliament and coordinated with other development programmes of the EC, the member states and donors. Its implementation is a complex process that involves several actors as well as decision-making across scales.

The programme is founded on two pillars:

1 The food security definition from the World Food Summit in Rome and the London Food Aid Convention (1999):
   ... food security, at the individual household, national, regional and global levels, is achieved when all people, at all times, have physical and economic access to sufficient, safe and nutritious food to meet their dietary needs and food preferences for an active and healthy life. (FAO, 1996)
2 The European Council Regulation (EC) no 1292/96 of 27 June 1996, and the European Commission's Communication COM (473) 2001 on the integration of the programme within the EC's development policy and

development cooperation, which strongly influence EC's food security policy.

Operationalization

The EC's food security policy aims to address the underlying structural causes of food insecurity and stresses the importance of focusing on three food security dimensions (European Commission, 2001):

1   food availability at the national level (food production and trade);
2   food access at the household level resulting from poverty;
3   food utilization and nutritional status at the individual level.

Food security operations, however, have focused on increasing food access since 'food insecurity is above all a problem of lack of access (economic and physical) to food' (EuronAid, 2006, p9).

The EC works with partners at several levels to achieve its goals through direct aid to governments and governmental institutions, such as the Ministry of Agriculture and Forestry (MAGFOR), the Consultative Group on International Agriculture Research (CGIAR), the United Nations World Food Programme (WFP), the International Committee of the Red Cross (ICRC) and other international and local non-governmental organizations (NGOs). The total programmed aid per year, from 2003 to 2006, was 440.34 million Euros, 419 million Euros, 412.65 million Euros and 441 million Euros, respectively (European Commission, DG Development, 2005) (see Table 14.1).

It is noteworthy that although the EC collaborates with the United Nations World Food Programme at the international level, this partnership was severed in Nicaragua during the mid 1990s (pers comm). The EC emphasizes the need for consistency of the EP-FAFS with the EC's country and regional support strategies, as well as coordination with the EC's Development Policy and

Table 14.1 *European Commission fund allocations in food security for 2003 and 2004*

EC fund allocations (million Euros)	2003	2004
Support to government programmes:	134.8	140.8
NGOs	25	22.5
Others: United Nations Food and Agriculture Organization (FAO) and Consultative Group on International Agricultural Research (CGIAR)	22.2	27.7
Total funding for structural food security:	*182*	*190.7*
United Nations World Food Programme (WFP)	120	105
NGOs	50	50
United Nations Relief and Works Agency (UNRWA)	15	13
International Committee of the Red Cross (ICRC)	8	8
Government	49.35	30
Total food aid through annual allocations:	*242.35*	*206*
Capacity-building, price contingencies, monitoring and reserve	15.99	22.3
Total	**440.34**	**419**

Source: European Commission, DG Development (2005)

European Commission Humanitarian Office (ECHO) interventions. The programme also seeks to build on national poverty reduction policies and to foster local partnerships. There is also an EC mandate for impact appraisal of the programme, particularly in terms of disaster preparedness, and specifically in linking relief, rehabilitation and development. Examples of the EP-FAFS's operations include food aid in kind or cash, support for research operations (e.g. field training) and a variety of food security actions, such as supply of tools and inputs for agricultural production, credits and marketing, early warning systems and capacity-building (European Council, 1996; European Commission, EuropeAid Cooperation Office, 2001).

Several projects have been executed under the EP-FAFS budget lines; but since 2000 there has not been a new open call for project proposals in Nicaragua. The last significant project was PRODELSA (pers comm). What is more, the food security budget line is closing.

Targeting: Selection of countries and vulnerable populations

The importance of identifying vulnerable groups properly is stressed throughout the EP-FAFS, which is critical not only for effective use of the funds, but also for a better apprehension of their coping strategies with the food risks (European Commission et al, 2006). Several targeting criteria have been used at different scales.

Country level

Economic criteria and indicators such as being a least-, low- or middle- income country are employed. A list of 127 eligible countries was provided in the Regulation 1292/96 Annex; from these, a list of priority countries is created every funding period. For example, the Food Aid/Food Security budget lines for 2005 to 2006 selected about 30 countries for funding (European Commission, DG Development, 2005). The selection changes from year to year depending upon the countries' food security situations relative to the global context (European Commission et al, 2006). In Latin America, priority countries between 2004 and 2006 included Bolivia, Ecuador, Guatemala, Honduras and Nicaragua.

Regional and project level

According to the EC's 2005 to 2006 food security work plan, 'lack of [food] access results from ... weak purchasing power. At project level ... international organizations and NGOs are adequate for [ensuring] an appropriate targeting' (EuronAid, 2006, p9). The targeting method at this level is driven by a call for proposals. These are launched by the EC field delegations, which, together with EC headquarters, draft the *Country Technical Document* defining the selection criteria, fields of interventions and geographic targeting (European Commission et al, 2006). Other international organizations such as the FAO, CGIAR, WFP and ICRC have annual funds allocations following negotiated procedures with the EC (see Table 14.1).

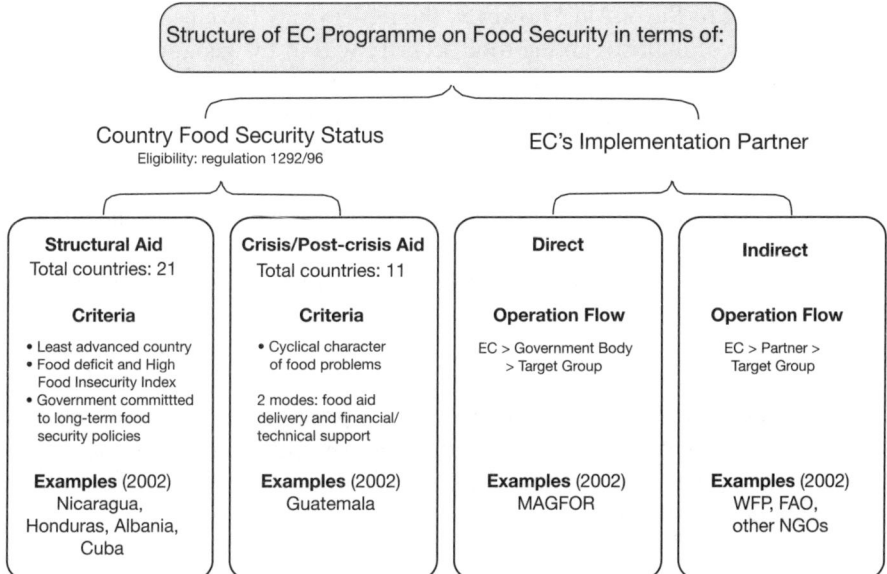

Figure 14.3 *EP-FAFS programme operation modes according to the food security states of the beneficiary countries and type of implementation partner*
Source: chapter author

Structure of operations

The programme operating modes depend upon the food security status of the countries and the EC's implementation partner (see Figure 14.3). In direct operations, which represent the majority of the funds allocations, the EC works directly with the government of the recipient country, while in indirect interventions an intermediary such as an NGO mediates. The indirect mode of operation has not been active in Nicaragua since 2000. The type of intervention adopted is critical because the sources of information, regional targeting, strategies and decision-making processes (i.e. integration of stakeholders) differ.

Looking ahead: The new EC Food Security Policy

Since 2004, the EP-FAFS has been undergoing a controversial reform. Regulation (EC) No 1292/96 was repealed by Regulation (EC) No 1905/2006 of the European Parliament and of the Council of 18 December 2006 establishing a financing instrument for development cooperation (*Official Journal* L 378 of 27 December 2006).

Profound changes in the EC food security policy are expected through a major policy reform. However, at the time of writing this chapter, the process had not been completed. For the EC, the aim of the reform is to emphasize the role of field actors and to foster non-state actors' participation through reduced top-down approaches (and more contextualized operations).

According to EuronAid,[1] however, the reform may have negative implications in terms of mainstreaming food security within EC geographical budget lines and the assignment of food aid management solely to ECHO, which may compromise the flexibility to review programming targeting as needed.

Two projected changes are relevant for this study:

1. the decision to transfer the emergency food aid programmes to the humanitarian instrument (ECHO); and
2. the fact that from 2007, food security country programmes will be financed through geographical instruments which may undermine the priority of food security actions *vis-à-vis* other priorities for the regions.

While the implications of these changes on the programme performance goals are yet to be seen, there are some specific concerns about the long-term viability of food security, since they would become part of other EC development programmes and related actions (pers comm).

The science/practice interface

The web of agents: Multiple levels of articulation

The science/practice interface has been characterized by scholars' understandings of knowledge flows among the actors within scientific and practical domains. These conceptualizations range from basic bridges to complex spider webs that serve as canals for knowledge flow (Cash and Moser, 2000; see also Chapter 4 in this volume). The metaphor of a spider web of actors (see Chapter 1) seems appropriate in this case, given the complexity of interactions that emerge from the actors involved at different operational levels of the programme, and the diversity of the targeting scales and implementation strategies of the EC's food security operations. These multiple levels of articulation result in several knowledge pathways and the linking of member states, private donors, international food programmes and other programmes of the EC, European and local NGOs, and governmental institutions of Nicaragua, among other actors.

Evaluating the individual contribution of the EP-FAFS to enhance food security is very complex because it is a component of the broader EC food security policy, and thus strongly linked to a variety of other EC budget lines through national and regional geographical programmes (e.g. the European Development Fund and the Asia and Latin America Programme). In fact, a problem for the official evaluators of the programme was to determine the added value of each food security instrument and its specific role in achieving food security (PARTICIP GmbH, 2004).

The two major actors of this programme are the EC's Cooperation Office, EuropeAid (Brussels) and the EC Delegation in Nicaragua.

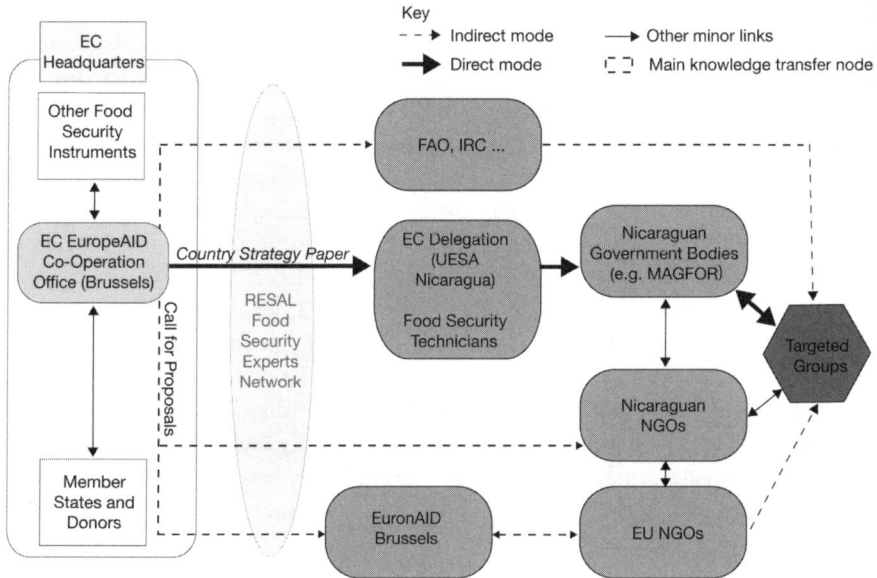

Figure 14.4 *The European Programme for Food Aid and Food Security in Nicaragua: Diagram of web actors and modes of operation*

Source: chapter author

The EC Delegation in Nicaragua is the nexus between EC headquarters and the EC's operations in Nicaragua. The EC Delegation is thus a key actor in both direct and indirect operating modes. In indirect operations, the delegation launches the call for proposals with support from EC headquarters. In direct operations, its role involves coordination and involvement with Nicaraguan and Central American institutions. Figure 14.4 describes the links between the actors of the programme according to the mode of operations (i.e. direct or indirect).

Operation mode and knowledge pathways

The EP-FAFS's mode of operation has critical implications for the questions addressed in this chapter. The *direct mode* is the preferred mode of operation. In this case the European Commission provides funding for government programmes via budget support, or else by giving direct support to private or public bodies at the local level. This type of aid requires a funding agreement between the beneficiary government and the European Commission, specifying the implementation arrangements as well as the conditions to which this aid is submitted (European Commission, EuropeAid Cooperation Office, 2001). The grounding for these actions is stated in the *Country Strategy Paper* drafted for Nicaragua, which describes the EC's agreement with Nicaragua's government on long-term food security strategies (European Commission, 2002). The establishment of priorities is thus a negotiating procedure. According to the programme's legal regulations (European Council, 1996), the policy should be set in coordination with other EC financial instruments and other major donors, and build on nationally owned policies and strategies for poverty reduction (European Commission, 2002).

Indirect operations are executed via partner organizations of the EC once the recipient countries have been targeted by the commission. These implementation partners include the following. United Nations Food and Agriculture Organization (FAO) interventions comprise a Food Insecurity and Vulnerability Information Mapping System (FIVIMS) and country-level projects (European Commission et al, 2006). The International Committee of the Red Cross (ICRC) receives an annual allocation from the EC for food aid distribution in countries at war or in high distress. The World Food Programme (WFP) is the privileged EC partner for distribution of food aid in serious food crises. Operations are based on yearly agreements with the EC at national levels. In the case of Nicaragua, however, it is noteworthy that collaboration with the WFP has been broken since the mid 1990s (pers comm). No reference to this has been found in the policy documents, however.

EuronAid is a European network of NGOs of food aid and security primarily funded by the EC. It offered services to NGOs implementing humanitarian aid programmes funded by the European Commission until 2008. EuronAid had an important role in facilitating the implementation of NGO programmes and promoting dialogue on food aid and food security policies with the EC, as well as with other interested parties (EuronAid, 2006). It offered professional services on accessing EC's food security budget lines and, when applicable, played an active role in allocations of food aid granted by the commission to NGOs (see Table 14.1) and the use of indicators of countries' performances.

EuronAid's capacity for networking and lobbying for food aid is very relevant to the questions of this study, as well as its role as a knowledge-transfer agent through training and information activities on food security (EuronAid, 2006). However, its involvement in Nicaragua is uncertain since its contract in the area ended in 2005 (pers comm). From a knowledge-transfer perspective, EuronAid could have served as a key feedback node between the NGOs and the EC, and as a means of communicating ground estimates for food aid to the EC.

Local NGOs are supported as a way of capitalizing on specific expertise via a call for proposals by the European Union. The PARTICIP GmbH report (2004) evaluated projects being implemented in 2004:

- Commercialization of Organic Produce (COPIBO): was successful in terms of commercializing organic produce to Europe.
- Intermón-Oxfam: offered legal services in rural areas to solve conflicts regarding land tenure; was successful, but had low impact compared to the investment.

- Acción Contra el Hambre (ACH): performed an agro-socio-economic study of a department in Nicaragua without much impact upon food security.
- Centro Regionale di Intervento per la Coperazione (CRIC): created a farming school for livestock production training; had good impact in local commerce, but very slow in its progress.
- Agency for Technical Cooperation and Development (ACTED): performed a project oriented towards production and technical training, but had low impact and low sustainability.

Other experts have emphasized the importance of the quality of NGO partnerships and intervention methodologies as key factors for success, and have cited the exemplary work of ACH in Madriz (a department in Nicaragua) in terms of capacity-building and natural resources conflicts (pers comm).

In addition to the actors mentioned above, others include expert networks such as the European Food Security Network (RESAL) and individual experts (see Figure 14.4).

RESAL was established in 1998 as a network to reinforce the capacity for analysing the local food security situation, helping with the decision-making process, and acting as a liaison between the EC and beneficiary countries with functions in the implementation of long-term food security policy. In principle, these attributions positioned RESAL as an ideal node of exchange of information between producers, users and the EC. Indeed, this network has been characterized as a positive asset of the programme. However, the network as such stopped in 2001 through an 'internationalization process' that integrated the experts within the delegations (pers comm).

Individual experts: since 2001 and up to 2004, the expertise gap left by RESAL was filled, to some extent, by contracted experts.

The Consultative Group on International Agricultural Research (CGIAR) is an international research network in food security and natural resources management, but the mechanism for channelling this research to the other actors of the programme is not clear. It is funded by the EC's food security budget line (23 million Euros annually); but at the time of this study did not seem to be operative in Nicaragua (pers comm).

Local experts: the programme is explicit in its intention of increasing participation and partnerships with local experts; but so far, no specific role has been made explicit for them within the programme. Their degree of involvement and interaction with other actors, particularly EC's experts, remains unclear.

The European Commission: A successful boundary organization?

Boundary-type organizations are critical players in the process of knowledge transfer between the policy and research arenas. These organizations have been formally defined as organizations mandated to act as intermediaries between the arenas of science and policy (Guston, 1999; Guston, 2001; Cash et al, 2003). Previous research has linked their effectiveness to their capacity for navigating (and benefiting from) the interaction between their science/practice communities, as well as the salience, credibility and legitimacy of their programmes' results (Cash et al, 2003).

A full evaluation of this question is beyond the scope of this chapter; instead, an initial reflection is offered here as a useful means to approach the knowledge transfer process within the European Programme for Food Aid and Food Security (EP-FAFS) in Nicaragua.

At the outset, the EC programme provides a common medium for knowledge exchange between the science and policy communities, and is accountable to both of them. However, as seems to be the case here, the existence of a common umbrella does not necessarily lead to an effective transfer of knowledge. The findings suggest deficiencies in the incorporation of experts' knowledge within the programme. So far, several configurations of expertise internalization have been used with limited success.

Characterizing the actors of the programme according to their role in the creation and usage of knowledge is a way of facilitating the understanding of its knowledge paths. This is a challenging process, however. In this case, assigning actors strictly to one of these categories does not seem adequate because their activities are not exclusive to one of the two groups. Instead, we distinguish between three categories: dominant knowledge producers, dominant knowledge users and hybrid. The latter case is particularly interesting for knowledge exchange since these hybrid actors may act as nodes of contact between user and producer communities.

Within the EP-FAFS, we may characterize the NGOs as 'dominant users', whereas FAO and EuronAid rest more in the hybrid domain. The knowledge producer communities are more clearly represented by the former European network of food security (RESAL), and the individual and local experts.

Discussion of Vulnerability and Resilience in Practice project (VARIP) questions

This section focuses on the implications of the EP-FAFS's science/practice interface in the programme's performance and the knowledge exchange among its actors in the context of the following VARIP questions.

Knowledge base articulation: How adequate is the knowledge base about vulnerability? Who has generated that knowledge? How is that knowledge distributed among the actors involved in this programme?

Insights about the knowledge base on vulnerability have been gained by assessing how the concepts of vulnerability, resilience and adaptation are articulated throughout the programme operation. This section focuses on the conceptualization of vulnerability to food insecurity. First, the association between vulnerability and poverty throughout the programme is described. Next, the programme's targeting process is examined by looking at the methodologies used to characterize vulnerable groups, followed by an analysis of the strategies used to reduce food insecurity.

Food security, vulnerability and poverty

Since its inception in the 1996 Food Summit, the EP-FAFS has been influenced by the evolution of food security conventions as well as by European development and cooperation programmes.

The result is a programme theoretically founded on a comprehensive food security vision that recognizes three key elements of food security (availability, access and utilization) and stresses the need to integrate its multiple spatial and temporal dimensions. This vision is much in line with the scholarly literature on the subject. The EC's policies are designed to ensure that all of these dimensions are addressed, particularly the structural causes of food insecurity.

Operationally, however, the EC's global goals and priorities, which centre on poverty reduction, determine food security policy. This seems also to be the case in other international organizations, as well as the Government of Nicaragua, for whom poverty reduction and the strengthening of the democratization process became a major focus after the near collapse of Nicaragua's economy during the 1980s (European Commission, 2002). The association between food security and poverty is explicit throughout the programme's regulation, particularly in its objectives (e.g. enhancing food security geared to alleviating poverty). At the national level, food insecurity is attributed to faltering development and a weak trade position, whereas, at the household level, poverty is identified as a major cause of food insecurity (European Council, 1996).

A review of the EC's poverty reduction policy as described in *Poverty Reduction Strategy* documents (Government of Nicaragua, 2000, 2001) and related development policy documents, such as the *Country Strategy Paper, Nicaragua* (European Commission, 2002), also encounter this theory/practice dichotomy. Although the multidimensional character of food security is emphasized (e.g. as a lack of access to healthcare and natural resources), the proposed instruments for food insecurity reduction centre on poverty, and are strongly based on income indicators. For instance, vulnerability is closely associated with poverty: 'poverty is often associated with ... high vulnerability to exogenous events, and lack of economic opportunities' (Government of Nicaragua, 2000, p4).

According to an interviewee, 'no official concept of vulnerability exists' (pers comm). Several usages of vulnerability, however, have been found in the *Poverty Reduction Strategy* documents:

- Susceptibility to suffering: according to these documents the poor are more likely to be negatively affected by a stressor(s) than the non-poor. Low food consumption and hunger are considered household vulnerabilities associ-

ated with low income. Other causes of vulnerability cited include lack of information, gender inequality and limited trust in formal organizations.
- Environmental vulnerability on its 'own right': Nicaragua has a high exposure to natural hazards, which has been shaping vulnerability perceptions and fostered a risk/hazards approach to vulnerability.
- Targeting and identifying 'vulnerable groups': one of the goals of the poverty reduction strategies is 'to ensure that society's most vulnerable groups have ... access to social and development programs' (Government of Nicaragua, 2001, pp3–4). This reference to vulnerable groups is used frequently, but often without specification of their defining characteristics. Further indications to targeting of the young, the elderly, the handicapped and the extremely poor give a clearer picture of the EC's characterization of vulnerable groups.

To summarize, the concept of poverty seems to capture vulnerability in relation to food security throughout the EP-FAFS, which, in turn, has fostered policies focused on long-term poverty eradication in the form of development initiatives. According to an evaluation report, however, neither the technicians from EC delegations nor many of the field cooperation organizations believe that Nicaragua's current strategy for growth is going to ameliorate the situation of the most poor in the short run (PARTICIP GmbH, 2004). In addition, the 2007 restructuring of the EC's policy as a cross-cutting theme has aggravated these concerns. The strong association between poverty and certain food insecurity outcomes (such as malnutrition or hunger) is widely acknowledged in the literature. However, given the complexity of food insecurity causes, it does not follow that efforts aimed at reducing poverty will necessarily address food security issues comprehensively.

This near equalization of poverty with food insecurity has important practical implications. On the one hand, it 'logically' leads to food security-mitigating instruments focused on promoting economic growth, while, on the other, it significantly reduces the dimensions of vulnerability to food insecurity being addressed, and, thus, potentially the effectiveness of the programme.

Methodologies for characterizing vulnerable groups

This study has not developed an integrative framework for targeting the beneficiaries of the EP-FAFS that integrates all criteria across spatial scales. The methods used for targeting follow mostly a top-down filtering approach in which the selection process starts at the country level, based on economic indicators such as income and food deficit status.

The *Country Strategy Paper, Nicaragua* guides interventions at a national level, in general terms, without explicit spatial delineations (European Commission, 2002). In this regard a field evaluation (PARTICIP GmbH, 2004) has stressed the need to redefine Regulation CR 1292/96 and its method of targeting populations at risk.

A systematic methodology has been developed specifically for targeting by the RESAL network (Laforge, 2001). However, no evidence of its application has been found in this programme beyond an attempt of validation in two municipalities of Nueva Segovia. Personal interviews suggest that the extent to which the method has informed regional EC priorities has been minimal, if at all, although at some point the document had the potential of being used by NGOs.

The information at hand suggests that, ultimately, the local targeting of recipient groups is contingent upon the type of EC partner. In the case of direct aid to the Government of Nicaragua, it is likely that the targeting will be driven by poverty indicators and field information, which are commonly used at the local level. Two NGOs operating in Nicaragua, Save the Children USA and Catholic Relief Services (CRS), based their targeting on poverty and extreme poverty maps, respectively, but were not receiving funding from the EC. Other organizations may rely on information from the United Nations World Food Programme–Vulnerability Analysis and Mapping (UNWFP–VAM, 2001), livelihoods approaches, and/or ground experiences of local organizations and practitioners' networks.

Strategies for reducing food insecurity

Despite a comprehensive food security policy geared towards addressing its several dimensions, the character of the programme operations suggests that its strategies are defined from a national standpoint, focused, for the most part, on reducing poverty with structural adjustments and market reforms. These are often based on rural development guidelines and the so-called 'key building blocks for improving food security': rapid economic growth (particularly in the agricultural sector) and the strengthening of effective social safety nets.

According to the evaluation reports, a wide variety of actions have been supported with mixed results in food security improvement. These were positive in some cases, but also localized and not viable in the long term. An interviewee expressed that a problem with some of the projects was their lack of adaptability to the ground and disproportionate focus on the means of production (e.g. technical tools) rather than capacity-building.

The European Commission's assessment and monitoring of its European Programme for Food Aid and Food Security

The examined field reports note communication deficiencies between the NGOs and the EC regarding the objectives for food security priorities and consistency in the selection of the projects being funded.

An additional perspective on the programme's knowledge base is derived from the EC's own monitoring and assessment of its food security policies. The EC has a strong interest in monitoring at all levels, from local projects to general food security goals. Desk and field reports are produced by external consultancies with later support from NGOs.

This monitoring process has informed policies and budget line revisions, such as the future programme's restructuring process and assessments of EC institutions' performance – for example, the EC delegation in Nicaragua. At the outset, there is an explicit aim for learning from past experiences, improving coordination among the EC's partners, and revising the programme constantly. However, the extent to which these objectives are materializing in the EC's actual learning and capacity-building requires further research with the practitioners. According to a field evaluation of the programme in Nicaragua (PARTICIP GmbH, 2004), the local results have contributed little to major food security goals for the most part, and their long-term sustainability has also been compromised due to resource limitations. For instance, the entire evaluation of the programme for Central America was funded for only four weeks, which included a workshop in Brussels, field visits and extensive interviews (PARTICIP GmbH, 2004). It is difficult to evaluate the progress made in indirect operations because this mode of operation has been terminated.

Evaluations at the programme level are concerned with the programme's overall relevance, effectiveness, impact and consistency. The type and character of indicators used to measure the progress in food security suggested by the EC are very much related to poverty measures, nutritional levels and indicators alike (European Commission et al, 2006). Further research is needed; but initial insights suggest that critical dimensions of vulnerability related to food security are possibly not being addressed.

So, how adequate is the knowledge base about vulnerability? Who has generated that knowledge?

The analysis has not been conclusive regarding this question; more research is needed.

The programme's policy design seems much in line with the vulnerability literature with respect to understanding the multidimensional character of food security and recognition of the need to address food security structural causes; but there are deficiencies in its operationalization because most of the strategies are focused on reducing poverty, and economic growth.

Another important aspect of the EC's knowledge for reducing food insecurity is the fit of its operations to the local context and needs. Insufficiencies in this respect have been mentioned in the context of the Hurricane Mitch operation, which was delayed and provided 'pre-packaged aid not adapted to the needs' (pers comm). The failure to set an early warning system for food security, such as the Food Insecurity and Vulnerability Information Mapping System (FIVIMS), is also an interesting example. In this case the cause of failure is complex and very much related to inter-agency coordination issues, as well as local grounding. It is noteworthy that Guatemala has such a system in place, although it has some operational issues (pers comm).

Regarding the knowledge transfer about food security to the targeted groups, there seems to be a mismatch between the budgeting of the operations and their actual result in terms of their capacity-building – that is, it has been reported that many resources are invested in technical training or machinery, but few in the capacity-building of the targeted groups (pers comm).

To summarize, some of the aspects of knowledge about food security that need to be strengthened in this programme are the methodology for targeting the vulnerable groups (PARTICIP GmbH, 2004), the economic impacts of food security operations (FAO field office report; pers comm), and food safety considerations, such as those relating to genetically modified crops (pers comm). For instance, it has been reported that a food security operation may have disincentives for internal production (pers comm).

To what extent do the actors make use of the knowledge available to them? How do the actors structure the vulnerability and resilience discussion in the transfer of knowledge? How relevant and pertinent is the knowledge to the needs of decision-makers and other actors?

The unavailability of relevant information about food security does not seem to be a major limiting factor to the success of the programme. Within EC delegations, at least, information is relevant and abundant about the EC's food security policies and programmes, as well as about other agencies and organizations operating in the country. It is the extent to which available information is used, and its influence on decision-makers, that is most important. In this respect, institutional practices promoting or restricting science/practice interactions and the existence of conflicts among the actors involved are critical issues.

The type of information that supports a decision depends upon both the nature of the decision and the operational level involved (e.g. budget versus desire for local projects). Additional factors include the leverage of the EC delegation in the decision, the nature and interactions among the delegation's staff members and the existence of experts. According to an interviewee, none of the staff or decision-makers' decisions are based at all on the academic literature, and there are deficiencies in integrating internal reports relevant to the decision-making process (e.g. reports on rural development, or water and land management).

The findings suggest that experts' perception of their influence in the decisions' outcome is minimal, despite the relevance of their analyses to the matter being considered. The scale of the decisions (i.e. national or local) is a key determinant on experts' influence in decision-making. For instance, experts typically have a low impact on 'high-level decisions' (e.g. programme budget allocations). Although experts are usually affiliated with field delegation offices, the negotiation process at the EC's headquarters is one in which political interests are very influential, and experts have very little say. They are more influential at the project level in decisions regarding the orientation of the projects and their monitoring (pers comm). But, again, their degree of influence is case specific, and in many instances relates to exogenous factors, such as the programme's structure and mode of operation. For example, one product of the experts' group RESAL was a methodology for the characterization of vulnerable groups. This is the only comprehensive methodology found in this study, but has not been used because it was tied to a programme's mode of operation (indirect) that has been discontinued. Therefore, even though the expertise and users' interest existed, the methodology was not applied because there was not an administrative procedure in place for its usage (pers comm).

What barriers and failures limit the knowledge exchange in the science/practice interface? How important are the intermediaries between science and practice and who are they?

The configuration of the EP-FAFS is a major factor in creating or smoothing barriers. A fragmented structure hinders communication and coordination among the actors despite the existence of coordination instruments. In this case, the barriers relate to:

- The high complexity of the programme due to its operational magnitude in terms of the numbers of actors involved, poses a coordination challenge. One such case is illustrated by the discrepancies between the EC's technicians and EC headquarters. The technicians feel that the EC's criteria to select projects within the call for proposals are not clear due to inconsistencies regarding the projects that they recommend and the ones they fund. Coordination deficiencies also exist at other levels, and are particularly acute at the interagency level, which has prevented efforts such as FIVIMS (pers comm).
- The programme operation, particularly the call for a proposals procedure which has been described as a slow bureaucratic process, is highly complicated and resource consuming (PARTICIP GmbH, 2004; pers comm). For example, once a project is selected, its signing with the NGOs could take up to 15 months, delaying its implementation. The European Commission Humanitarian Office (ECHO) also issues a call for proposals, but allows for the renegotiation of the projects in view of local conditions. However, no explicit coordination has been observed between the two processes despite their overlapping objectives.
- EC/NGO communication deficiencies and difficulties in implementing the EC's guidelines on the ground: these issues have compromised consistency in the application of EC goals at regional levels and posed tensions in the usage of regional methodologies that require local adjustments.
- Strict project implementation rules: these are often based on *ex-ante* conditions and hinder adjustments along the life of the project. A field evaluation has mentioned coordination deficiencies between EuronAid and the EC delegation in Nicaragua regarding food security issues and the implementation of field project procedures. In addition, the inflexibility is

such that NGOs have reported an extraordinary administrative burden that slows down the delivery of products, and is compromising the credibility and efficiency of the programme.
- Short-term orientation of initiatives, which discourages long-term partnerships and undermines long-term sustainability of the projects: the case of the experts' network RESAL, an initiative that was terminated after three years, is a good example. Other short-lived efforts aimed at retaining expertise in the EC have followed, but none seem to have been successful in the long term. The short period allowed for project completion, as stated in the call for proposals (48 months maximum), has also been an issue.
- Availability and allocation of resources: for some experts, part of the effort put into answering the call for proposals may have been better spent in learning from the local expertise of NGOs (pers comm). Additionally, some reports point to shortages of the administrative and human resources necessary to implement the future restructuring of the programme. It has also been noted that the Technical Assistance Unit of the EP-FAFS has more administrative personnel than technicians.

Role of intermediaries

The role of intermediaries has been sufficiently stressed throughout this study. The EC seeks partnerships to capitalize on their expertise and to fulfil the goals of the programme. The type of EC partner (e.g. governmental versus NGO) determines the programme's operation mode, which in turn has strong implications for the food security approaches adopted and the targeting of vulnerable groups.

It is interesting to compare the performance of direct and indirect modes of operation. Arguably, policies implemented by the state are more likely to be affected by national-level priorities and political influences (e.g. big producers' pressures) than indirect operations. On the other hand, indirect operations, such as those carried out by NGOs, have the advantage of being tied to local realities and have a stronger impact at this scale; but their long-term viability is often compromised. EC collaborations with local institutions also bring the added value of the learning experience and increased capabilities for the involved governmental organizations (e.g. the creation of a food security unit in MAGFOR).

Perhaps the most important conflicts among actors involved in the EP-FAFS are posed by political interests, which frequently operate to the detriment of social justice objectives and the mitigation of the structural causes of food insecurity. One example of this was the coincidence of the food security crisis in the Rio Coco in 2005 and the regional elections (pers comm). It has also been noted that the EC's policies are often tied to structural adjustments and commercial agreements that may not improve the food security situation, at least in the short run. Political pressures can guide the type of information used in negotiating processes. Indeed, some interviewees have observed that information is not really used to inform the decisions, but rather to justify *a priori* political arrangements, and that lobbying is of major importance in decisions (pers comm).

How does the nature of institutions shape the science/practice interface? To what extent is institutional fragmentation a problem? How permeable are the boundaries of institutions to new information?

The EP-FAFS is hosted in an institution of enormous complexity. It consists of several nation states and is heavily influenced by international relations, particularly those between EC headquarters and Nicaragua. This structure poses a major challenge in terms of the coordination of all the instruments and actors involved, which counters the benefits that their synergy can bring.

Regarding institutional configurations and barriers to the knowledge transfer, there are deficiencies regarding how the experts are integrated within the programme and the use of their capabilities. Several configurations of knowledge transfer have existed throughout this programme, but none seem to have achieved the full integration of expertise in the long term. Whether this can be attributed specifically to the intrinsic nature of the EC is difficult to assess. For instance, a key factor in this question relates to the position of the experts within the programme and the resources available to them (e.g. having full contract, vehicles for fieldwork, or supporting staff proportional to the magnitude of their projects). The case of the RESAL network is illustrative in this respect. Initially, the EC funded this network; but its components were not EC employees, which arguably, in many cases, facilitated fresher interactions and contributions to the programme. When the experts became part of the EC delegations staff, additional administrative responsibilities and adhesion to standard compliance limited their capabilities to contribute cutting-edge expertise. After this adhesion process, a significant level of expertise in the programme was lost (pers comm).

The fragmentation of resources allocation can also be problematic. This is the case of the food security technicians who, once a compact unit, were distributed into several EC delegations, a process that decreased their capability to prioritize food security operations in decision-making processes.

The future restructuring of the programme is a major issue for several actors who are concerned about the implications of addressing food security as a transversal theme, and the strong focus that will be placed on the programme's humanitarian aspect (i.e. food aid). For some, these changes would compromise the priority of the food security element by conditioning it to economic development and poverty reduction strategies.

How can the science/practice interface best be improved?

Regarding the programme's methodology, there is a lack of integration across the scales at which the programme operates (e.g. targeting and local project implementation). Since different aspects of vulnerability are revealed at different scales, the identification of vulnerability groups is also likely to be scale dependent. A limited cross-scalar approach may impair the programme's capability to effectively address the multiple dimensions of food insecurity (i.e. from structural causes to the household factors) and restrict the range of instruments visible for improving food security.

Recommendations

Some specific recommendations follow:

- Improving the articulation of the programme actions across scales would be facilitated by adopting an overarching framework that brings bottom-up information into high-level decision-making without compromising the overall consistency of the programme. Despite the emphasis on decentralized operations, the intervention priorities are largely based on a top-down approach that relies, for the most part, on poverty indicators. The EC values the experience of field-based organizations; but their contribution is often determined by the EC's implementation guidelines in areas selected from a regional perspective that may obscure food security issues at local scales. A top-down oriented approach may be 'safer' to ensure regional consistency, but may undermine the efficiency of the programme both in terms of reaching the vulnerable groups and taking full advantage of the expertise of the field organizations.
- Political will and institutional capacities are critical for the success of the programme, and particularly for institutional projects such as PRODELSA. Both of these capacities need to be strengthened if some real progress is to be made. The fact that more than half of the population is food insecure in one of the biggest, most productive, yet less populated countries of the region is outrageous (pers comm).
- With respect to the programme configuration, it would be beneficial to increase operational flexibility in order to improve its adaptability to decision-makers' and the population's needs.
- More resources need to be allocated to revitalize the involvement of local organizations and civil society for achieving food security via a procedure that encourages collaboration between the two.
- Taking full advantage of local expertise is the key for success; thus, it is suggested that more effort should be placed on learning from the past experiences of local partners. At the time of writing, a database of the location of all activities and the targeted populations was not found.
- A promising avenue has been the creation of the Regional Programme on Food Security which may address some of the issues mentioned, with respect to the *Strengthened Growth and Poverty Reduction Strategy* (SGPRS) and the *Country Strategy Paper* (CSP). This new strategy is attractive to some evaluators of the programme because it targets areas selected by the local actors, would be supported by an information system,

and would have strengthened analytical capabilities based on experts' coordination. Furthermore, its design has brought together regional institutions concerned with food security for the first time. A negative point is the lack of elements for discussion and coordination with other related programmes and donors, which has left aside NGOs who have worked with the EC in previous calls for proposals. However, it is not clear how this institution will interact with the 2007 EC thematic line in food security. At the time of writing, this regional programme has not been reviewed positively, as per an interviewee.
- Centring the discussion about the process of the knowledge transfer on the type and scale of the decision being made rather than the decision-maker offers additional insights. It is also important to consider the differences between the type of knowledge that is being used in both arenas.
- The lack of coordination among the actors of the programme has been attributed to their different food security/poverty conceptualizations and operational strategies. Nevertheless, food security issues seem to be a good focal point to bring several institutions together, as is suggested by the case of the Regional Programme on Food Security.

Conclusions

This chapter has reviewed the European Programme for Food Aid and Security in the context of the knowledge-transfer process among research and practice areas. The objective was ambitious and the resources very limited, which made a thorough interview process of the people involved impossible. Conceptually, the questions posed by this study are very challenging; attempting to assign actors to the science or practice arena has, in many instances, proven questionable. A more complete picture would be obtained by reviewing other programmes operating in the area.

Despite the difficulties, this study offers a set of significant insights:

- There is no explicit overarching framework for identifying highly vulnerable populations; economic indicators are often the key.
- Spider webs describe the interface between the science and practice of this programme better than the notion of bridges.
- In some respects, the EC appears to function as a boundary organization.
- There is not a sharp (i.e. mutually exclusive) divide between the science/practice categories; many actors in this programme are better characterized as dominant scientists, dominant practitioners and hybrids.
- Food insecurity is often equated with poverty, which prompts poverty reduction strategies.
- The targeting of vulnerable groups is contingent upon the type of EC partner; however, it is important to highlight that direct aid can end up being implemented by private organizations if the government chooses to

do so. Monitoring has played a key role in reshaping policies and budgets; but restructuring procedures need to be more flexible in order to adapt actions to the realities more efficiently.
- The lack of an adequate early warning system (such as the FIVIMS) needs to be addressed.
- At the time of writing, capacity-building among targeted groups was weak.
- The use of expertise has been observed to be stronger at project than at programme levels.
- Political barriers have been a major obstacle in pursuing structural changes in this programme.

Note

1 EuronAid was a network of European NGOs that was very active on food security at the time of this study. The EuronAid Secretariat closed on 31 March 2008 after a decision taken by the General Assembly of member NGOs in May 2007 to cease all operations by December 2007. See http://www.euronaid.net/index.html?id=020200000065.

References

Argote, L. and Ingram, P. (2000) 'Knowledge transfer: A basis for competitive advantage in firms', *Organizational Behavior and Human Decision Processes*, vol 82, no 1, pp150–169

Cash, D. W. and Moser, S. C. (2000) 'Linking global and local scales: Designing dynamic assessment and management processes', *Global Environmental Change*, vol 10, pp109–120

Cash, D. W., Clark, W. C., Alcock, F., Dickson, N. M., Eckley, N., Guston, D. H., Jäger, J. and Mitchell, R. B. (2003) 'Knowledge systems for sustainable development', *Proceedings of the National Academy of Sciences of the United States of America*, vol 100, no 14, pp8086–8091

Davenport, T. H. and Prusak. L. (2000) *Working Knowledge: How Organizations Manage What They Know*, Harvard Business School Press, Boston, MA

EuronAid (2006) *EuropeAid Annual Work Programme 2006 for Grants Food Security and Food Aid*, www.euronaid.net, accessed 9 April 2006

European Commission (2001) *Communication from the Commission to the European Parliament and the Council – Evaluation and Future Orientation of Council Regulation (EC) No 1292/96 on Food Aid Policy and Food Aid Management and Special Operations in Support of Food Security*, COM/2001/0473 final/2, Brussels, 12/09/2001, http://eur-lex.europa.eu/LexUriServ/LexUriServ.do?uri=COM:2001: 0473:FIN:EN:PDF

European Commission (2002) *Country Strategy Paper, Nicaragua, 2002–2006*, http://ec.europa.eu/external_relations/nicaragua/csp/02_06_en.pdf

European Commission, DG Development (2001) *Bi-Annual Report 2000/2001, Food Security at the Heart of Poverty Reduction*, EC Food Aid and Food Security Programme, Brussels

European Commission, DG Development (2005) *Food Aid/Food Security Budget Lines*, 21 02 01 – 21 02 02 – 2101040 (DEV/5459/05-EN), Brussels

European Commission, EuropeAid Cooperation Office (2001) *Staff Working Document: Report on the Implementation of the European Commission's External Assistance, Situation at 01/01/01*, EuropeAid Cooperation Office, Brussels, D(2001) 32947, http://ec.europa.eu/europeaid/multimedia/publications/documents/annual-reports/europeaid_annual_report_2000_en.pdf

European Commission, EuropeAid Cooperation Office, Food Security/Food Aid Sector, (2006) *Annual Work Programme 2006 for Grants, Food Security and Food Aid* (AIDCO/EN PE/2006/1103), Brussels

European Council (1996) European Council Regulation (EC) No 1292/96 of 27 June 1996 on Food-Aid Policy and Food-Aid Management, and Special Operations in Support of Food Security, Official Journal No. L 166, 5 May 1996, pp0001-0011

European Parliament and of the Council (2006) Regulation (EC) No 1905/2006 of the European Parliament and of the Council of 18 December 2006 Establishing a Financing Instrument for Development Cooperation, Official Journal L 378 of 27 December 2006

FAO (United Nations Food and Agriculture Organization) (1996) *Food Report of the World Food Summit*, 13–17 November 1996, www.fao.org/docrep/003/w3548e/w3548e00.htm, accessed 5 April 2005

FAO (2006) *Food Security Statistics 2006*, www.fao.org/faostat/foodsecurity/, accessed 1 June 2006

Government of Nicaragua (2000) *A Strengthened Growth and Poverty Reduction Strategy (SGPRS)*, www.imf.org/external/np/prsp/2000/nic/01/nic/01/nic/pdf

Government of Nicaragua (2001) *A Strengthened Growth and Poverty Reduction Strategy (SGPRS)*, Coordination and Strategy Secretariat of the Presidency, www.imf.org/external/np/prsp/2000/nic/01/nic/01/nic/pdf, accessed 2010

Guston, D. H. (1999) 'Stabilizing the boundary between politics and science: The role of the office of technology transfer as a boundary organization', *Social Studies of Science*, vol 29, no 1, pp87–112

Guston, D. H. (2001) 'Boundary organizations in environmental policy and science: An introduction', *Science, Technology & Human Values*, vol 26, no 4, pp399–408

IMF (International Monetary Fund) (2000) 'IMF and World Bank open way for US$4.5 billion debt relief for Nicaragua under the enhanced HIPC initiative', Press release No 00/78, 21 December 2000, www.imf.org/external/np/sec/pr/2000/pr0078

Laforge, M. (2001) *Metodología de caracterización de grupos vulnerables*, Institut de Recherche et d'Application des Méthodes de Développement (IRAM), Red Europea de Seguridad Alimentaria (RESAL), Nicaragua

NOAA (National Oceanic and Atmospheric Administration) (2005) 'Mitch: The deadliest Atlantic hurricane since 1780', NOAA Satellite and Information Service, National Climatic Data Center, www.ncdc.noaa.gov/oa/reports/mitch/mitch.html, accessed 9 April 2006

PARTICIP GmbH (2004) *Thematic Evaluation of Food Aid Policy and Food Aid Management and Special Operations in Support of Food Security*, Synthesis Report – Country Reports, final version, vol 4, Evaluation for the European Commission, EC-DG Development Documents Repository, http://ec.europa.eu/development/icenter/repository/FA-FS_volume_4_country_reports_EC_en.pdf, accessed 9 June 2006

UNWFP-VAM (United Nations World Food Programme–Vulnerability Analysis and Mapping) (2001) *Food Security and Livelihoods Survey in the Autonomous Atlantic Regions of Nicaragua*, http://documents.wfp.org/stellent/groups/public/documents/vam/wfp073961.pdf

World Bank (2000) *Data and Research*, www.worldbank.org/

15

Land in Transition: Coping with Market Forces in Managing Rangelands in Mongolia

Togtohyn Chuluun and Dennis Ojima

Introduction

The Mongolian rangelands have a diversity of ecosystems, ranging from forest steppe in the north, to the Gobi Desert in the south, and the steppe ecosystem dispersed in between. The Altai Mountains in the southwest and the Khangai Mountains and Khentii Mountains in the north-central part of the country add to the diversity of landscape, habitat and resource availability. The Mongolian nomadic pastoral cultures occur as an emergent feature of the variable ecosystem dynamics of the arid and semi-arid systems (Chuluun, 2006; Fernández-Giménez, 2006). These pastoral systems have adapted to variable environmental conditions, responding to variation in resource availability. The emergence of hierarchical pastoral networks or cooperative groups based on family relationships or common locations of grazing, as a complex adaptive system, has increased the resilience of these systems to climate variability.

The Mongolian nomadic pastoral system exhibits oscillatory movements in regions where the climate and rangeland production dynamics are relatively more predictable, and can accommodate the need to move only between summer and winter camps (see Table 15.1). The forest steppe areas are typical of this oscillatory pastoral movement. More frequent seasonal oscillatory movements with more than one movement during the summer season occur in the mountain steppe, forest steppe and in the wetter regions of the steppe. In

Table 15.1 *Changes in land-use patterns, land-use regulation and land tenure from the pre-*negdel *to post-*negdel *period: *Negdel *refers to pastoral collective*

Attributes	Pre-negdel period (until late 1950s)	Negdel period (1960–1990)	Post-negdel period (since 1990)
Land-use patterns	Nomadic movements with ecological conditions Short-term, long-distance moves of livestock	Less frequent and more distant movements, but often with conservation of cultural landscapes Enforced *otor* (individual migrations) Many shelters and wells built	Further reduced distance and frequency of moves Less *otor* Enforced *otor*: short-term, long-distance moves of livestock Year-round use of riparian and reserve pastures Animals' concentration near towns and roads
Regulatory institutions	Traditional pastoral networks (little formal regulation)	*Negdel*	None (few newly emerged *hot ail* and new co-operatives)
Land-use regulation	No enforced formal regulation of movement Neighbourhood groups migrate together using animal carts	*Negdel* enforces seasonal moves and *otor* Machinery provided by *negdel* for transportation and hay-making Species specialization by kind, age and sex	No formal regulation or enforcement Little coordination of seasonal movements by *hot ail* (traditional network of households sharing resources within a particular region) Diverse species composition
Land tenure and legal framework	Customary rights within administrative units	*Negdel* allocates pasture, often along customary lines All property state owned Disputes resolved by brigades and *negdel hural* (elected representatives)	Customary rights are weak Livestock, shelters and wells are privatized Disputes are resolved by local governments (*hural bag* and *sum*)

Source: Bazargur et al (1993); Fernández-Giménez (1999), as modified by Chuluun and Ojima (2002)

regions with relatively higher climate variability and increased uncertainty, such as the Gobi region, pastoral movements tend to be more chaotic and follow more opportunistic strategies to secure forage. These movements are associated with drier parts of the steppe and the Gobi Desert steppe and desert areas where non-equilibrium ecosystem dynamics are observed (Ellis and Chuluun, 1993; Fernández-Giménez, 1999; Chuluun, 2000; Bedunah and Schmidt, 2004). The herders from these regions move to places where better rangeland conditions exist – especially during the summer season. Currently, their movements are constrained to administrative boundaries, rather than to the ecological extent of suitable rangelands. However, the herders still use pastures located in other administrative units, but they cause pasture conflicts. With privatization of livestock and talk of land privatization (GISL, 1997;

Fernández-Giménez, 2006), it is becoming more difficult to move livestock across the landscape to access water and forage.

Before the formation of pastoral collectives or *negdels* during the late 1950s, movement of the Mongolian herders still incorporated traditional pastoral management concepts associated with using forage and water resources within a landscape context and emphasizing community-based movement decisions. During the collectivization or *negdel* period between 1960 and 1990 (see Table 15.1), herders were moving less frequently and across longer distances with mechanized transport vehicles (e.g. trucks and tractors). These collectives or *negdels* took control over these land-use regulations, enforcing seasonal moves and *otor* (individual migrations). *Negdel* herders were specialized according to the class of livestock herded, characterized by livestock species, sex and age class. The government promoted *otor* (short- or long-distance movements of livestock) as a strategy to fatten the stock in summer and autumn and to avoid drought and *zud* (blizzards and bitter cold) conditions. Shelter construction, well-building and hay-making provided some mechanisms to reduce the vulnerability of these pastoral systems to climatic extremes. *Negdels* or collectives were dissolved with the privatization of livestock during the early 1990s.

Since 1990, Mongolia has shifted to a free-market economy, which has led to changes in the livestock sector (see Table 15.1). During this transition period, unemployment increased because of the stagnation of enterprises in the capital city and other civic centres; in addition, poverty increased, in part due to accelerating inflation. Accompanying these trends was a decline in the social services available in the civic centres. Numerous small administrative units or villages also became less viable due to a lack of economic and resource support from the central government. Although livestock was privatized in rural areas and the number of livestock increased, the livestock industry has suffered due to degradation of pasture and unfavourable climate conditions in past years (1999 to 2002). As a result, livestock, the source of the livelihoods of rural citizens, has shrunk in size and poverty has grown, in general. People's migration from rural areas to three major cities (Ulaanbaatar, Darhan and Erdenet) has increased (see Figures 15.1 and 15.2).

Environmental degradation

Environmental degradation has expanded markedly in Mongolia and is associated with increased livestock numbers. Since approximately 1995, the area of highly degraded land has increased 1.8 times (MNE, 2001), and desertification in the arid and semi-arid region of Mongolia has increased by 3.4 per cent during 1990 to 2004 (MNE, 2006). Acceleration in desertification has occurred, in part, because of human influences and partly from changing climate. As the number of livestock has exceeded pasture-carrying capacity, a number of impacts have been observed. These include pastureland degradation, declines in plant production, breakdowns in ecosystem, and a shift of grasslands to more desert-like conditions. In some areas, soil deterioration and

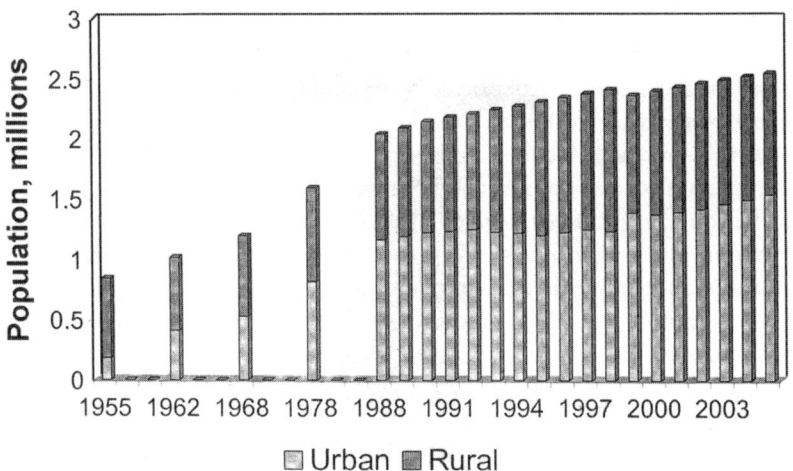

Figure 15.1 *Population dynamics of Mongolia*

Source: chapter authors

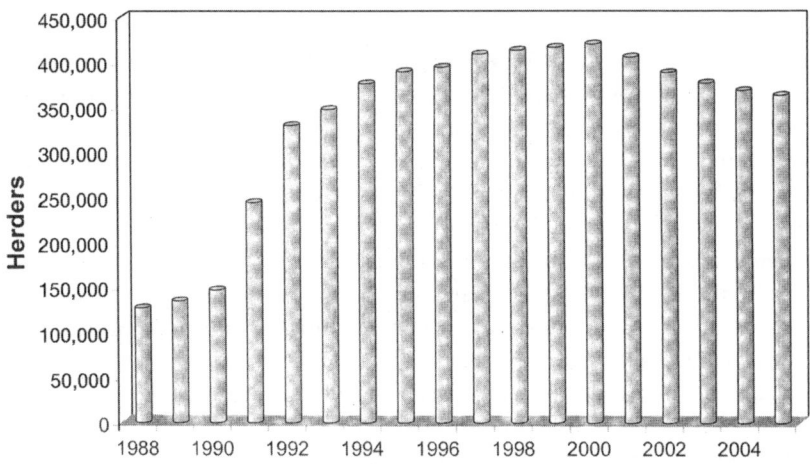

Figure 15.2 *Number of herders in Mongolia*

Source: chapter authors

sand migration have increased as bushes and trees, which are the main factor for arresting sand migration, have been cut for local use.

According to 2000 statistical information (National Statistical Office, Mongolia, 2001), the total number of water points in Mongolia was slightly less than 31,000, with approximately 8000 being mechanical wells and the rest hand wells. Of the approximately 21,000 water points located in pastureland, about 25 per cent are dysfunctional. Between 1990 and 2000, the number of water points decreased by 28 per cent with about three-quarters of these being mechanical wells. Of the wells constructed between 1960 and 1990, only about 40 per cent are still operational today (UNDP, 2005).

Because the number of water points continues to decline from year to year, traditional nomadic pastoral patterns of seasonal grazing have been disrupted. Increased grazing pressure around the remaining water points has resulted in the overgrazing of pastures. In addition to these grazing effects, a warmer and drier climate has created conditions promoting the expansion of deserts. Since the beginning of the 20th century, global warming has intensified in the northern latitudes, and the temperature in Mongolia has increased by 1.8°C since 1940 (Batima et al, 2005). The forage availability determined by remote-sensing data from 1982 to 2002 in the central parts of Mongolia has been affected by these climate effects (Ojima et al, 2004). Ellis et al (2002) have shown that the steppe area adjacent to the Gobi region is especially vulnerable to climate changes; with increased grazing pressures during the past decade, these changes have led to desertification.

From 1960 to 1990 many aspects of the traditional nomadic culture were replaced by socialistic practices. Herdsmen were commonly organized into collectives and were allowed only a small number of animals for private ownership (Sneath, 2003). The provincial and national governments established a strategy for short-term, long-distance moves (*otor*) to safeguard against drought and rangeland overuse. This strategy also served as a mechanism to fatten the stock during the summer and autumn seasons. The collectives made all the decisions over allocations of animals and specialization in tasks and species. Although these functions may have been subsumed by the collectives, the customary institutions did not disappear altogether, as is demonstrated by the fact that many *hot ails* (traditional network of households sharing resources within a particular region) have re-emerged during the transition to a market economy (Schmidt et al, 2002; Janzen 2005; Reading et al, 2006; Schmidt 2006).

Since de-collectivization started in the early 1990s, pastoral movements have become less frequent and shorter due to the lack of subsidies to maintain transportation for long-distance travel and preference to stay closer to settled areas (Janzen, 2005; Fernández-Giménez, 2006; Reading et al, 2006). The higher concentration of livestock near settled areas and year-round use of riparian zones have led to deterioration of the rangelands (Ojima et al, 2004; Chuluun et al, 2005a; Janzen, 2005).

Recent changes affecting land use and tenure

Mongolia's collectives began to dissolve during the summer of 1991. Under the legislation on privatization, individual collectives were allowed a free hand in deciding how they should privatize (Mearns, 1993). This has resulted in a diversity of organizational structures utilized today. During this initial phase of privatization, the number of herders doubled. These conditions have increased the vulnerability of Mongolian pastoral systems since the transition to a market economy in 1990 and are related to climate change and an increased density of livestock (Ojima et al, 2004). The termination of state-organized fodder production and hay-making has also reduced the capacity of pastoral communities to sustain prolonged droughts and severe winters.

During the early transition period, land tenure did not change and was still in the domain of the state and administered through county-level administrators. However, the privatization of livestock was the major policy objective and this was aided by liberalizing prices and the distribution of livestock among the population. The expectation of these policies was that market processes would act to stimulate the livestock industry and a market-driven system would provide an economic base for the country. However, the impact of these policies on steppe ecosystems, the degradation of these ecosystems under changing grazing pressures, and the decline in watering points were not considered relative to changes in the way in which livestock are seasonally moved.

Livestock privatization provided incentives for increasing livestock numbers. Without state subsidies, having a greater number of animals in one's herd became a dominant risk-management strategy as 'insurance' against future uncertainties and climate variability. The number of livestock increased from 25.2 million in 1993 to 33.6 million in 1999 (see Figure 15.3). However, livestock species composition shifted during this decade – the number of camels halved; cattle decreased by 27 per cent; sheep decreased by 21 percent; while the proportion of horses remained constant. The increase of livestock numbers was driven by the number of goats, which was a response to access to the cashmere market in China and the rest of the world. Goat numbers doubled from 1990 to 1998, and even tripled up to 2006, reaching 15 million goats nation-wide. The combined drought and *zud* (severe for livestock winter conditions) during 1999 to 2002 resulted in heavy livestock losses (see Figure 15.3).

The vulnerability of herders to climate variability has increased since 1990 with the dissolution of the *negdels* and the partial disruption of traditional resilient pastoral networks. Since the dissolution of *negdels*, no formal regulation or enforcement of land use has been installed. During the transition to capitalism, the livestock, shelters and wells were privatized and customary rights to certain pasture lands became weak or unclear, especially in central Mongolia. Now local governments are responsible for resolving land-use disputes rather than using the former policies of the *negdel* period; but current policies still have some vestiges of the former customary herding rights.

In planning de-collectivization, emphasis was placed on the transfer of assets from the state into the hands of private individuals, but at a pace that would not substantially disrupt production or the herding population itself. Consequently, many intermediate forms of organization between collective and private ownership of livestock have appeared. In some areas, they have persisted and will continue to persist; in others, they were only a short-term solution and have been dissolved. The most common form is the livestock company. Of the 255 collectives in Mongolia which were privatized, 80 still exist in the form of joint stock companies. In structure, these are very similar to the former collectives, although the relationship between company and member herder is significantly different. Typically companies have retained

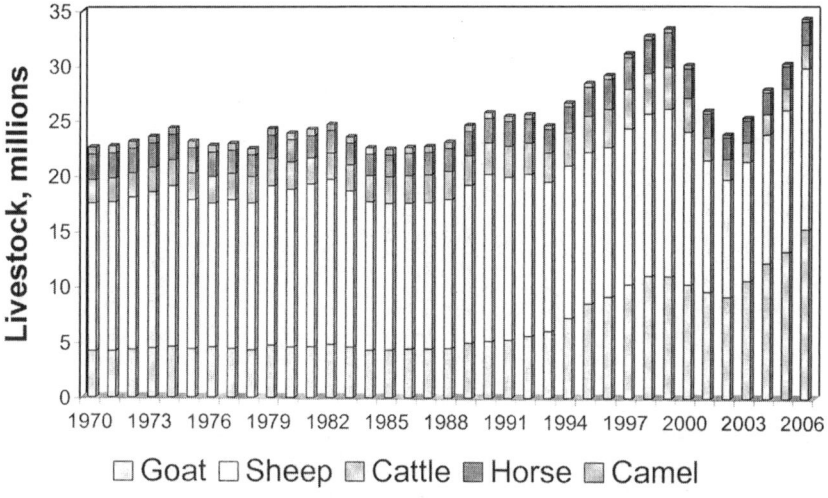

Figure 15.3 *Livestock dynamics in Mongolia*

Source: chapter authors

ownership of large numbers of animals which are leased to their members, although the terms of the lease vary a great deal between companies. A small number of companies continue to pay their herders a salary and retain full ownership over the animals.

Livestock dynamics and human well-being

Livestock numbers were relatively stable until 1990, oscillating between 20 and 25 million (see Figure 15.3). However, since the privatization of livestock and entry into the open-market system, the numbers increased to 34 million by 1999 and were up to 35 million in 2005 (National Statistical Office, Mongolia, 2006). A dramatic increase in goat numbers was experienced with a near tripling in number due to increased access to the cashmere market. Opening Mongolia to the global market has had a drastic effect on herders, especially with the cashmere export (see Table 15.2) opportunity to China (World Bank, 2003).

Since the privatization of livestock, moderate and wealthy herders have emerged and have been gradually increasing in numbers until 1999 (see Figure

Table 15.2 *Exported cashmere during 2005 and 2006*

	Raw Cashmere (tonnes)	Total price (US$ millions)	Processed cashmere (tonnes)	Total price (US$ millions)
2005	381.3	13.4459	919.2	52.7937
2006	1675.7	62.814	1388.2	79.4635

Source: chapter authors

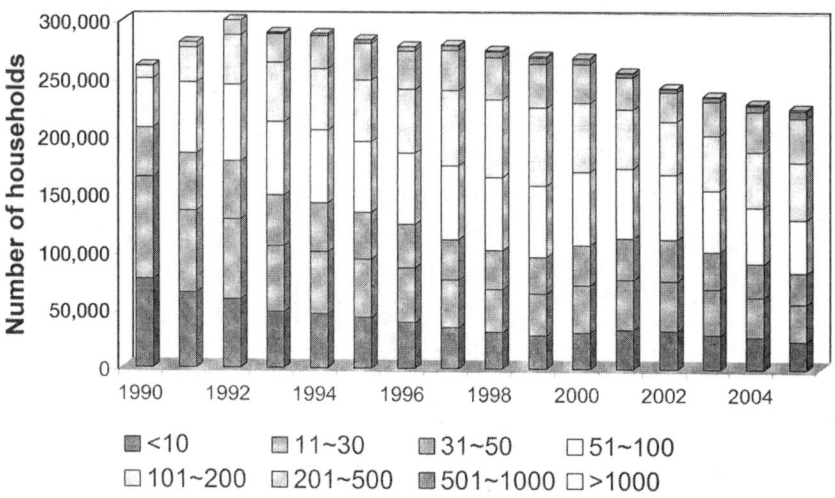

Figure 15.4 *Herders' household wealth distribution based on privately owned livestock numbers*

Source: chapter authors

15.4). Since 2000, moderate-income families have accounted for approximately 40 per cent of total households and wealthy households about 2.4 per cent. However, extreme climatic events have reduced the number of wealthy and moderate households and increased the proportion of poor households (i.e. households owning less than 100 animals), which now account for two-thirds of total households. An increased number of herding households failed during and after the extreme events of 1999 to 2002. Since then, however, the number of poor households has declined overall. Today, Mongolia has about 250,000 herding families, about 60 per cent of whom own less than 100 livestock (National Statistical Office, Mongolia, 2006). Now wealthy and moderate herders are larger in numbers than in 1999.

Cashmere and meat are the main economic factors for the increase in livestock numbers, especially for goats. However, there is a lack of direct correlation between the export of cashmere (see Figure 15.5) and the increase in goat numbers. Raw cashmere exports from Mongolia have increased during the past decade (National Statistical Office, Mongolia, 2006), and between 2005 and 2006 exports quadrupled. Currently, Mongolia contributes 25 to 30 per cent of the world cashmere market, the rest coming from China. Meat production in Mongolia, however, is not experiencing the same increase as increasing livestock numbers would imply (see Figure 15.6).

Changes in the Mongolian rangelands

Recent policy changes have been proposed to modify the major administrative boundaries of Mongolia to allow greater access to natural resources and

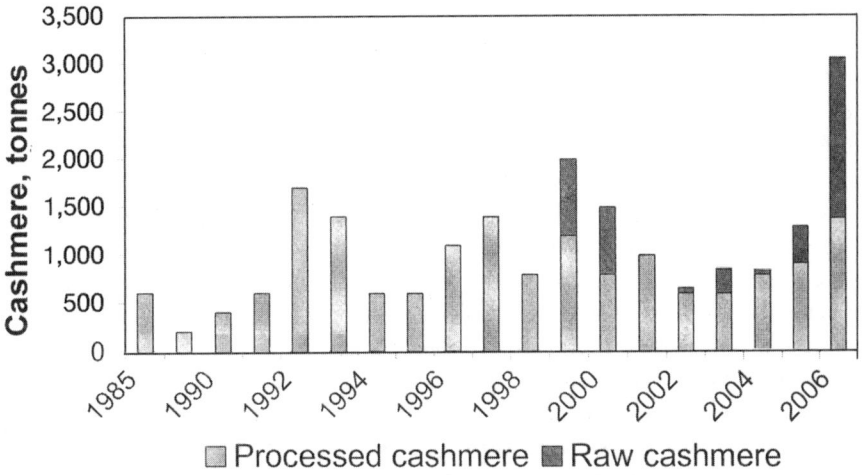

Figure 15.5 *Cashmere export of Mongolia*

Source: chapter authors

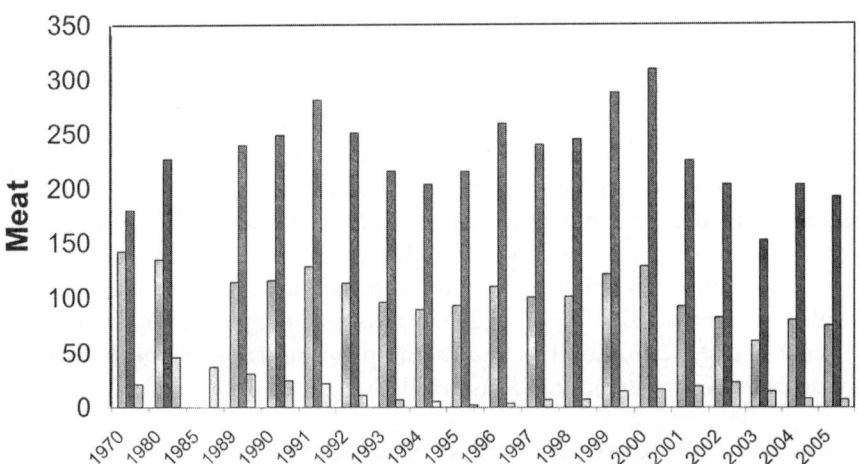

Figure 15.6 *Meat production and export*

Source: chapter authors

seasonal grazing lands, better sustain the pastoral livelihood, and maintain ecosystem integrity (Chuluun et al, 2005b). The policy changes are designed to develop a settlement pattern that reduces the concentration of population around major civic centres and promotes the usage of resources associated with rural areas of the country. In Mongolia, about half of the county-level administrative units (*sums*) are too small to include seasonal grazing lands and the land cover is more homogeneous. Because of this, pastures have degraded due to continuous overuse throughout the year. These new administrative and territorial units have been organized to enhance socio-economic optimality, environmental sustainability, and historical and cultural acceptability by citizens.

Changes in land-use patterns, land-use regulations and land tenure from the pre-collective or *negdel* period to the post-*negdel* period are summarized in Table 15.1. An example of land-use change from the pre-*negdel* to the post-*negdel* period is shown for the region covering Uyanga, Nariin Teel and Baruunbayan-Ulaan *sums* of Ovorhangai aimag (Bazargur, 2005). This region covers the Hangai Mountains in the north, steppe in the middle and Gobi Desert in the south. The connectivity between landscapes through pastoral movements has been reduced and cultural landscape fragmentation has grown from the pre-*negdel* or the *negdel* to the post-*negdel* period. There were three common pastures/meeting places for different herders' communities during the pre-*negdel* period. However, these common pastures were reduced to two during the *negdel* period and only one was located within a *sum*. These common meeting places served a critical role for exchange of livestock and for social interactions. The loss of connectivity resulted in animal live-weight reduction and pasture degradation, according to Bazargur (2005). Such a modification of range management and restricted pasture movement can lead to ecosystem degradation, forage decline and reduced livestock production if prolonged for an extended period of time.

Since 1990, increased overstocking on common properties and violation of traditional grazing management have begun to destroy key resources. Animal losses have become widespread due to overgrazing around animal watering points and urban centres, trespassing onto reserve pastures used during natural calamities such as severe winter storms (*zud*), and drought by herds grazing without permission. During 2000 to 2002, severe weather events (comprised of three consecutive years of drought and severe *zud*) led to decreases in livestock numbers to 1990 levels. These large animal losses exemplified the vulnerability of these pastoral communities and the need for appropriate mechanisms to buffer them from such natural disasters.

Proposal for restoration of cultural landscapes at administrative level

Enforcing the policy of cultural landscape is an optimal way of linking the scientific knowledge of ecosystems and landscape dynamics with social considerations to improve the capacity to overcome natural disasters, such as droughts and *zud*. According to the definition by the United Nations Educational, Scientific and Cultural Organization–World Heritage Convention (UNESCO–WHC), the cultural landscape is the agro-ecosystem that has survived the industrialization era (UNESCO–WHC, 2005). UNESCO is calling for the increased preservation and development of these cultural landscapes. Their enhancement could lead to ecosystem sustainability and be integral to sustainable development strategies (Dagvadorj et al, 2003). The cultural landscape is fashioned from a natural landscape by cultural activities, such as nomadic pastoral land use in the case of Mongolia. In the Mongolian context, the complex of four seasonal pastures, long-distance use of pastures to fatten animals, hay-making land, reserve pastures used during the *zud* and drought, and sacred land that local people worship constitute the Mongolian cultural landscape (see Figure 15.7). This cultural landscape with its traditional pastoral community has emerged as a dissipative structure (Chuluun, 2000) in a highly variable climate with a diverse landscape and has evolved to become a sustainable coupled social/environmental system.

Fragmentation of arid and semi-arid regions reduces bio-complexity through the reduction of its connectivity (Galvin et al, 2008). Fragmentation of the cultural landscape through administrative division fractioning increases the vulnerability of coupled socio-environmental systems to global perturbations. If the cultural landscape is revived with its traditional community, it will require a minimum level of investment and will be less vulnerable and more resilient. In this case, the pastoral system can be considered a complex adaptive system and will be more stable, and the biodiversity and health of the system will be improved.

Historically, Mongolia had provinces, or *aimags*, oriented across environmental gradients which occurred in a north–south orientation, providing

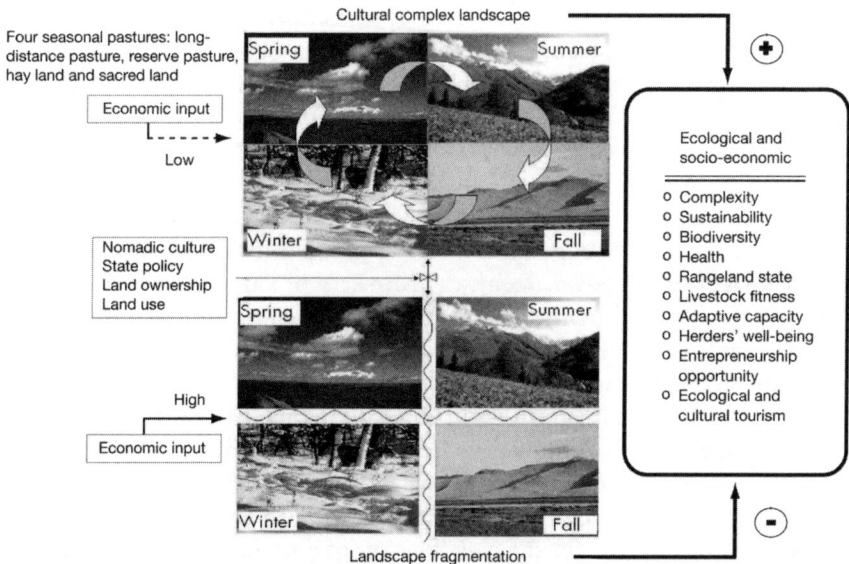

Figure 15.7 *Mongolian cultural landscape incorporates four seasonal pastures: Otor (a long distance from the main camp pasture), reserve pastures (used during extreme events such as drought and zud), hay-making lands and sacred lands (worshipped by local people due to their religious, aesthetic and biodiversity values)*

Source: modified by Chuluun, after Galvin et al (2008)

access to a greater diversity of ecosystems, including forest steppes (Mongolian Academy of Sciences, 1990). These provinces were organized for political and environmental reasons to maintain grazing lands to support a mobile military force and operated for many centuries (Baabar, 1999).

Currently, there are 21 provinces, or *aimags*, 330 counties, or *sums*, and 1671 commonwealth areas, or *bags*, within the administrative units in Mongolia. The fragmentation of the cultural landscapes began during the late 1800s and continued under the period of Soviet influence (Sonomdagva, 1967; Bawden, 1968; Sonomdagva, 1998). An outcome of the reduced size of many of these *sums* is that almost half of the 330 *sums* lack one or two seasonal pastures (Bazargur et al, 1993; Bazargur, 2005). Because of this, conflicts of pasture use have become more common. Since the privatization of livestock, greater pressure has been exerted to produce more pastoral products.

The Mongolian government has approved a regional development plan (Ikh Khural (Parliament) of Mongolia, 2001). According to this development plan, Mongolia will establish four economic regions: Western, Khangai, Central and Eastern regions. The Government of Mongolia has promoted development planning through this goal by maintaining essential public services delivered to citizens, such as health, education and telecommunication, and has resolved the issue of amending the current administrative structure and organization in the government action programme for 2004 to 2008 (Ikh Khural (Parliament) of Mongolia, 2004). The basic principle for developing the new division of administrative and territorial units of Mongolia is to carry it out in a complex manner consistent with historical traditions and current needs, on the basis of a thorough study, and with the aim of improving the quality of public administration and services and expanding the rights and opportunities of citizens to participate in the creation of social wealth (Ikh Khural (Parliament) of Mongolia, 2004). By the resolution of the president and the prime minister of Mongolia, a working group and sub-groups, including representatives of the government, the Ministry of Construction and Urban Development, the Academy of Sciences and other respected professional organizations, have been working to develop a proposal to amend the administrative and territorial division.

As Mongolia's administrative and territorial units are enlarged, the capacity for self-sufficient development under market conditions, human resource utilization, and efficiency of investments and budgets will improve, risks will be reduced, and the capacity to adapt to globalization and climate change will increase. The main goals for the new division of administrative and territorial units are its socio-economic optimality, environmental sustainability, and cultural and historical acceptability by the citizens. A balanced division of population, land, natural resources and industries are of special significance. The new socially and economically optimal administrative and territorial division of Mongolia will be the main impetus to improve the state of unemployment, poverty, inbreeding, quality and access to public services, and the promotion of local self-sufficient development.

In the government proposal, there will be 68 proposed administrative/territorial units (*hoshuus*) and each will contain several ecosystems in contrast to many of the current sums, which have only a single ecological zone. *Hoshuus* in the Gobi Desert, desert steppe and dry steppe regions often have larger territories to offset higher climate variability. Plant onset trends were used to evaluate the natural resource capacity. The trends were studied (Ellis et al, 2002) using satellite data to determine the areas with vulnerable pastures. In the border territory between Gobi and the steppes zones, there is a trend towards a one-month delay in the beginning of green-up or growth of vegetation. So, many *sums* with territories entirely in this zone are vulnerable to climate change. Territory vulnerable to climate change comprises a small part of the entire territory of the proposed *hoshuus* (enlarged administrative/territorial units), so enlarging administrative/territorial units will serve as a factor to mitigate the vulnerability to climate change (Chuluun and Altanbagana, 2005).

Introducing desirable futures for the Mongolian rangelands in the new millennium

Ironically, the tragedy of the commons was not as widespread in the Mongolian rangelands as it is now during the transition period from a centralized to a market economy. Land degradation due to overgrazing has intensified near cities, areas of concentrated infrastructure and water resources since the early 1990s. Traditional land-use strategies that operated across larger spatial scales (cultural landscape) are in danger of being lost without access to larger cultural landscape elements which allow for seasonal grazing. A reduction in the spatial scale of the cultural landscape has led to reduced resilience and increased vulnerability of the coupled human/environmental systems.

The key to sustainability lies in enhancing the resilience of communities (Walker and Salt, 2006). Resilience is the ability to absorb change and still retain basic functions and structure. Interestingly, co-operatives based on traditional pastoral networks are re-emerging in Mongolia in response to the drought and extreme winter conditions of 1999 to 2002. The comparative study of pastoral communities shows that a pastoral community with a co-operative based on traditional networks has lost less livestock compared with other communities during the recent climate disaster (Chuluun and Enkh-Amgalan, 2003). Today the Mongolian rangelands are at the bifurcation stage between two pathways: predominantly private landownership or traditional land-use culture operated by traditional pastoral resilient networks. It appears that enhancement of co-operatives based on traditional resilient pastoral networks will lead to a reduction in the future vulnerability of Mongolian pastoral systems.

Finally, the development of the modern cultural landscape of the Mongolian rangelands includes the strengthening of pastoral traditional networks with the introduction of modern technologies such as wireless communication, renewable energy resources, access to appropriate livestock breeding, use of healthy veterinary practices, and access to markets of more finished products. Development of early warning systems with the use of integrated technologies such as remote sensing and modelling and distribution through wireless technologies can also reduce risk in these very vulnerable but productive systems. Given the high literacy rate (97 per cent) of the Mongolian people, there is a good probability of success if such innovative approaches to sustainable development of these cultural landscapes can be implemented and lead to conservation of the critical cultural landscapes of the region.

Acknowledgements

Research for this chapter was supported by the Analysis of Integrated Assessment of Climate Change Project of START and funding to the Natural Resource Ecology Laboratory at Colorado State University on Northern Eurasian C-land use–climate interactions in the semi-arid regions, supported by NASA Project NNG05GA33G.

References

Baabar, B. (1999) *History of Mongolia*, NEPKO Publishing, Ulaanbaatar, p354
Batima, P., Natsagdorj, L., Gomboluudev, P. and Erdenetsetseg, B. (2005) *Observed Climate Change in Mongolia*, AIACC Working Paper 13, START Secretariat, Washington, DC
Bawden, C. R. (1968) *The Modern History of Mongolia,* Praeger, New York, NY
Bazargur, D. (2005) *Strategy of Administrative-Territorial Division and Socio-Economic Development of Mongolia*, Ulaanbaatar, Mongolia
Bazargur, D., Shiirevad'ya, C. and Chinbat, B. (1993) *Territorial Organization of Mongolian Pastoral Livestock Husbandry in the Transition to a Market Economy*, PALD Research Project No 1, University of Sussex, Institute of Development Studies, Policy Alternatives for Livestock Development in Mongolia, Brighton, UK
Bedunah, D. J. and Schmidt, S. M. (2004) 'Pastoralism and protected area management in Mongolia: The case of Gobi Gurvan Saikan National Park', *Development and Change*, vol 35, pp167–191
Chuluun, T. (2000) 'Climate variability, nomadic society and turbulent history: A Mongolian case study', *Update*, Newsletter of the International Human Dimensions Programme on Global Environmental Change, issue 1/2000
Chuluun, T. (2006) 'A new administrative-territorial division of Mongolia as a mechanism to increase adaptive capacity to climate change', Paper presented to the Sixth Open Meeting of the International Human Dimensions of Global Environmental Change Research Community, Global Environmental Change, Globalization and International Security: New Challenges for the 21st Century, Bonn, Germany, 9–13 October 2005, Conference Book 436
Chuluun, T. and Altanbagana, M. (2005) 'Use of remote sensing information in administrative-territorial reform of Mongolia', in *The First National Conference on Remote Sensing and Geographic Information System Applications*, Ulaanbaatar, Mongolia, pp71–76 (in Mongolian)

Chuluun, T. and Enkh-Amgalan, A. (2003) 'Tragedy of commons during transition to market economy and alternative future for the Mongolian rangelands', *African Journal of Range and Forage Science*, vol 20, no 2, p115

Chuluun, T. and Ojima, D. S. (eds) (2002) 'Fundamental issues affecting sustainability of the Mongolian steppe', in *Proceedings of the Open Symposium on Change and Sustainability of Pastoral Land Use Systems in Temperate and Central Asia*, Ulaanbaatar, Mongolia, 28 June–1 July 2001, Interpress Publishing and Printing, Ulaanbaatar

Chuluun, T., Altanbagana, M. and Sarantuya, G. (2005a) 'Vulnerability and adaptation assessment of the Mongolian pastoral systems to climate and land use changes', *The Proceedings of the Department of Biology*, School of Natural Sciences, Mongolian Education University, vol 4, pp182–189

Chuluun, T., Nergui, P., Davaanyam, S., Altanabagana, M. and Ariunmunkh, B. (2005b) 'Proposal of new administrative-territorial division of Mongolia', submitted to the President of Mongolia by the Division of Regional Settlement Planning of the Ministry of Construction and Urban Development

Dagvadorj, D., Nyamtseren, L., Chuluun, T., Tsedendamba, L., Suhbaatar, Ts. and Enebish, T. (2003) 'A concept on sustainable development of Mongolia and regional development', *Development Studies of Mongolia*, Institute of National Development, Mongolian Academy of Sciences, vol 1, pp31–38 (in Mongolian)

Ellis, J. and Chuluun, T. (1993) *Cross-Country Survey of Climate, Ecology and Land-Use among Mongolian Pastoralists*, Report to the Project on Policy Alternatives for Livestock Development (PALD) in Mongolia, Institute of Development Studies, University of Sussex, Brighton, UK

Ellis, J., Price, K., Boone, R., Yu, F., Chuluun, T. and Yu, M. (2002) 'Integrated assessment of climate change effects on vegetation in Mongolia and Inner Mongolia', in T. Chuluun and D. Ojima (eds) *Fundamental Issues Affecting Sustainability of the Mongolian Steppe*, Interpress Publishing and Printing, Ulaanbaatar, pp26–34

Fernández-Giménez, M. E. (1999) 'Sustaining the steppes: A geographical history of pastoral land-use in Mongolia', *Geographical Review*, vol 89, pp315–342

Fernández-Giménez, M. E. (2002) 'Spatial and social boundaries and the paradox of pastoral land tenure: A case study from post socialist Mongolia', *Human Ecology*, vol 30, pp49–78

Fernández-Giménez, M. E. (2006) 'Land use and land tenure in Mongolia: A brief history and current issues', in D. J. Bedunah, D. E. McArthur and M. Fernández-Giménez (eds) *Rangelands of Central Asia: Proceedings of the Conference on Tranformations, Issues, and Future Challenges*, US Department of Agriculture, Forest Service Proceedings RMRS-P-39, Washington, DC, pp30–36

Galvin, K. A., Reid, R. S., Behnke, R. H. Jr. and Hobbs, N. T. (eds) (2008) *Fragmentation in Semi-Arid and Arid Landscapes: Consequences for Human and Natural Systems*, Springer-Verlag, New York, NY

GISL (1997) *Strengthening of Land Use Policies in Mongolia*, ADB-TA No. 2458-MON, Final Report Phases I and II, GISL, Hants, UK

Ikh Khural (Parliament) of Mongolia (2001) *Regional Development Concept, Resolution No 57*, Mongolia

Ikh Khural (Parliament) of Mongolia (2004) *Program of the Mongolian Government for 2004–2008*, Mongolia

Janzen, J. (2005) 'Changing political regime and mobile livestock-keeping in Mongolia', *Geography Research Forum*, vol 25, pp62–82

Mearns, R. (1993) 'Territoriality and land tenure among Mongolian pastoralists: Variation, continuity and change', *Nomadic Peoples*, vol 33, pp73–103

MNE (Ministry of Nature and Environment, Mongolia) (2001) *State of the Environment: Mongolia*, United Nations Environment Programme, Klong Luang, Thailand

MNE (2006) *Mongolia: State of the Environment, 2004–2005*, Ulaanbaatar, Mongolia

Mongolian Academy of Sciences (ed) (1990) *Information Mongolia: The Comprehensive Reference Source of the People's Republic of Mongolia*, Pergamon Press, Oxford

National Statistical Office, Mongolia (2001) *Mongolian Statistical Yearbook 2000*, National Statistical Office, Ulaanbaatar, Mongolia

National Statistical Office, Mongolia (2003) *Mongolian Statistical Yearbook 2002*, National Statistical Office, Ulaanbaatar, Mongolia

National Statistical Office, Mongolia (2006) *Mongolian Statistical Yearbook 2005*, National Statistical Office, Ulaanbaatar, Mongolia

Ojima, D. S., Chuluun, T., Bolortsetseg, B., Tucker, C. J. and Hicke, J. (2004) 'Eurasian land use impacts on rangeland productivity', in R. DeFries and G. P. Asner (eds) *Ecosystem Interactions with Land Use Change*, Geophysical Monograph Series, vol 53, American Geophysical Union, Washington, DC, pp293–301

Reading, R. P., Bedunah, D. J. and Amgalanbaatar, S. (2006) 'Conserving biodiversity on Mongolian rangelands: Implications for protected area development and pastoral uses', in D. J. Bedunah, D. E. McArthur and M. Fernández-Giménez (eds) *Rangelands of Central Asia: Proceedings of the Conference on Tranformations, Issues, and Future Challenges*, US Department of Agriculture, Forest Service Proceedings RMRS-P-39, Washington, DC, pp1–18

Schmidt, S. M. (2006) 'Pastoral community organization, livelihoods and biodiversity conservation in Mongolia's Southern Gobi Region', in D. J. Bedunah, D. E. McArthur and M. Fernández-Giménez (eds) *Rangelands of Central Asia: Proceedings of the Conference on Tranformations, Issues, and Future Challenges*, US Department of Agriculture, Forest Service Proceedings RMRS-P-39, Washington, DC, pp18–29

Schmidt, S. M., Gansukh, G., Kamal, K. and Swenson, K. (2002) 'Community organization: A key step towards sustainable livelihoods and co-management of natural resources in Mongolia', *Policy Matters*, vol 10, pp71–74

Sneath, D. (2003) 'Land use, the environment and development in post-socialist Mongolia', *Oxford Development Studies*, vol 31, pp441–459

Sonomdagva, Ts. (1967) *Changes in the Administrative-Territorial Division of Mongolia (1921–1965)*, Ulaanbaatar, Mongolia

Sonomdagva, Ts. (1998) *Changes and Reform of the Administrative-Territorial Division of Mongolia (1691–1997)*, Ulaanbaatar, Mongolia

UNDP (United Nations Development Programme) (2005) *Economic and Ecological Vulnerabilities and Human Security in Mongolia*, Interpress Publishing and Printing, Mongolia

UNESCO–WHC (United Nations Educational, Scientific and Cultural Organization–World Heritage Convention) (2005) *Operational Guidelines for the Implementation of the World Heritage Convention*, WHC 05/2, UNESCO World Heritage Centre, Paris

Walker, B. and Salt, D. (2006) *Resilience Thinking: Sustaining Ecosystems and People in a Changing World*, Island Press, Washington, DC

World Bank (2003) *From Goats to Goats: Institutional Reform in Mongolia's Cashmere Sector*, Report 26240-MOG, Washington, DC

16

Managing Floods and Scarcity in a Monsoon Climate

Louis Lebel, Po Garden and Phimphakan Lebel

The Ping

Life in northern Thailand has long been organized around the monsoon. Nearly 90 per cent of the annual rainfall falls between May and October. Stream flows rise several months after the start of the wet season, but decline rapidly after the rains end, underlining low storage capacities in soils and groundwater in the mountain catchments (Alford, 1992; Sharma et al, 2007). Water managers, thus must deal with periods of both excess and shortage each year. The Ping River drains much of the mountainous northern region of Thailand, flowing through a wide inter-montane valley that has a history of more than 700 years of communal and state-managed irrigation (Surarerks, 1986, 2006; Cohen and Pearson, 1998). During the wet season farmers divert water to grow rice, and fish in the floodplains and wetlands. In the dry season only a few farmers with access to main river waters or large all-season tributaries can grow crops. The rest repair canals, build baskets and traps, and harvest products from community-managed forests.

Several decades of economic growth in the inter-montane basin around Chiang Mai and Lamphun have transformed livelihoods and the landscape (Rigg and Nattapoolwat, 2001; Glassman and Sneddon, 2003; Lebel et al, 2009b). The first key transition was towards year-round agriculture. This was made possible by storage reservoirs and the extension of irrigation infrastructure by the state (Cohen and Pearson, 1998). A second transition was urbanization that accompanied the growth of manufacturing industries and tourism-sector development (Tran Hung, 1998; Pearson, 1999). Wetlands and irrigated areas were sliced and paved for housing and industrial estates (Tan-kim-yong et al, 2005; Lebel et al, 2007b; Lebel et al, 2009b). Capture fisheries became restricted to largely recreational and rare subsistence needs, having been replaced by an expanding pond and river-cage aquaculture industry (Chaibu et al, 2004; Lebel et al, 2007a). Economic development in the Ping River Basin may be making livelihoods and ecosystems more vulnerable to, and at greater risk of, exposure to dry-season shortages. Further supply augmentation and integrated water resources management (IWRM) are seen by many government officials as keys to dealing with vulnerabilities associated with seasonal water scarcity.

At the same time, wet season floods wreak more havoc and damage each year as infrastructure encroaches across floodplains (Manuta et al, 2006). The flood events in 2005 caused substantial economic damage, despite largely effective early-warning systems (Lebel et al, 2009c). Flood protection promises are increasingly made, but often without much consideration for the side effects of new infrastructure that also redistribute risks to already vulnerable places and people (Lebel and Sinh, 2007; Lebel and Sinh, 2009; Lebel et al, 2009a). Current patterns of urban construction in the Ping Basin may be making residents and businesses more vulnerable to flooding. Most officials look to quick technical fixes, hoping to manage new risks in concrete; a few reflect and argue for more emphasis on securing and maintaining ecological resilience.

In upper tributary areas, households barely integrated within the market economy sit next to luxurious resorts and orchards with sprinkler irrigation systems. In some valleys farmers are under an institutional barrage to change practices or move to make way for expanding protected areas (Laungaramsri, 2000; Walker, 2004). In other valleys competition for stream-water flows in the dry season are significant (Walker, 2003; Neef et al, 2006). Conflicts and disputes over causes of impacts and appropriate ways to develop, manage or conserve land and water are frequent (Charoenmuang, 1994). Better management of land uses in upper-tributary watersheds is seen by most government officials, and much of the wider public, as crucial to maintaining the ecological and hydrological services to lowlands. A few voices point to livelihood dependencies and needs for development opportunities for upland farmers (Walker, 2004; Lebel et al, 2008; Thomas et al, 2008).

Poor water quality is a recurrent issue in the more densely settled and industrialized parts of the basin. Surveys by the Pollution Control Department during 1997 to 2001 found surface water quality deteriorated moving downstream in the Mae Kuang tributary from Sansai district to Muang district of Lamphun (Panya Consultants and Sigma Hydro Consultant, 2003). Contamination due to phosphorus, nitrate/nitrogen and ammonia is higher than in other river sub-basins in the Ping watershed. Water quality indicators often exceed pollution standards. Pollution is caused by many sources, including municipal and industrial effluents, as well as intensive cropland and pig and dairy farms. Other sources of pollution come from small-scale and cottage industries, as well as other industries not covered by typical regulations applying to large factories (Thomas, 2006a). The reported high quantities of ammonia and nitrogen are a result of effluents released directly from communities without treatment. Organic pollution results in very low dissolved oxygen all year round, posing threats to river organisms. Every dry season finds reports of dead fish found floating in the river near industrial plants and dense urban settlements. There is a history of conflicts over water pollution among farmers, urban water users, and industrial and service operators.

Conflicts among water and land uses have grown more frequent and more difficult to solve, despite improvement of technical skills, information-in-hand and the financial resources of managers. Marginalized groups, both in remote rural locations as well as on the urban fringe, remain vulnerable to various combinations of floods, seasonal scarcity, loss of watershed services and poor water quality. In the age of enhanced information technology, the bureaucracy, community and political leaders access and use different kinds of knowledge to criticize, propose and justify their respective agendas and call for policy changes (e.g. CMU, 2004; Thomas, 2006a). This includes ignoring as well as exaggerating vulnerabilities – for example, it is common to 'talk up' normal seasonal dry periods as abnormal 'drought' crises with the aim of gaining public support and budgetary allocations for new water storage and diversion infrastructure.

Knowledge is clearly very important to improving how various water-related vulnerabilities and risks are managed in the Ping Basin. What remains far from clear is which organizational forms and institutional arrangements would best support the continuing evolution towards a more effective knowledge system (see van Kerkhoff and Lebel, 2006). This is the key problem with which this chapter grapples. Would it be preferable to combine many of these functions and house them in a one-stop-shop government agency, accountable to the public constituency that it serves and a huge array of government parent bodies in which it would no doubt have formal links? Or is some form of consortium, with even a competitive component, more desirable? Even more fundamentally, should the design of the knowledge system be left to bureaucrats, consultants, water users or researchers? Or should the knowledge system be allowed to self-organize and, therefore, remain only loosely integrated?

In this chapter we analyse the last decade or so of water and land use, as well as disaster management policy, and discuss how they have been affected by public administration reforms. We focus on the demands being made on the knowledge system for management of water-related vulnerabilities in the Upper Ping River Basin, northern Thailand. Our ultimate goal is to help shape the emerging water management knowledge system in ways that would support more equitable and sustainable negotiations around integrated basin management.

As we proceed, we will explain in more detail the structure and sources of four high-priority water-related vulnerabilities and the sources of resilience which may be managed to address them. We then introduce the key actors in our analysis, looking at the contributions and demands that they make of the knowledge system. Next we explore in more depth four specific arenas to illustrate key opportunities and barriers to improving knowledge/action linkages. Finally, we reflect on the evidence presented previously to draw broader conclusions about limitations of the current knowledge system, what might be done about them, and insights that might be of wider relevance.

Vulnerabilities and sources of resilience

Structure of vulnerabilities

In managing water in the monsoon climate in northern Thailand, four vulnerabilities stand out (see Table 16.1). First, there is the elevated risk of insufficient water for irrigation (and possibly also other uses) at the end of each dry season. After six months of low rainfall (see Figure 16.1a), many local water-storage schemes and small tributaries and streams are dry. Farmers with dry season crops in the field in downstream ends of irrigation schemes are particularly vulnerable to these recurrent seasonal shortages. Second, there is the increasing risk of significant economic costs from flooding in the wet season. Cyclonic depressions each year can bring widespread and prolonged episodes of rainfall, leading to major increases in river discharges that exceed natural bank and levee heights. Vulnerabilities are spatially heterogeneous with some low-lying poor urban areas inundated frequently and for longer periods. Third, there are the risks associated with degradation of upper tributary watershed landscapes that provide ecosystem and hydrological services at several spatial levels. The direction and extent of changes in upland environments, and, consequently, changes to risks, are highly place specific. Fourth, there are risks to human health and aquatic life from poor water quality, arising from a combination of agricultural, residential and industrial pollution sources. Low water quality also restricts wider uses and increases economic costs for water treatment.

These four vulnerabilities each arise from combinations of longer-term changes in important slow variables and more immediate changes in fast variables. For example, dry-season shortages vary substantially from year to year depending upon the vagaries in the onset of the monsoon, and this interacts with long-term trends of increasing demands for water so that demands on storage and scarcity are becoming more and more likely (see Table 16.1). Inter-annual variability in total rainfall is moderately high (coefficient of variation = 20 per cent). At the main city observation station there has been a long-term decline in annual rainfall of about 3.3mm per year, or 0.28 per cent per year (analysis by authors; see Figure 16.1b). The inflow into Bhumipol Dam at the end of the Upper Ping River Basin is a good integrative signal of changes in rainfall and water-use by vegetation (agriculture, gardens and natural). The main pattern of highs and lows is similar to that for rainfall, but with higher inter-annual variability (coefficient of variation = 31 per cent). There is also a

Table 16.1 *Water management-related vulnerabilities in the Upper Ping River Basin*

Structural Variables	Water management-related vulnerabilities			
	Scarcity	Floods	Services	Quality
	Insufficient water for irrigation (and other uses) at the end of each dry season	Increasing value of infrastructure and agricultural activities at risk from floods during the wet season	Declining quality of ecosystem and hydrological services from upper tributary watersheds	Declining water quality
Affected stakeholders	Irrigated rice, field crop and orchard growers in some locations	Urban residents, aquaculture, flood-sensitive farming activities	Upland and lowland farmers	Aquaculture and downstream users (higher treatment costs)
Sources of vulnerability: slow variables	Increasing water consumption (users and uses)	Building in the floodplain River narrowing Wetland conversion	Decreasing areas of native forest cover (some catchments)	Increasing number of point sources (domestic, agricultural and industrial) Degrading aquatic ecosystems
Sources of vulnerability: fast variables	Late start to the monsoon, extending the dry season period	Prolonged cyclonic depression rains Flash flooding Poor runoff water quality from sewerage, agricultural and industrial sources	Intense rainfall events, causing landslides Poorly built mountain roads	Seasonally low flows Untreated wastewater releases from factories
Knowledge uncertainties (causes, drivers)	Land-use change impacts upon water balances Trends in agricultural markets	Traditional weirs in urbanized areas Impacts of climate change on cyclones, intense rainfall events and, ultimately, flood regimes	Effects of complex landscape-level changes on hydrological functions	Assimilation capacities and sensitivities of natural ecosystems
Knowledge uncertainties (solutions)	Water-saving and efficiency measures Social acceptability of pricing, supply augmentation and other policies	Diversions and wetland retention Flood walls and dykes Living with flood architecture Effective early-warning systems to reduce moveable property losses	Low predictability of flash flooding Assessing and acting on landslide risk Impacts of plantations and orchards	Wastewater treatment (industrial, domestic) Reducing wasteful fertilizer application
Governance challenges	Water-sharing arrangements Allocation priorities among sectors and places Fair distribution of risks, burdens and benefits	Fair distribution of risks and compensation	Multi-use versus exclusive protected area and watershed policies Land, forest and water access and use rights	Effective regulatory procedures Better practice incentives

Source: chapter authors

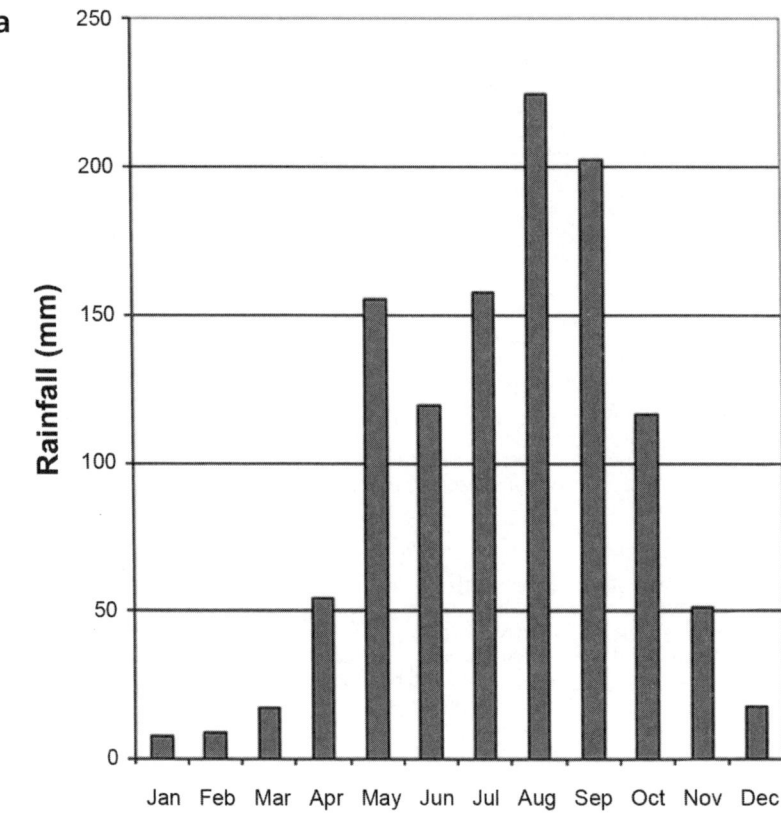
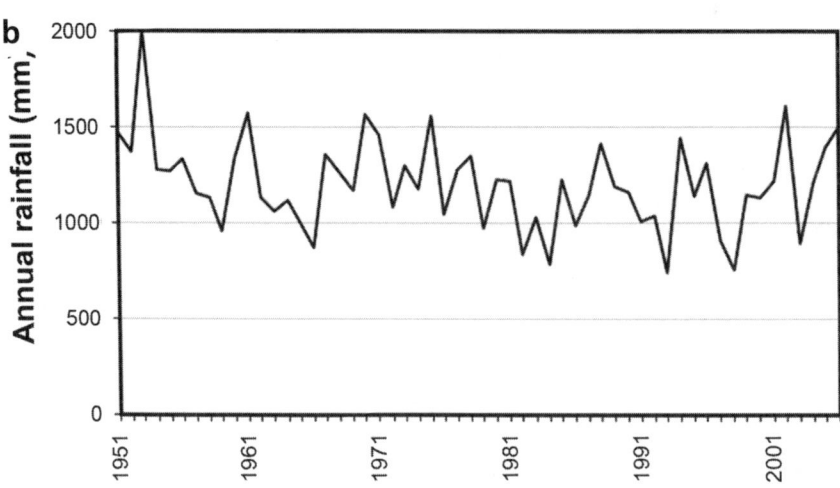

Figure 16.1 *(a) Monthly variation in rainfall for Chiang Mai City based on monthly mean, 1971–2000; (b) inter-annual variability in total annual rainfall for Chiang Mai, 1951–2006*

Source: chapter authors

significant long-term decline, as for rainfall, of 0.47 per cent per year relative to the long-term average (Sharma et al, 2007).

Interactions between fast and slow processes are also important to dynamic vulnerabilities from floods. Longer-term patterns of human settlement encroaching on wetlands and irrigated fields constrain floodplain areas, which, in turn, can interact with high runoff from individual storm events within cities or from surrounding mountains to increase flood water heights.

Vulnerabilities can be multilevel. The dry season flow allocation challenge is not only about setting priorities and negotiations among farming, industrial and urban uses within the Upper Ping River Basin upstream of Bhumipol Dam; it is also about trade-offs with generating electricity at Bhumipol Dam and supplying irrigation water to the lower Chao-Phraya Basin above Bangkok.

Vulnerabilities may be interconnected. For instance, operational decisions for Mae Ngad and Mae Kuang dams upstream from Chiang Mai must balance opposite risks. Major depressions near the end of the wet season, when rivers are already high and soils are moist, can create major flooding. In anticipation of such events, dams may release waters a few days in advance to help moderate flood levels, based on information from satellite imagery and weather forecasts. On the other hand, if rainfall does not arrive as expected, storage at the onset of the dry season will be much lower than normal. Within residential areas, irrigation canals can be a source of unwanted flood waters during the wet season. Chiang Mai residents and farmers have argued for years on the merits and otherwise of removing several traditional weirs and associated canals to reduce flood-related risks. Obviously, doing so would likely have adverse implications for dry-season water access for farmers. In both these examples reducing vulnerabilities to scarcity tends to increase risks to floods, and vice versa. The impacts of organic water pollution are often more serious during periods of low river flows when dissolved oxygen levels may fall below critical levels, threatening aquaculture and native aquatic life. Flood spates, on the other hand, may result in waste treatment ponds overflowing or create longer-lasting pools of water that are dangerous for human health.

Vulnerabilities are associated with uncertainties in knowledge about both causes and solutions (see Table 16.1). Some of these uncertainties are fundamental and others are more about linking scientific insights and practical or experience-based forms of knowledge. Thus, there are still some important constraints in the understanding of complex landscape changes on hydrological functions (e.g. dry-season flows) and watershed services (e.g. sediment delivery to streams). Supply augmentation decisions also need an understanding of infrastructure performance, costs and impacts, and these are often heatedly contested by experts using different models and assumptions. Much more analysis is needed, for example, about the potential impacts of climate change on flood regimes in this region, especially how they might interact with major changes in land use already under way (Lebel et al, 2009a). At the same time, not enough is understood about the side effects of introducing more flood protection infrastructure on floodplain functions and people's livelihoods.

Better technological and social intervention options are needed to control water pollutants from the large numbers of farms, homes and smaller firms which lie along the banks of the Ping River. Conventional regulatory and monitoring approaches that are applied to large factories cannot be applied so easily to these other water user groups. Likewise, and with both quality and quantity implications, groundwater use by agricultural, residential and commercial users, particularly in the Lamphun area, needs much more careful scrutiny as expansion has occurred with little regard for longer-term sustainability (Margane and Tatong, 1999). An estimate made several years ago suggests groundwater withdrawals were around 7.9 million cubic metres per year, which is close to the safe yield (Panya Consultants and Sigma Hydro Consultant, 2003). Many more wells have been dug in the last decade. Ultimately, these short-term 'adaptive' practices may be reducing overall resilience of the water system in the Upper Ping.

Vulnerabilities are, in part, social constructs. They are forwarded by actors to achieve other objectives, such as having construction projects funded or drawing attention to unfairness in how risks and burdens are distributed. As a consequence, competing claims about the needs to reduce specific vulnerabilities, or not, are a normal part of any efforts aimed at reducing vulnerability and building resilience. Knowledge, power and interests are never entirely separable; it is prudent, therefore, to ask how vulnerability and resilience are governed (Lebel et al, 2006).

Sources of resilience

The Ping Basin is a coupled social–ecological system with several important sources of resilience that could be managed to reduce water-related vulnerabilities. Sources of resilience are multilevel and spatially heterogeneous. In Table 16.2 we have disaggregated sources of resilience by zone at the household-plot level as an illustration of a finer layer of heterogeneity. Households in each zone face different challenges and have different strategies to maintain resilience. In addition, the key set of actors with responsibilities for managing vulnerabilities or with related interests are somewhat zone specific.

Upland landscapes and biological diversity are very important to farmers for maintaining the resilience of their livelihoods. Access to a portfolio of products from forests and cultivated and aquatic systems (rivers, paddies) provides insurance and safety mechanisms for households that have very few convertible assets (such as cash or savings). Ethnic, cultural and institutional diversity is a very significant feature of the uplands in the region (McCaskill and Kampe, 1997). The main ethnic minority groups in the Upper Northern Region of Thailand include Pwo Karen, Sgaw Karen, H'mong, Lahu, Yao, Akha, Lisu and Lua. Within and across groups there is a large diversity of languages and traditional institutional arrangements for managing both private and common pool resources. There is a variety of land-use systems, with many variants of shifting or rotational systems as well as small-scale irrigation systems (Thomas et al, 2002). These land-use systems are dynamic and rapidly modernizing, with improved linkages to regional markets (Thomas et al, 2008). This diversity produces a landscape mosaic likely to be a source of social and ecological resilience.

On the other hand, several important ecological changes over the past several decades imply that some 'conventional' sources of resilience are being undermined. For upland communities, the option to depend upon hunting is disappearing because of illegality and the reductions in populations of most large mammal species (as well as birds and fish). At the larger scale, the general trend has been declining forest cover, although this transformation is ending and has already reversed in some locations. Restrictions on access to higher-quality forest areas to meet national conservation objectives have created significant hardships and vulnerabilities for some households. Accordingly, people are increasingly capable and willing to work elsewhere, and where there are locational advantages (Thomas et al, 2008) they try new crops or even tourism-related livelihoods (see Table 16.2).

In contrast to upland areas, lowland forest ecosystems have almost completely disappeared. At best they persist in small patches in foothills and valleys as 'community forest' areas where useful products are extracted, mostly for domestic consumption. In a few cases they are maintained to support the arts/crafts industry; but even these are increasingly becoming private tree-crop plantations. As a consequence, the importance of biodiversity *per se* for lowland communities is less obvious and direct. Lowland northern Thai, or Khon Muang, still often identify themselves with traditions from the *Lanna* Kingdom (literally, million rice fields). For lowland livelihoods and economic sectors, maintenance of watershed functions in the uplands as well as in the lowland floodplains is the main issue. Several key wetland services are probably maintained in the long-lasting traditional irrigation schemes, and even the larger diversion dam-based systems that developed during the 1960s and 1970s. The degree to which 'diversity' (landscape and biological) is important for these various functions is open to debate and an area requiring much more research.

Lowland plots are largely privately owned. Some consolidation of ownership has taken place as rice areas have declined in the main valley. Share-cropping arrangements are significant. In both lowlands and uplands, knowledge about farming techniques, varieties and market opportunities appears to come through social networks in which the private sector, agribusinesses and traders play a key role (Thomas et al, 2008). Over the past two decades most innovations have been associated either with intensifying production systems and/or reducing labour inputs. In the lowlands a common strategy has been to switch to fruit trees.

The urban system has an unusual relationship with biodiversity. On the one hand, it would seem that there is almost zero dependence for daily activities. On the other, the importance of tourism for the regional economy in the

Table 16.2 *Examples of sources of resilience to challenges of water scarcity, floods, altered hydrological services and quality at different levels in the social–ecological system; at the smallest level we have also disaggregated by broad zones*

Level in the social/ ecological system		Sources of resilience to challenge of:			
		Scarcity	Floods	Services	Quality
Household (plot)	Upland	Emphasize rain-fed crops Store rainwater for consumption Wild animal populations as alternative food	Some fields in different locations	Longer fallow rotation Water source, steep slope and riparian community forests	Maintain headwater forests for purification services and secure drinking water supplies
	Lowland/ rural	Off-farm dry season livelihoods Shallow wells Water-sharing agreements	Some flood-insensitive livelihood activities	Water-sharing agreements	Alternative water supplies Pond ecosystem-based pre-treatment of water for non-consumptive uses
	Lowland/ urban	Low water-use gardens and fields (allow seasonal 'browning') Shallow wells	Houses on stilts or two levels		
Village to sub-district (landscape)		Irrigation water-user groups with strong institutions Wetlands as retention and groundwater recharge areas Rainfall storage for drinking and residential use (uplands)	Preserved and restored wetlands and key floodplains Effective early-warning systems Community-based aid networks and safety nets (e.g. temples) Water diversion capabilities; resources to protect key infrastructure	Rice fields as sediment traps Riparian buffer strips of vegetation	Functioning wetlands and in-stream ecosystems assimilating nutrients
Ping Basin to Northern Region (region)		Water-use planning	Functioning early-warning and disaster management systems	Land-use planning	Land-use planning Water pollution regulations and enforcement

Source: chapter authors

Ping River Basin should not be underestimated, and at least part of this tourism depends upon the perceptions of diverse cultural and natural landscapes (Kaosa-ard, 2001b; Shamshub, 2010).

Knowledge producers and action takers

Many actors are involved in the management of water-related vulnerabilities. Some are crucial to knowledge system functions, whereas others are more indirect pushers and shapers. The same individual or organization can be a knowledge producer or an action taker in different circumstances or relationships. In this section we discuss each of the major groups, shown in Figure 16.2. Our focus is on the contributions and demands that each makes on the knowledge system. Where possible, we also offer a preliminary assessment of the adequacy of the knowledge content for efforts to reduce vulnerabilities and build resilience.

Bureaucrats

The Royal Irrigation Department (RID), which sits in the Ministry of Agriculture and Co-operatives (MOAC), is probably the most conspicuous part of the technical bureaucracy in the Ping Basin today and a powerful agency in the overall national administration. RID has a long history of applying science and engineering to water problems. It originally, and still primarily, focuses on water provision to farms; but increasingly it is also responsible for flood management. It views itself primarily as both a knowledge producer and

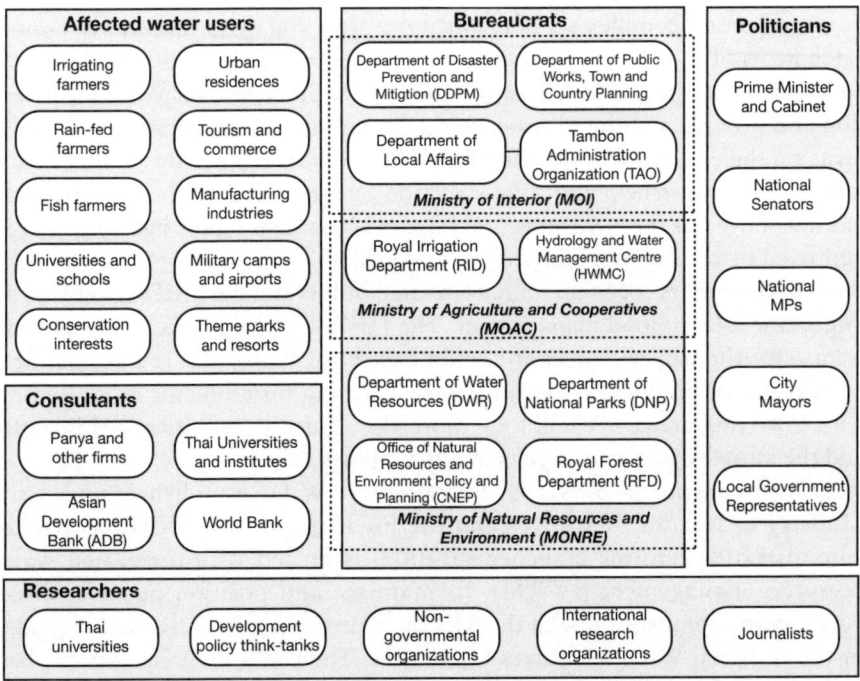

Figure 16.2 *Some of the actors involved in the emerging knowledge system for the Upper Ping River Basin*

Source: chapter authors

action taker. The RID is involved in planning, developing (building), managing and regulating water resources. In northern Thailand, many of its activities in rural areas were overlaid on local pre-existing water infrastructure and institutions (Cohen and Pearson, 1998; Tan-kim-yong et al, 2005). The outcome in the field is complex and remarkably robust because there is always, literally, another channel for information, water and resolving a dispute.

The pattern of growth in demand for water by industry, commerce and residential users in the Ping Basin (see Table 16.1) is well understood by bureaucracy, but not talked about openly. RID manages the Mae Kuan and Mae Ngat dams. It must plan water releases to irrigation districts and face the tough task of reducing dry-season allocation and planting areas through a 'negotiation' process with farmers, while priority must continue to be given to rising demand from urban residences and industry in Chiang Mai and its neighbouring Lamphun Province. Supply augmentation is the main strategy to delay having to make tougher decisions on rural water uses. One recent example is the Bhumipol reservoir inflow augmentation project which involves an inter-basin transfer from the Salween Basin across the border in Burma/Myanmar to the Ping Basin through a tunnel under Om Koi district, Chiang Mai Province, to flow into the upstream area of the Bhumipol Dam. This would have benefits for hydropower generation and perhaps also dry-season irrigation in the lower reaches of the Chao Phraya River.

Another is a complex set of planned transfers and diversion between upper catchments of the Ping, involving both tunnels and canals. This would provide additional water to the Mae Kuang storage dam that could be useful for irrigation and provide additional water to the Lamphun industrial estate and urban areas on the east bank of the Ping River. Public hearings for this controversial mega-project were held and a construction budget prepared; but construction has not started as the environmental impact assessment report has not yet been approved by the National Environment Board.

Apart from its focus on scarcity-related vulnerabilities, RID also plays an important role in flood management. The Hydrology and Water Management Centre for the Upper Northern Region based in Chiang Mai is a key actor; it has developed practical systems for early warning for all major urban centres with low-lying areas. We will look more closely at the activities of this centre and the knowledge-action arenas that it supports below.

The Department of Water Resources (DWR), established under the Ministry of Natural Resources and the Environment (MONRE) as part of administrative reforms launched in 2002, is tasked with integrated water resources management (IWRM). Its mandate and position put it in direct bureaucratic competition with the RID on many issues. Its activities are guided by the National Water Resources Committee. The DWR prepares master plans and policy advice. It promotes and supports river basin and sub-river basin committees or organizations. River basin organizations (RBOs) are viewed as the implementation tool for integrated water resources management. The latter is seen by some other actors, however, as a euphemism for introducing wider and higher charges for water under the guise of reallocating flows to more efficient uses. Lobbying by farmers with concerns over water pricing has been an important factor in the institutional stalemate over a national water law that would give administrative, policy-making and regulatory teeth to river basin organizations, the Department of Water Resources and the National Water Resources Committee. As of late 2010, the Water Law still had not been passed. We will look more closely below at RBO experiences in the Ping River Basin.

In Thailand, natural resources in the uplands and lowlands are managed separately (see Table 16.2). In the lowland areas, the RID has dominated water issues. The upper tributary watersheds have historically been under the jurisdiction of the Royal Forest Department, but now also the newer Department of National Parks. Both departments are in MONRE. The Pollution Control Department is another key agency dealing with water quality issues. In these areas local government or Tambon administration organizations (TAOs) often have little scope for their formal 'control' (Garden et al, 2010). For several decades the Ministry of Interior (MOI) has often pursued an independent agenda, providing roads, schools and health services, even to the many villages which are not recognized by the Royal Forest Department because they are considered to be on 'their' land.

Although it is conventional to analyse the policies, politics and knowledge foundations of the MOAC and MONRE when it comes to water and land issues in northern Thailand, the central role of the MOI has often been neglected. Many of the oldest and more powerful parts of the Thai bureaucracy remain under the MOI. Its influence is through its coordination mandates under administrative laws. The MOI decides when it is a drought or flood and is central to decisions on large infrastructure projects. Although it would be unconventional to say so, in many ways they are the *primary* water resource developers.

One way this influence is felt is through the administrative hierarchy. Integrative plans at the provincial office bring together the proposals of other agencies with the aim of avoiding duplication. Individual line agencies, however, largely continue to plan for themselves (Garden et al, 2010). Provincial governors are powerful players in the Thai administrative system and play an important role in finalizing these plans. Larger water-resource development projects cut across ministries; in these cases, cabinet decrees can help to solve coordination problems. Public consultation has traditionally come after cabinet resolutions – as a social-marketing exercise in gaining public acceptance for ideas that come from experts in the bureaucracy.

Another way of influencing the future which may become more important is through new structures. The MOI is modernizing and restating direct claims on the management of vulnerabilities. The Department of Disaster Prevention and Mitigation (DDPM), established in October 2002 as the principal government agency on disaster prevention and mitigation, ranks disaster risk from floods as its highest priority (DDPM, 2006). It replaced the earlier Civil

Defence Division, also within the MOI. It is organized through a hierarchy comprising a national committee (a system of 12 regional centres and the local Civil Defence Committees) as part of the normal bureaucratic structure of the Ministry of Interior. The introduction of the DDPM has not been an unqualified success; the agency has struggled to define a role for itself among line agencies with overlapping areas of responsibility (Manuta et al, 2006). Chiang Mai was not one of the 12 regional centres, as might be expected, making DDPM's relevance within the Ping even less than might otherwise have been the case. Significant disconnects among agencies responsible for preventing and preparing for disasters, on the one hand, and those more oriented towards relief and emergency measures, on the other, persist (Lebel et al, 2010b). The August 2007 Disaster Prevention and Mitigation Act was expected to strengthen the department's legal status; but this has not yet had much impact upon division of labour or disaster management planning practices within the Upper Ping Basin.

In summary, the longstanding competition among different ministries for roles in water management is likely to remain. Different departments and ministries stake claims on areas of expertise and place demands on the wider knowledge system for complementary research inputs.

Consultants

The main body of research is done by private consulting firms or universities in response to requests from government agencies or multilateral banks. A handful of consulting firms do most of the water-related work commissioned by the Thai government. There are usually close links between the larger firms and particular agencies and ministries. Decisions about projects are made by a panel whose composition depends upon what jurisdictions and mandates are crossed. Who chairs the panel is very important. In general, large project concepts arise from within bureaucracy, but are developed through consultant-led feasibility studies.

Panya Consultants, Bangkok-based and a relatively large and professional organization by Thai standards with close links to government departments, has been particularly important in carrying out assessments and feasibility studies for water infrastructure. It has also been commissioned to coordinate participatory planning activities. The company's sources of knowledge, understanding and experience are clearly an important part of current decisions. Panya Consultants maintains, for example, data on stream flow, groundwater and irrigated agriculture areas that have been acquired from working with various government agencies in the past. A typical example is the preliminary study of the Mae Ngat-Mae Kuang tunnel, which would divert water from the Mae Taeng River to the Mae Ngat Reservoir in Chiang Mai (Sutiwanich et al, 2006). Panya Consultants has been actively involved in the preparatory phases for structuring river basin and river sub-basin organizations as well.

According to the Public Debt Management Office in the Ministry of Finance, whose regulations oversee the hiring of consultants and which maintains a database of acceptable ones, there are three primary reasons why government agencies might use consultants (Government of Thailand, 2002). First, they may need the expertise. Second, they may need an opinion 'outside the bureaucratic mindset'. And, third, they may not have enough manpower to do the work. The database contains names of several hundred consultants per category, such as water supply, sanitation, construction, and agricultural and rural development. Consultants are hired through a competitive bidding process and direct hiring. Direct hiring is frequently used for major projects through the allowable mechanism of cabinet decrees, a potential avenue for corruption and non-transparent, less-than-independent advice.

The World Bank, the Asian Development Bank (ADB) and the Japan Bank for International Cooperation (JBIC) have all had important influences on the emerging knowledge system in the Upper Ping Basin (see Figure 16.3). Their influence has been through two primary channels: driving institutional changes at the national level, and supporting technical assessments and reviews. They both provide and commission the work of consultants. The World Bank has been more active in the latter areas, providing key support to the influential study of the Office of Natural Resources and Environmental Policy and Planning (ONEP) on organizational models for river sub-basin organizations (Thomas, 2005), as well as other technical activities.

The ADB and the JBIC leverage the Thai government's water policies using loan conditionalities to the agricultural sector (Abonyi, 2005; Molle, 2005). These include guidelines and requirements for most of the recent institutional changes in the water sector, from promotion of integrated water resources management concepts to the introduction of river basin committees or organizations and the drafting of new water laws and their creation of so-called apex regulatory authorities (Lebel et al, 2009a). The banks' interests are, of course, in making loans and seeing that they are repaid, and that more opportunities for loans arise when the water resource sector is further opened up to foreign investment.

Apichart Anukularmphai, widely acknowledged to have played an important role in introducing integrated water resources management and RBO concepts to the Thai bureaucracy and public (Lebel et al, 2009a), underlines the importance of viewing change as an evolving participatory process, with increasingly greater engagement of stakeholders. Many of these ideas are part of a global discourse on IWRM (Molle, 2008a) and shared in this region through the Network of Asian River Basin Organizations, which is supported by the ADB. The Upper Ping River Basin Organization, through its president, Supaporn Thornpook, expressed interest to join the Network at the third general meeting in February 2008, but at the time of writing had yet to formally do so.

Researchers

The boundary between consultants, researchers and advocates is fuzzy. Consultants carry out research on demand, whereas the knowledge and commentary from researchers are often uninvited. Some organizations and individuals may take on different actor roles at different times. Researchers often have substantial difficulties in getting their views listened to and are regularly criticized for findings that lack practical or operational-level relevance.

Most university groups have been adept at maintaining some critical independence from donors, but are restrained by the agreed terms of reference. Non-governmental organizations, media outlets and university researchers also carry out independent studies without requests from the state or banks. Chiang Mai University is the main source of research consultancies, guest speakers, task and assessment positions, and conductors of independent research. Among the academic community a few key individuals and their groups stand out because their analysis and activities affect public policy debates.

Wasan Jompakdee, for example, is an engineer at Chiang Mai University. For a decade or longer, he has been active in mobilizing Chiang Mai residents about ecologically inappropriate riparian developments. He appeals to traditional culture and beliefs more than 'research papers' when campaigning. At the same time, he spearheaded a major study by a Chiang Mai University consortium (CMU, 2004, 2005) on the condition, threats and vulnerabilities to water resources in the Upper Ping River Basin. Another key figure has been David Thomas, who headed the International Centre for Research in Agroforestry (ICRAF) office in Chiang Mai. He has been able to pursue participatory watershed management activities and partnerships with local government and the Royal Forest Department (Thomas et al, 2000; Thomas et al, 2004). He was also a primary consultant to ONEP in forming the new river sub-basin organizations (Thomas, 2006a). The ICRAF office has also contributed significantly to the improvement and sharing of regional land cover and other datasets with Thai government agencies.

The development economist Mingsarn Kaosa-ard was head of the Thai government policy think-tank Thailand Development Research Institute in Bangkok, and now leads the Social Research Institute at Chiang Mai University. She publishes articles, reports and books regularly, mostly in Thai, on environmental and development policy issues (Kaosa-ard and Rerkasem, 2000; Kaosa-ard, 2001a). She sits on various boards and committees and gives talks at major water-related events. Her primary contributions have been in institutional design.

Some of the more prominent research communities with regard to water management in northern Thailand are not necessarily based within the basin. Some key international organizations and domestic policy think-tanks have located their offices hundreds of kilometres downstream in Bangkok, where they can be close to key bureaucrats. Key domestic organizations include the

Thailand Development Research Institute (TDRI) and the Thailand Environment Institute (TEI) (Kaosa-ard, 2001a). Important international organizations include the World Conservation Union (IUCN), the International Water Management Institute (IWMI) and the World Wide Fund for Nature (WWF). A significant amount of the research on water-related vulnerabilities has been carried out by non-Thai citizens; but because it is not available in the Thai language, it does not always influence experts and advisers in the bureaucracy as it otherwise might.

Independent research tends to be much more critical of current practices and policies than work carried out by consultants for the state. Irrigation infrastructure development and related water policies have come under scrutiny for being too quick to adopt institutional and organizational models from elsewhere without due attention to local physical and social contexts (Pearson, 1999; Molle, 2008a; Tan-kim-yong et al, 2005). Women, for example, continue to be excluded from new basin-organization structures, repeating gender biases present in earlier irrigation management structures (Resurreccion et al, 2004). Studies of flood and disaster management suggest both coordination and equity problems (Manuta et al, 2006; Lebel and Sinh, 2007). These studies also show an over-reliance on technical planning expertise, without much understanding of behavioural responses or social impacts. Side effects and instances of risk redistribution are ignored. Likewise, there is a substantial body of research studies that is critical of land and conservation policies in upland watersheds. Many of these focus on the arbitrariness or unfairness of zoning schemes and the lack of recognition of use and access rights of minorities and other poor farmers (Laungaramsri, 2002a, 2002b; Vandergeest, 2003; Daniel, 2005). Claims about upland forest loss are widely touted as leading to loss of hydrological services, inducing both water shortages and floods (e.g. Krairapanond and Atkinson, 1998); but much of the knowledge base is incomplete or unconvincing, and best understood as a selective use of science (Forsyth, 1996, 1998; Walker, 2003; Bruijnzeel, 2004; Lebel et al, 2004; Ziegler et al, 2004).

Journalists are grouped here under the category of researchers (see Figure 16.2) because they have played an important role in drawing public attention to water management issues and the interests of key actors involved in decision-making (e.g. Garden and Nance, 2007). In northern Thailand they have, at times, acted as conduits for the positions of the bureaucracy and politicians; but on other occasions they have also highlighted alternative and minority viewpoints effectively, creating space for deliberation and contesting knowledge claims.

Water users

Water users are often an under-recognized foundation of the knowledge system. Their local practical experience is crucial to reducing vulnerabilities. The most well established, but still often ignored in recent government reforms, is the long history of community-based organizations concerned with managing irrigation water (Surarerks, 1986; Cohen and Pearson, 1998; Tan-kim-yong et al, 2005). Some of these are linked to RID activities, especially when they use water from canals run by that department; but others are still relatively independent of direct state management.

The Thai government has been trying to get rid of traditional communally managed weirs because of their putative impacts upon urban flooding and to replace them with a system of water gates (Anon, 2006). The manager of the removal project based at the RID office in Chiang Mai argued that:

> *The Ping River has been encroached upon and is now too narrow in key places. In addition, weirs prevent water from flowing freely. We have to clear the river. Water gates should replace weirs so that water can flow freely in the rainy season, but still be used for agriculture in dry season.* (Skol Harnpitakul, pers comm, 2006)

The proposed plan for a 6.5m high water gate was approved by the cabinet on 25 July 2006. The project is justified as coming from the king through a special committee to coordinate royal-initiated projects headed by Julnop Snitwong Na Ayudhaya. This argument was disputed by some; the head of a traditional weir committee said that the king personally opened the weir in the Commercial Banking Act BE 2505, a claim denied by the RID.

There are also unresolved disputes about whether an environmental impact assessment carried out for a similar project in 1997 by the RID needs to be redone. A senior official with the RID believed another environmental impact assessment was not required, whereas the deputy Chiang Mai governor strongly urged that a new study be conducted, regardless of laws, because conditions could have changed. The shift from traditional weirs to new gates would have major repercussions for control of irrigation waters; local committees are worried that they will lose their function and capacity to sort out water management problems. The argument is, in part, one between traditional knowledge that is part of the package deal with traditional infrastructure and institutions, and, conversely, 'modern scientific knowledge' that brings a shift in control with the introduction of new technology. Put simply, it is knowledge about managing water for irrigation versus flood protection.

In upper tributaries, water-user groups are usually labelled watershed networks and deal more explicitly with land and forest management issues, as well as more direct water supply and allocation issues. As in the lowlands, many of these networks predate the river sub-basin initiatives of the Department of Water Resources.

Farmers are not the only stakeholders with significant expertise important to water management. The practices of commerce and industry are also important. Although rarely discussed, these stakeholders are significant and growing users of water in the Ping Basin (Pearson, 1999; Glassman and Sneddon,

2003). They often have expertise in wastewater treatment and recycling and thus are key knowledge holders and action takers with respect to water quality-related vulnerabilities.

Politicians

Politicians are very active in the knowledge game (see Figure 16.3). They make promises of flood protection, security from drought, forest conservation and clean rivers that, in turn, highlight uncertainties in understanding and create demand for new knowledge (or, in some cases, its suppression).

Local government agencies, including both Tambon administrative organizations at the sub-district level and municipalities, can have substantial local-level experience and access to much more experience through local networks. Elected councils demand better data and information to justify plans and projects. Some local government agencies are beginning to collect and organize datasets, not just in response to higher levels in the administrative hierarchy, but also to help with planning and developing better proposals for local development projects.

Large water-hungry projects that increase the vulnerabilities of farmers to dry-season shortages continue to be proposed and implemented because their proponents are able to draw effectively on cultural symbols or political concerns (Molle, 2008b). Thus, investors with political influence were able to secure land and water for the construction of a controversial, but high-profile, tourist attraction, Night Safari, on the outskirts of Chiang Mai (Khuenkaew, 2007). Next door, and almost at the same time, the Royal Floral Expo grounds, Ratchapreuk, were developed in tribute to His Majesty the King. Both sites became popular with foreign and Thai tourists, but also created substantial tensions with other water users who reportedly are 'contained' by the state (*Nation*, 2006b). Proposals to extend the Night Safari with a large elephant park have been opposed by residents and others (Khuenkaew, 2007; *Nation*, 2007). Both projects were initiated by the government of former Prime Minister Thaksin Shinawatra as part of the plan to turn Chiang Mai into a major tourism destination. Farmers in the area were worried about increasing vulnerability to dry-season shortages and were asked to reduce dry-season cropping (*Bangkok Post*, 2006; *Nation*, 2006b).

Floods create political opportunities, such as new budget lines to be captured or directed. Many of the projects for protecting Chiang Mai from floods that resurfaced in late 2005 and 2006, as the flood waters receded, had already been included in consultant or RID reports years earlier. The then deputy prime minister, Suwat Liptapanlop, who had been put in charge of flood prevention for Chiang Mai, quickly obtained approval for projects worth a total of 13 billion baht (Lebel et al, 2009d). Other projects were led by Newin Chidchob, an influential politician from the northeastern region. Almost immediately local communities organized opposition to the larger dyke- and weir-levelling projects (Phanayanggoor, 2006). The opposition was largely not against more infrastructure of any sort per se, but on the way in which such decisions were being made, which options were being considered, who should be responsible for management, and how they would be maintained (Lebel et al, 2009d). National political crises, including a military coup in 2006, stalled the various projects (*Nation*, 2006a). Even so, several months later, the mid 2007 municipal elections for Chiang Mai were notable for a campaign in which posters of most candidates pictured them standing waste deep in flood waters, illustrating how management of water vulnerabilities remained a central political issue.

Linking knowledge and action

The various actors producing and using knowledge for water management are linked to each other in various formal and informal networks and hierarchies, some relatively well established and others new. These institutionalized relationships provide something of the 'infrastructure' upon which a knowledge system can evolve – for example, towards sustainable water-resource management goals. We next look at four institutional or organizational initiatives to better link knowledge and action for the management of water. Each example focuses on one of the three water-related vulnerabilities we identified earlier (see Table 16.1) and highlights increasing degrees of engagement and power-sharing (van Kerkhoff and Lebel, 2006).

Communication for preparedness and warnings

The Hydrology and Water Management Centre (HWMC) of the RID maintains hydrological data on rivers in northern Thailand and has developed flood prediction models for major rivers in northern Thailand (Lebel et al, 2009c). These models can usefully predict floods caused by rainfall upstream of Chiang Mai about eight hours in advance. Timely warnings can save lives and property. Early-warning systems are based on both water levels and rainfall data. The HWMC is cautious about over-automating early warning systems as it needs to consider the possibility of electricity failures and maintenance costs.

Public awareness campaigns using leaflets and the mass media are used to prepare the wider population in each wet season. In Chiang Mai, public awareness campaigns are conducted in both English and Thai, as a large number of foreign tourists visit the city each year (Lebel et al, 2009c). The HWMC is well set up to give briefings to government officials and the mass media about flood threats, status and management options. It has a specially designed 'media-briefing' room filled with visual aids. There are maps upon which water levels and rainfall data are regularly updated using simple manual cards. There are rule-of-thumb-type calculation charts about when to expect different peak flood heights and issuing warnings. The director of the centre, Thada Sukhapunnaphan, maintains an impressive set of Power Point presentations with video clips and animations to explain the causes of different kinds of floods, how early-warning systems work, and the impacts of previous floods. The seminar and briefing rooms at the HWMC are important knowledge-action arenas where scientific information is condensed and communicated in readily understood ways to a range of government and community stakeholders.

What we find unusual in an expert bureaucracy is that the HWMC has spent considerable effort in coming up with practical solutions, as opposed to pursuing the more typical goals of 'highest available technology' approaches to communication. In our view this has largely worked. At the same time, the organization is continually exploring ways of improving its observation networks (e.g. using weather radar) and its communication tools (e.g. through computer-based animations) while still using and improving physical display boards for use in the field. The HWMC also has an informative functional website.

Early-warning systems and the associated flood-control operations that they trigger to reduce vulnerabilities in and around built-up areas of Chiang Mai are not, however, foolproof. Our fieldwork during the series of costly floods which struck Chiang Mai in 2005 (Wood and Ziegler, 2007) underlined the importance of redundancy in communication channels. During several events that year, community radio stations played an unexpectedly important role (Lebel et al, 2010b). Affected people called in with requests for assistance and situation reports and public officials joined in and used the stations to organize assistance. The radio stations also exposed limitations with the formal early-warning systems.

The improved levels of preparedness possible with good early-warning systems can have unintended consequences. Preparedness was high for the last expected flood event of the 2005 wet season in Chiang Mai (Lebel et al, 2010b). In September 2005, as Cyclone Damray made its way across the Gulf of Tonkin near Vietnam, large hotels and firms in Chiang Mai's central business district protected themselves with sandbags and water pumps. The flood waters arrived and were deflected and redistributed; when those on the receiving end removed the barriers to let the waters flow, conflicts ensued. A lot of warning and experience gave local authorities, businesses and communities time to prepare, and they did, but often by redistributing risks and burdens to others. The redistribution of risks is a common political dimension of efforts to reduce vulnerabilities (Lebel and Sinh, 2009d) and likewise raises social justice issues with respect to larger-scale and longer-term adaptation measures (Lebel et al, 2009a).

In comparison to riverbank overflow floods, flash floods in mountains around Chiang Mai remain much harder to prepare for and to issue early warnings. Landslides, debris flows and flash floods in these areas continue to cause serious losses of property and life (Asian Disaster Preparedness Centre, 2006; Yumuang, 2006). Radar can provide generic warnings about the likelihoods of intense rainfall events over broad areas, but needs to be complemented by local monitoring and communication systems to be of any practical use (Lebel et al, 2010b). People living in areas of risk must learn how to recognize warning signs (e.g. intense prolonged rainfall, and stream and river quality indicators). Appropriate channels of communication to inform others of the need to take precautions are required. These must overcome social constraints – like a history of top-down approaches focused on emergency response and relief that has left many communities complacent about risks while unaware of their own capabilities and roles in disaster preparedness (Lebel et al, 2010b).

A shared understanding among stakeholders, from users and affected groups through to managers and policy-shapers, is not easy to achieve in any domain. In some situations, such as early warnings for floods, the problem of communication is a relatively straightforward one of 'passing the message on'. But in many other situations it is a much deeper one of 'sharing frames of understanding' – for example, in the allocation of rights to water for crops. The latter situations clearly require much greater two-way information exchanges than the former.

Integration for managing scarcity

The Upper Ping River Basin Committee was established in 1999 but reconstituted with broader stakeholder membership and new selection procedures in 2001 (Anukularmphai, 2004). Later, the committee became known as one of the pilot river basin organizations (RBOs) promoted by the DWR and has since gone through several changes in structure and membership.

Different agencies have different ways of explaining and drawing their relationships with each other. From the perspective of the Department of Water Resources, 'local' administrations (in the MOI) are nested inside and below 'river basin organizations'. The much more powerful MOI *knows* the world works quite differently (Garden et al, 2010). Most of the skills and authority in planning and coordination exist within the MOI. It supplies administrative staff to the Upper Ping River Basin secretariat. Most of the water-related data, infrastructure planning and engineering skills, on the other hand, lie within the RID. But there are links. A working group on information in the early years of the Upper Ping RBO, for example, was chaired by the director of the HWMC, as discussed above.

On paper, the mandate of the Upper Ping River Basin Organization is huge and includes advising the Ministry of Natural Resources and Environment on water policy, planning and projects, formulating river basin management plans, gathering data relevant to water resources management, dispute resolution, solving water management problems, and cooperating with other government offices and river basin organizations. In practice, most of the work in 2005 was public relations and planning. We directly observed underinvestment in public consultation exercises where major 'discussions' were squeezed into narrow timeslots in a hectic schedule. The only conclusion which can be drawn is that the exercise was instrumental, so that 'participation' could be checked off the list. Moreover, the existence of several parallel planning processes has proven difficult for us as researchers to untangle and, not surprisingly, mystifies many of the participants.

The interaction between basin and more conventional area-based jurisdictions within the MOI is a source of both opportunities and challenges. Although the mandate of the Upper Ping River Basin Organization is not strongly backed by a new water law, some RBO representatives don't see this as a drawback, claiming that a water law is unnecessary as there is already authority through local government, such as the Tambon administration organization (TAO) and line agency representatives.

The introduction of the RBO structures has created a demand for knowledge at multiple levels. This demand for both level-dependent information and cross-level understanding is not yet met adequately by current data and information systems. Many feel that that the Upper Ping River Basin Organization has not done enough to justify its role as a knowledge broker or integrator of basin planning activities. The RBO's capacity to develop relationships with the line agencies, which they have been told to 'bring together', let alone the private sector and other agricultural water users, is still limited. Most of the 'integration' has been little more than 'compilation'.

An alternative may be to recognize the distributed expertise relevant to sustainable water resources management and to solve integration challenges. Intermediary bodies such as river basin and river sub-basin organizations might be seen as arena creators or places where multi-stakeholder dialogues can take place in a way that has some fairly well-defined means of linking to more formal assessment processes when needed and to decision-making processes as a whole – that is, places where dialogue among different parties can take place, but without giving them any significant power to set policy or make plans. Rather than trying to re-staff and re-tool the Upper Ping River Basin Organization in this knowledge management role, we think that there are better prospects in expanding its role as a facilitator of the emerging multi-level platforms for deliberation on water resources development and management interventions.

Learning to manage services

Many watershed networks have emerged, adapted from traditional local-level institutions or created out of local needs to manage forest, agricultural land and water allocation problems and disputes (Na Ayutthaya, 1996; Wittayapak and Dearden, 1999; Lebel et al, 2008). Some of these have a substantial history of organizing information and drawing on external knowledge sources and expertise.

The experiences of ICRAF in the Mae Chaem watershed in developing participatory science or evidence-based methods of monitoring are a good example (Thomas et al, 2004; Saipothong et al, 2006; Thomas, 2006b). In a reflective interview we heard how this work 'showed that the whole basis for land use was that people needed to negotiate among themselves the upstream/downstream issues; we showed that the sub-watershed unit was a very useful management unit at the local level – a lot of local names are based on the streams at the sub-watershed level' (David Thomas, pers comm., 3 August 2007). The Mae Chaem River Sub-Basin Group is made up of at least 25 local sub-watershed networks that merged together. We helped researchers working with ICRAF by putting the maps and data and time series on the table and lifting the standard of debate.

Several other river sub-basin organizations were formed with support from the Office of National Resources and Environmental Policy and Planning and a grant from the World Bank. They are entering at an intermediate spatial level, usually larger than pre-existing watershed networks or communal irrigation systems, but at a much smaller and manageable level than the Upper Ping RBO, discussed above. There are prospects for much more bottom-up rather than more typical top-down design of river basin organizations at this level (Thomas 2005, 2006a).

Watershed networks can draw on formal expertise to learn better how to manage hydrological and ecological services – for example, through participatory mapping, as done by ICRAF in Chiang Mai (Lebel et al, 2010a). The need to understand better spatial variation and distribution of land uses and covers became obvious early in their research programme so that more mapping and remote sensing was carried out. This led to rethinking the categories of land use through to the dynamics of change and its impacts upon people and the environment, and reframing agro-forestry issues at landscape rather than just at individual plot levels. Maps also became useful boundary objects in negotiations over land use in watershed management (Saipothong and Thomas, 2007). But many village-level mapping exercises are too small in extent for use in watershed management, a problem overcome by capturing the information about land-use and land-cover maps in digital form and using geographic information system (GIS) techniques to allow flexible presentation and a higher, more standardized, quality output with government and other agencies (Saipothong et al, 2005).

In this example ICRAF's influence has been diffuse, important and consistent with its mission: strengthening evidence-based reasoning and deliberation (Lebel et al, 2010a). David Thomas, then head of the ICRAF office in Chiang Mai, told us: 'It was not up to us as a research organization to say what was right and what was wrong, but we were saying that this is the evidence of what is, what has been, and what's going on. Now go compare this to what people say and make your own judgments' (Lebel et al, 2010a).

Planning for quality

The selection of two sub-basins (Ping Part 1, Mae Kuang) from the Upper Ping for initial further pilot organizational development took place at the water forum meeting in Chiang Mai in March 2005. Non-governmental representatives made up at least half of the 23 and 24 members, respectively, of the two sub-basin working groups, and officials from MONRE about one third (Thomas, 2005). During their first year the river sub-basin organizations (RBSOs) focused on developing action plans. These drew strongly on earlier Ping Basin programmes, in part because of time pressures, but also reflecting the interests of the government agencies involved. The plans appear to be evolving in ways that largely reproduce the goals and plans of individual government agencies. In contrast to the policy attention given to upland ecosystems as providers of valued services and a source of resilience, the functional value of ecosystems in the lowlands in dealing with water quality issues has largely been ignored. Both river RSBOs were active in the years immediately following their formation, but gaining agreement on organizational structures and planning procedures has proven time consuming.

Larger political events have had repercussions for the Mae Kuang RSBO and its parent body, the Upper Ping RBO. During 2007 to 2009, changes in governments caused multiple changes in RSB and RSBO committees with new rounds of selection. A new Ping River Basin Committee was appointed in August 2008 and other committees were reappointed in 2009. After August 2008, the Mae Kuang RSBO was to have 36 members distributed widely across stakeholder groups. Representation of non-central government stakeholders was substantially better than in organizations and committees at higher and lower spatial administrative levels. As of March 2010, selection for the new Mae Kuang River Sub-Basin Working Group was still ongoing.

Engaging stakeholders in the Mae Kuang River Sub-Basin near Chiang Mai has been particularly challenging because of the huge diversity of land uses and interests that stretch from upland minority villages, through irrigated farms, to peri-urban and industrial users of water.

Several platforms co-exist in the Mae Kuang River watershed. Some focus on interests of agricultural water users connected to the Mae Kuang Dam; others are based on traditional *muang fai* irrigation schemes; and yet others are related to industrial water users and upper tributary watershed networks. The DWR's efforts to introduce an RSBO met with little initial success for a variety of reasons, including bureaucratic competition, lack of resources, and a failure to adequately take into account these pre-existing platforms and institutions. Powerful local coalitions continue to support and work through alternative platforms and channels. The persistence of cross-sectoral allocation and water quality management problems suggests that to be effective as a multi-stakeholder platform the Mae Kuang River Sub-Basin Organization should restart, working from existing capacities and organizations.

A consequence of the interaction between decentralization reforms that have produced planning demands and capacities at the TAO level with the current action-planning procedures is a proliferation of proposals for a huge number of small water-related projects, many of which could conceivably be compiled into much more efficient river sub-basin or even basin-wide initiatives (e.g. when related to training and capacity-building requests).

Construction-oriented projects dominate and strongly reflect the conservative knowledge and conventional interests of line agencies active in the various basins. Total budgets requested across many small projects appear to be high relative to the kinds of funds which might eventually be channelled through Ping Basin mechanisms (Thomas, 2006a). Nevertheless, individual agencies work to get working groups to forward their projects onto the table and into planning documents. Most stakeholders' input that reaches the plans is bureaucratic, fragmented and still reflects organizational interests rather than what might have been expected from greater stakeholder engagement and local knowledge inputs. This leaves a huge challenge for setting priorities. Transparent and evidence-based procedures, as well as monitoring and evaluation systems need to be much further developed.

Knowledge, power and vulnerability

The knowledge system emerging to manage water and water-related vulnerabilities in the Ping Basin in northern Thailand has a few strengths and several weaknesses. We have explored above the knowledge/action arenas created or supported by selected organizational initiatives in four issue areas. As can be seen in Figure 16.3, there are some connections between these and the vulnerabilities they manage. Some links are direct through shared actors engaged, whereas others are more at the level of knowledge demands and content. For instance, ICRAF has been an important actor in dealing with watershed services issues through support for upland networks and participatory planning processes, as well as dealing with the formation of RSBOs in lowland areas where water scarcity and quality issues are much more important. The HWMC of the RID has provided important data and understanding on water balances in various platforms, including the Upper Ping RBO with its concerns for allocation issues, and to other actors in dealing with flood management issues.

From providing timely warnings of floods through to negotiating water and land uses to stave off threats of scarcity or ecological collapse, several recurrent tensions and gaps are evident, as well as openings or opportunities to address them. In this section we summarize the challenges, first, and the possibilities for addressing them, second.

Tensions

The first tension arises from the predominance of belief-centred rather than evidence-based practices. In exploring the knowledge system supporting decision-making, we identified two recurrent information problems: no data and limited analysis. The problem of no data came in two forms. Either none of the needed observations had yet been made, or they had but were not being shared among agencies. The problem of no data has to be addressed by measurement and observation networks and by agreements to document and

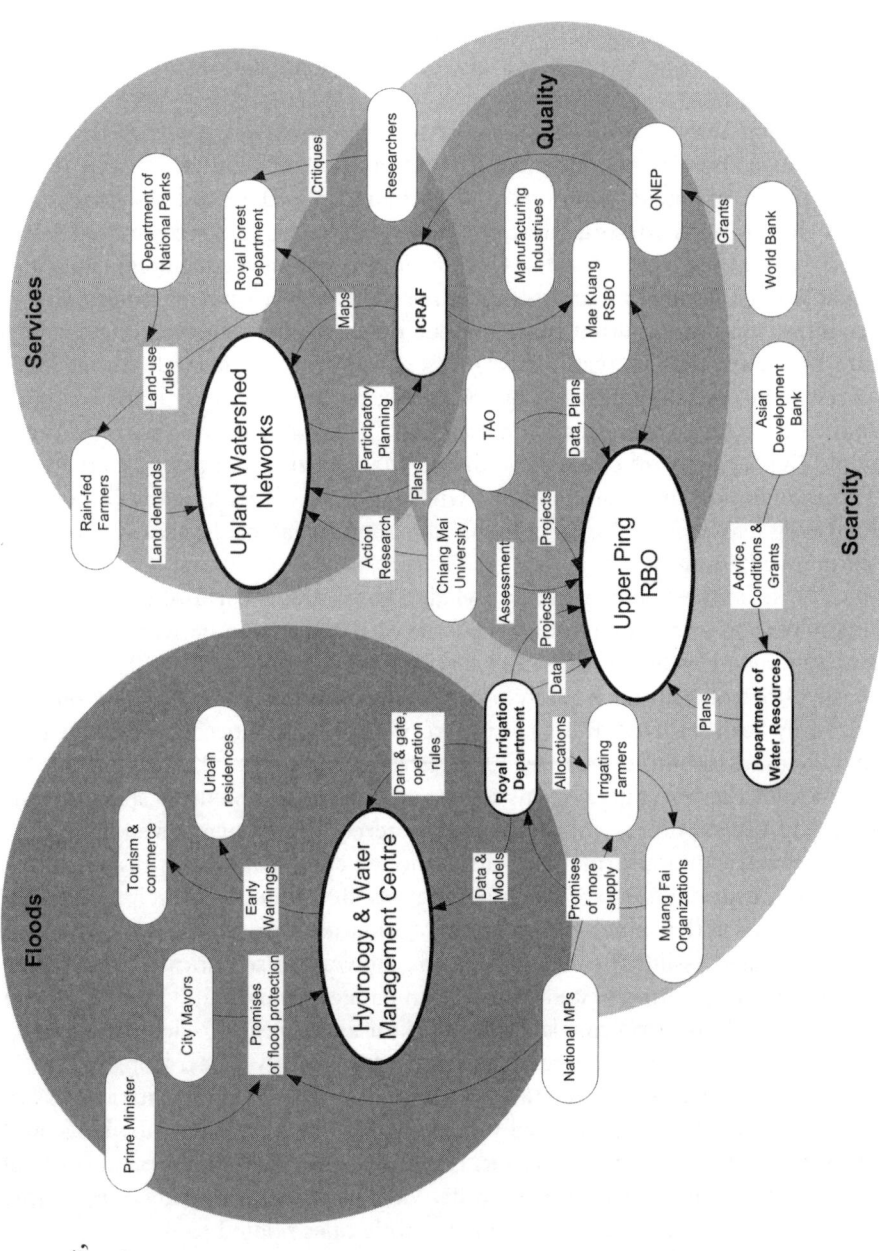

Figure 16.3 The emerging knowledge system around floods, scarcity, services and quality issues in the Upper Ping River Basin

Source: chapter authors

share datasets across agencies. The problem of no analysis was not expected when we started our research, but obvious when justifications for projects and their critiques were scrutinized.

One explanation is that proponents and opponents don't need to do analysis because there is no demand: reasoning is never challenged and initial positions are assumed to be fixed and unchangeable. Data and plans are made to fit expectations and organizational interests. Research has been used selectively to reinforce conventional wisdoms in the management bureaucracy and in the non-governmental anti-water infrastructure development community. A second explanation is that there may be no capacity to collate, analyse and interpret data that is available. This is a frequent claim made about local government and community organizations, but can be easily self-serving to central bureaucracy interests. Lack of capacity is addressable with training, while lack of demand can only really be changed through public challenges. This is an important tension because while some beliefs are valuable, others are hindering learning in what is a highly dynamic social–ecological system with shifting vulnerabilities.

The second tension is the knowledge gaps and uncertainties created by fragmentation and competition. The issue of coordination between watershed and administratively defined scales is raised in David Thomas's (2006a) report to the Office of Natural Resources and Environmental Policy Planning but has not really been resolved. Rather, it is part of ongoing 'politics' among ministries that continue to unfold after the 2002 restructuring of the environment portfolio. Within the bureaucracy this is an outstanding problem, despite various reforms. Information is not shared widely across departments even within the same ministry; and informal networks among people with direct responsibility and interest in datasets are often much more developed and reciprocal. One consequence is that much of the basic information that exists (somewhere) is effectively inaccessible (by most). There are, however, also critical gaps in observation and monitoring systems for stream gauges, rainfall and pumping from irrigation and diversion canals. Procedures for assessment of scientific information are poorly developed, largely *ad hoc*, and frequently lack transparency.

Another form of fragmentation is that arising from differential access to knowledge within bureaucracy by those outside it. Access to knowledge and information flows is uneven across classes in society. It is an especially critical problem for minorities in the uplands. Various factors, starting with simple ones such as language, and other more subtle ones related to power relations and political influence (*amnat*), constrain access to normal channels of public information. Competition can be healthy if it leads to pursuit of excellence in knowledge making or action taking, but, on the other hand, lack of cooperation in sharing data means that each actor is working from unnecessarily incomplete understanding in efforts to reduce vulnerabilities. This tension also undermines administrative integration efforts.

The third tension is the variation in standards and procedures in reaching decisions on smaller, intermediate and larger projects. The assessment and decision-making processes for large supply projects remain the least transparent. The convention of holding off stakeholder engagement until after key decisions are already made, and the process can be turned into one of social marketing and gaining public acceptance for done deals, continues. To date, most of the assessment in the water sector has been conducted by infrastructure builders and their consultants, not users. The Upper Ping River Basin Organization remains completely silent and sidelined on larger projects. Independent assessments of science in areas of contested public policy are not happening by themselves. Lack of acceptable assessments can be addressed either in championing an independent body or making a legitimate, credible process that is open to inputs and challenge. It is a fundamental problem with institutional development.

Engagement with a broader range of stakeholders as projects or policies of larger scope are considered tends to bring in more sources of knowledge and this creates challenges for cross-validation. The final issue of 'integrated river basin management' is a relatively new one for the Upper Ping Basin, but increasingly hard to ignore. Representative local government provides some avenues for people to express their preferences about how knowledge is being used in decision-making, but much less opportunity for shaping what knowledge is sought or assessing what is really needed. Research and observation agendas are still largely shaped by the preferences and beliefs of state agencies.

The fourth tension is the complexity and relevance of planning procedures. Bureaucracy is good at making plans. But plans rarely reflect practical realities except when they are literally about rewriting history. A substantial amount of effort in the past decade of administrative reforms associated with political decentralization has been directed at planning procedures that have become more complex. Information-intensive planning processes have made it almost impossible for an ordinary citizen to track the route of project approval and budget processes. Water resources development and management issues easily cut across the mandates and jurisdictions of multiple authorities and levels, often making coordination a central problem. The knowledge system faces many of the same difficulties, but with an added one that bringing together different sources and kinds of knowledge might also require 'integration' at the level of understanding. Complexity in the structures and sources of vulnerabilities in the social–ecological system do not have to be mimicked by planning procedures to handle them.

Openings

New knowledge and approaches to better link knowledge and action are needed both at the levels of policy and practice. Tensions lead to cracks and create openings. The above review of four case studies and of the individual capacities and roles of actors suggests that there are opportunities to improve the effectiveness of the current knowledge system.

First, ways to meet the greater demand for knowledge at different levels of governance are needed. Many of these problems have emerged with the wider engagement of stakeholders directly and through local representative bodies. Tambon administration organizations, river basin organizations, and river sub-basin organizations are demanding more and better information for planning purposes. Much of this still comes from externally supported consultancies and is of variable quality. But, and at the same time, there are valuable networks of collaboration forming around common issues that cut across departmental boundaries and which could be the basis of high-quality information support to political deliberations and more administrative-style planning processes.

Second, effective integration and informed deliberation may not require centralized authority, just better coordination. As efforts are made to increase the level of engagement among vulnerable stakeholders, responsible practitioners and researchers, the challenges of sharing power and understanding become more closely intertwined. The contest among agencies means that often multiple partnerships and channels are open to solve each particular problem. It means that a lot gets done without the need for highly centralized coordination. It means that adaptive management can take place at the level of self-organizing relationships among agencies, rather than as the prerogative of a central master agency for water. We suggest that more attention should be given to strengthening the consortium approaches, where each party brings its data and understanding to the table because it anticipates some real benefits by doing so.

Third, an optimal configuration of actors and knowledge sources does not exist to allow institutional flexibility for adaptive governance. The introduction of river basin organizations with vague mandates for integrated water resources management (IWRM) brought either laughter or tears to most people whom we consulted. For those who are well intentioned but impatient with the dreams of multi-stakeholder processes leading to better coordination of water resource development across sectors, the process has been rushed and frustratingly under-resourced. For those who know better where power and decision-making lie, it looks as if it were designed to fail. For the hopeful, these are small steps in the right direction towards more bottom-up deliberative planning and project implementation processes, especially at the sub-basin level. We are most sympathetic with the third and last view. The introduction of river basin organizations should be seen as an experiment, testing the waters among agencies that have not really been able to work together very well in the past. It is rightly process oriented, opening up possibilities for new information and new players to get involved in different ways as it unfolds.

Fourth, individual agency, not just organizations and institutionalized relationships, is important. In all major collective actors or organizations, and in the three knowledge/action arenas we studied in more detail, there were particular individuals who had key roles in the knowledge system. Where appropriate, we have named them and noted their positions in our analysis above. Many of these key individuals know each other, even across the specific water-related vulnerabilities in which they have the largest roles. Indeed, this capacity to link otherwise disparate geographical or issue-oriented knowledge/action arenas (see Figure 16.3) is one of the critical contributions of these individuals above and beyond the specific knowledge inputs that they make.

Fifth, and finally, the needs and interests of marginalized groups should be prioritized in assessing vulnerabilities and risk management strategies. Dry-season water provision, flood protection measures and land-use planning regulations to maintain watershed services each require consideration of both current burdens and risks, as well as how these are altered by adaptation interventions. Public investments aimed at reducing vulnerability for wealthy peoples and places should be a lower priority than for the poor and other vulnerable groups. New approaches to linking knowledge and action should start from acknowledging that modes of engagement and how power is shared are crucial features and that this can, to a certain useful extent, be shaped by appropriate institutional forms. In the case of the Ping Basin, for the most part, these are forms which bring the concerns and interests of vulnerable groups to the fore, and open up decision exploring and decision-making to much wider and continuous public scrutiny. It is not good enough to consult; engagement activities should also empower.

Conclusions

A knowledge system is emerging for the Ping Basin that can support dialogue, negotiation and decision-making around the development of water resources and the management of water-related risks. After decades of data-sparse and analysis-free decision-making about water, there are, for the first time, prospects for agenda setting, exploration of institutional options, scrutiny of project rationales, and conduct of impact assessments to be informed by quality-controlled data, reproducible evidence and shared experiences. Although nobody expects the bureaucratic culture of secrecy around the planning of large water infrastructure projects and major water policies to disappear overnight, there is now sufficient commitment within the bureaucracy and demand from outside it for stakeholder engagement in the production and use of knowledge to become much more commonplace.

Acknowledgements

Several researchers, assistants and interns over the past few years have carried out fieldwork that has helped us to better understand water resource management issues in the Upper Ping Basin. In particular, we thank Sophie Carton, Supaporn Khrutmuang, Julie Poncet, Songphonsak Ratanawilailak, Patcharawalai Sriyasak and Nutthawat Subsin. Thanks also to Santita Ganjanapan, Francoise Molle, Thada Sukhapunnaphan and David Thomas for helping us to understand much more deeply many of the issues discussed in this chapter. Thanks to Rajesh Daniel for his constructive comments on earlier drafts of the chapter. This work was supported by separate grants from the Packard Foundation, the Rockefeller Foundation, the International Fund for Agricultural Research, Echel Eau and the Challenge Programme on Water and Food (Project PN50).

References

Abonyi, G. (2005) *Policy Reform in Thailand and the Asian Development Bank's Agricultural Sector Program Loan*, ERD Working Paper Series No 71, Economics and Research Department, Asian Development Bank, Manila

Alford, D. (1992) 'Streamflow and sediment transport from mountain watersheds of the Chao Phraya Basin, northern Thailand: A reconnaissance study', *Mountain Research and Development*, vol 12, pp257–268

Anon (2006) 'How and why government tries to get rid of traditional weirs' [in Thai], *Northern Citizen*, vol 241, pp12–13

Anukularmphai, A. (2004) 'River basin committees development in Thailand: An evolving participatory process (EPP)', Paper for training Programme on Integrated Water Resources Management and Strengthening of River Basin Committees, 26 July–6 August 2004, Bangkok and Chiang Mai, Thailand, Network of Asian River Basin Organizations (NARBO), Asian Development Bank, Manila

Asian Disaster Preparedness Centre (2006) 'Flashflood and landslide disaster in the provinces of Uttaradit and Sukhothai, northern Thailand, May 2006', *Rapid Assessment*, Asian Disaster Preparedness Centre, Bangkok

Bangkok Post (2006) 'Farmers asked to cut back on off-season crops', 6 November 2006, *Bangkok Post*, Bangkok

Bruijnzeel, L. A. (2004) 'Hydrological functions of tropical forests: Not seeing the soil for the trees', *Agriculture, Ecosystems and Environment*, vol 104, pp185–228

Chaibu, P., Ungsethaphand, T. and Maneesri, S. (2004) 'The costs and returns of Tilapia and Tubtim (red tilapia) cage culture in Chiang Mai Province, Thailand', [in Thai], *Journal of Fisheries*, vol 57, pp244–250

Charoenmuang, T. (1994) 'The governance of water allocation problems in Thailand: Four case studies from the upper northern region', in Thailand Development Research Institute (ed) *Water Conflicts*, Thailand Development Research Institute (TDRI), Bangkok, pp111–147

CMU (Chiang Mai University) (2004) *Project to Develop a Master Plan and Implementation Plan for Conservation and Development of Environmental and Water Quality of the Ping River and Its Tributaries*, Final Report submitted to the Office of Natural Resources and Environmental Policy and Planning, Ministry of Natural Resources and the Environment [in Thai], Chiang Mai University, Chiang Mai

CMU (2005) *Project on Monitoring and Evaluation of Environmental Quality and Implementation of Restoration of Natural Resources and Environment in the Ping Basin*, Interim Report submitted to the Office of Natural Resources and Environmental Policy and Planning, Ministry of Natural Resources and Environment [in Thai], Chiang Mai University, Chiang Mai

Cohen, P. T., and Pearson, R. E. (1998) 'Communal irrigation, state, and capital in the Chiang Mai Valley (Northern Thailand): Twentieth-century transformations', *Journal of Southeast Asian Studies*, vol 29, pp86–110

Daniel, R. (ed) (2005) *After the Logging Ban*, Foundation for Ecological Recovery (FER), Bangkok
DDPM (Department of Disaster Prevention and Mitigation) (2006) *Thailand Country Report*, Department of Disaster Prevention and Mitigation, Ministry of Interior, Thailand
Forsyth, T. (1996) 'Science, myth and knowledge: Testing Himalayan environmental degradation in Thailand', *Geoforum*, vol 27, pp275–292
Forsyth, T. (1998) 'Mountain myths revisited: Integrating natural and social environmental science', *Mountain Research and Development*, vol 18, pp126–139
Garden, P., and Nance, S. L. (2007) 'Forums and flows: Emerging media trends', in L. Lebel, J. Dore, R. Daniel and Y. S. Koma (eds) *Democratizing Water Governance in the Mekong Region*, Mekong Press, Chiang Mai, pp157–175
Garden, P., Lebel, L. and Chirangworapat, C. (2010) 'Local government reforms as work in progress', in C. Wittayapak and P. Vandergeest (eds) *The Politics of Decentralization: Natural Resource Management in Asia*, Mekong Press, Chiang Mai, pp137–160
Glassman, J., and Sneddon, C. (2003) 'Chiang Mai and Khon Kaen as growth poles: Regional industrial development in Thailand and its implications for urban sustainability', *Annals of the American Academy of Political and Social Science*, vol 590, pp93–115
Government of Thailand (2002) Office of the Prime Minister Regulation on Procurement, Section 3, Hiring Consultants, BE 2535 (BE 2545 amendment), Public Debt Management Office, Ministry of Finance, Bangkok
Kaosa-ard, M. (2001a) *Framework for Water Management Policy for Thailand*, Thailand Development Research Institute, Bangkok
Kaosa-ard, M. (2001b) *Tourism Master Plan for Thailand*, Thailand Development Research Institute, Bangkok
Kaosa-ard, M. and Rerkasem, B. (2000) *The Growth and Sustainability of Agriculture in Asia*, Oxford University Press for the Asian Development Bank, Manila
Khuenkaew, S. (2007) 'Call to axe two megaprojects', 8 May 2007, *Bangkok Post*, Bangkok
Krairapanond, N. and Atkinson, A. (1998) 'Watershed management in Thailand: Concepts, problems and implementation', *Regulated Rivers: Research and Management*, vol 14, pp485–498
Laungaramsri, P. (2000) 'The ambiguity of "watershed": The politics of people and conservation in northern Thailand', *Sojourn*, vol 15, pp52–75
Laungaramsri, P. (2002a) 'On the politics of nature conservation in Thailand', *Kyoto Review*, October, www.kyotoreview.cseas.kyoto-u.ac.jp/issue1/article_196.html
Laungaramsri, P. (2002b) *Redefining Nature: Karen Ecological Knowledge and the Challenge to the Modern Conservation Paradigm*, Earthworm Books, Chennai
Lebel, L. and Sinh, B. T. (2007) 'Politics of floods and disasters', in L. Lebel, J. Dore, R. Daniel and Y. S. Koma (eds) *Democratizing Water Governance in the Mekong Region*, Mekong Press, Chiang Mai, pp37–54
Lebel, L. and Sinh, B. T. (2009) 'Risk reduction or redistribution? Flood management in the Mekong region', *Asian Journal of Environment and Disaster Management*, vol 1, pp23–39
Lebel, L., Contreras, A., Pasong, S. and Garden, P. (2004) 'Nobody knows best: Alternative perspectives on forest management and governance in Southeast Asia:

Politics, law and economics', *International Environment Agreements*, vol 4, pp111–127

Lebel, L., Anderies, J. M., Campbell, B., Folke, C., Hatfield-Dodds, S., Hughes, T. and Wilson, J. (2006) 'Governance and the capacity to manage resilience in regional social-ecological systems', *Ecology and Society*, vol 11, p19

Lebel, P., Leudpasuk, S., Lebel, L. and Chaibu, P. (2007a) 'Fish cage culture in upper part of Ping river' [in Thai], *Journal of Fisheries Technology*, vol 1, pp160–170

Lebel, L., Thaitakoo, D., Somporn, S. and Huaisai, D. (2007b) 'Views of Chiang Mai: The contributions of remote sensing to urban governance and sustainability', in M. Netzband, W. L. Stefanov and C. Redman (eds) *Applied Remote Sensing for Urban Planning, Governance and Sustainability*, Springer-Verlag, Berlin, pp221–247

Lebel, L., Daniel, R., Badenoch, N. and Garden, P. (2008) 'A multi-level perspective on conserving with communities: Experiences from upper tributary watersheds in montane mainland southeast Asia', *International Journal of the Commons*, vol 1, pp127–154

Lebel, L., Foran, T., Garden, P. and Manuta, J. (2009a) 'Adaptation to climate change and social justice: Challenges for flood and disaster management in Thailand', in F. Ludwig, P. Kabat, H. van Schaik and M. van der Valk (eds) *Climate Change Adaptation in the Water Sector*, Earthscan, London, pp125–141

Lebel, L., Garden, P., Subsin, N. and Na Nan, S. (2009b) 'Averted crises, contested transitions: Water management in the Upper Ping River basin, northern Thailand', in D. Huitema and S. Meijerink (eds) *Water Policy Entrepreneurs: A Research Companion to Water Transitions around the Globe*, Edward Elgar, Cheltenham, UK, pp135–157

Lebel, L., Perez, R. T., Sukhapunnaphan, T., Hien, B. V., Vinh, N. and Garden, P. (2009c) 'Reducing vulnerability of urban communities to flooding', in L. Lebel, A. Snidvongs, C.-T. A. Chen and R. Daniel (eds) *Critical States: Environmental Challenges to Development in Monsoon Asia*, Strategic Information and Research Development Centre, Selangor, Malaysia, pp381–399

Lebel, L., Sinh, B. T., Garden, P., Seng, S., Tuan, L. A. and Truc, D. V. (2009d) 'The promise of flood protection: Dykes and dams, drains and diversions', in F. Molle, T. Foran and J. Kakonen (eds) *Contested Waterscapes in the Mekong Region*, Earthscan, London, pp283–306

Lebel, L., Daniel, R. and Tipraqsa, P. (2010a) 'Mission-integrated research for development and the environment in the uplands of northern Thailand', in W. Proctor, S. H. Dodds and L. van Kerkhoff (eds) *Integrated Mission-Directed Research: Experiences in Environmental and Natural Resource Management*, CSIRO Publishing, Collingwood, Australia, pp71–94

Lebel, L., Manuta, B. J. and Garden, P. (2010b) 'Institutional traps and vulnerability to changes in climate and flood regimes in Thailand', *Regional Environmental Change*, doi:10.1007/s10113-010-0118-4

Manuta, J., Khrutmuang, S., Huaisai, D. and Lebel, L. (2006) 'Institutionalized incapacities and practice in flood disaster management in Thailand', *Science and Culture*, vol 72, pp10–22

Margane, A. and Tatong, T. (1999) 'Aspects of the hydrogeology of the Chiang Mai-Lamphun Basin, Thailand that are important for the groundwater management', *Z. angew. Geol.*, vol 45, pp188–197

McCaskill, D., and Kampe, K. (1997) *Development or Domestication: Indigenous Peoples of Southeast Asia*, Silkworm Books, Chiang Mai

Molle, F. (2005) *Irrigation and Water Policies in the Mekong Region: Current Discourses and Practices*, International Water Management Institute, Colombo, Sri Lanka

Molle, F. (2008a) 'Nirvana concepts, narratives and policy models: Insights from the water sector', *Water Alternatives*, vol 1, pp23–40

Molle, F. (2008b) 'Why enough is never enough: The societal determinants of river basin closure', *Water Resources Development*, vol 24, pp217–226

Na Ayutthaya, P. N. (1996) 'Community forestry and watershed networks in Northern Thailand', in P. Hirsch (ed) *Seeing Forests for Trees: Environment and Environmentalism in Thailand*, Silkworm Books, Chiang Mai, pp116–146

Nation (2006a) 'Political crisis 'crippled flood-prevention work', 2 August 2006, *Nation*, Bangkok

Nation (2006b) 'Villagers vow "mob rule" if expo leads to lack of water', 6 November 2006, *Nation*, Bangkok

Nation (2007) 'Local groups together against elephant park', 5 May 2007, *Nation*, Bangkok

Neef, A., Chamsai, L. and Sangkapitux, C. (2006) 'Water tenure in highland watersheds of northern Thailand: Managing legal pluralism and stakeholder complexity', in L. Lebel, X. Jianchu and A. Contreras (eds) *Institutional Dynamics and Stasis: How Crises Alter the Way Common Pool Resources are Perceived*, Regional Centre for Social Science and Sustainable Development, Chiang Mai University, Chiang Mai, pp64–88

Panya Consultants and Sigma Hydroconsultant (2003) *Project to Develop Integrated Water Resources Management Plan in Ping Watershed* [in Thai], Department of Water Resources, Ministry of Natural Resources and Environment, Bangkok

Pearson, R. (1999) *A Political Economy Analysis of the Impact of Agrarian Change and Urbanisation on Communal Irrigation Systems in the Chiang Mai Valley, Northern Thailand*, PhD thesis, Department of Anthropology, Macquarie University, Sydney

Phanayanggoor, P. (2006) 'Local against dyke-levelling', 6 February 2006, *Bangkok Post*, Bangkok

Resurreccion, B., Real, M.-J. and Pantana, P. (2004) 'Officialising strategies: Participatory processes and gender in Thailand's water resources sector', *Development in Practice*, vol 14, pp521–532

Rigg, J. and Nattapoolwat, S. (2001) 'Embracing the global in Thailand: Activism and pragmatism in an era of deagrarianization', *World Development*, vol 29, pp945–960

Saipothong, P. and Thomas, D. (2007) 'Spatial information tools for land use management networks in montane mainland Southeast Asia', *Information Development*, vol 23, pp129–136

Saipothong, P., Kojornrungrot, W. and Thomas, D. (2005) 'Comparative study of participatory mapping processes in northern Thailand', in J. Fox, K. Suyanata and P. Hershock (eds) *Mapping Communities: Ethics, Values, Practice*, East–West Center, Chiang Mai, pp11–18

Saipothong, P., Preechapanya, P., Promduang, T., Kaewpoka, N. and Thomas, D. E. (2006) 'Community-based watershed monitoring and management', *Northern Thailand Mountain Research and Development*, vol 26, pp289–291

Shamshub, H. (2010) 'Enhancing sustainable tourism in Thailand: A policy perspective', in L. Lebel, S. Lorek and R. Daniel (eds) *Sustainable Production Consumption Systems: Knowledge, Engagement and Practice*, Springer, Dordrecht, pp211–235

Sharma, D., Gupta, A. D. and Babel, M. S. (2007) 'Spatial disaggregation of bias-corrected GCM precipitation for improved hydrologic simulation: Ping River Basin, Thailand', *Hydrology and Earth System Sciences Discussions*, vol 4, pp35–74

Surarerks, V. (1986) *Historical Development and Management of Irrigation Systems in Northern Thailand*, Chareonwit Printing, Bangkok

Surarerks, V. (2006) 'Muang fai communities in Northern Thailand: People's experiences and wisdom in irrigation management', *Journal of Developments in Sustainable Agriculture*, vol 1, pp44–52

Sutiwanich, C., Phienwej, N., Harnpattanapanich, T., Fasching, A. and Laubblishler, K. (2006) *Investigation and Preliminary Design of Mae Ngat-Mae Kuang Tunnel Chiang Mai Province, Thailand*, Gan-prachum Wichargarn 1 PS 2549, Panya Consultants Co. Ltd., Thailand

Tan-kim-yong, U., Bruns, P. C. and Bruns, B. R. (2005) 'The emergence of polycentric water governance in northern Thailand', in G. P. Shivakoti, D. L. Vermillion, W.-F. Lam, E. Ostrom, U. Pradhan and R. Yoder (eds) *Asian Irrigation in Transition: Responding to Challenges*, Sage Publications, London, pp226–252

Thomas, D. E. (2005) *Developing Organizational Models in Pilot Sub-Basins: An Interim Report to Office of Natural Resources and Environmental Policy and Planning*, Ministry of Natural Resources and Environment, World Agroforestry Centre, Chiang Mai

Thomas, D. E. (2006a) *Participatory Watershed Management in Ping Watershed: Final Report*, Office of Natural Resources and Environmental Policy and Planning, Ministry of Natural Resources and Environment, Thailand, Bangkok

Thomas, D. E. (2006b) *Results Measurement Framework for Pilot Sub-Basins*, Office of Natural Resources and Environment Policy and Planning, Ministry of Natural Resources and Environment, Bangkok

Thomas, D. E., Weyerhaeuser, H., Saipothong, P. and Onraphai, T. (2000) 'Negotiated land use patterns to meet local and societal needs', in X. Jianchu, A. Wilkes, H. Tilmann, M. Salas, T. Grinter and Y. Shaoting (eds) *Links Between Cultures and Biodiversity: Proceedings of the Cultures and Biodiversity Congress 2000, 20–23 July 2000*, Yunnan Science and Technology Press, Yunnan, People's Republic of China, pp414–433

Thomas, D. E., Preechapany, P. and Saipothong, P. (2002) 'Landscape agroforestry in upper tributary watersheds of northern Thailand', *Journal of Agriculture* (Thailand), vol 18, Supplement 1, ppS255–S302

Thomas, D. E., Preechapany, P. and Saipothong, P. (2004) *Developing Science-Based Tools for Participatory Watershed Management in Montane Mainland Southeast Asia*, Final report to the Rockefeller Foundation for Grant No 2000 GI 086, World Agroforestry Centre, Chiang Mai

Thomas, D. E., Ekhasing, B., Ekhasing, M., Lebel, L., Ha, H. M., Ediger, L., Thongmanivong, S., Jianchu, X., Saengchayosawat, C. and Nyberg, Y. (2008) *Comparative Assessment of Resource and Market Access of the Poor in Upland Zones of the Greater Mekong Region*, Report submitted to the Rockefeller Foundation under Grant No 2004 SE 024, World Agroforestry Centre, Chiang Mai

Tran Hung (1998) *Integrating GIS with Spatial Data Analysis to Study the Development Impacts of Urbanization and Industrialization: A Case Study of Chiang Mai-Lamphun Area, Thailand*, PhD thesis, School of Environment, Resources and Development, Asian Institute of Technology, Bangkok van Kerkhoff, L. and Lebel, L. (2006) 'Linking knowledge and action for sustainable development', *Annual Review of Environment and Resources*, vol 31, pp445–477

Vandergeest, P. (2003) 'Racialization and citizenship in Thai forest politics', *Society and Natural Resources*, vol 16, pp19–37

Walker, A. (2003) 'Agricultural transformation and the politics of hydrology in northern Thailand', *Development and Change*, vol 34, pp941–964

Walker, A. (2004) 'Seeing farmers for the trees: Community forestry and the arborealisation of agriculture in northern Thailand', *Asia Pacific Viewpoint*, vol 45, pp311–324

Wittayapak, C. and Dearden, P. (1999) 'Decision-making arrangements in community-based watershed management in Northern Thailand', *Society and Natural Resources*, vol 12, pp673–691

Wood, S. and Ziegler, A. D. (2007) 'Floodplain sediment from a 30-year-recurrence flood in 2005 of the Ping River in northern Thailand', *Hydrol. Earth Syst. Sci. Discuss.*, vol 4, pp3839–3868

Yumuang, S. (2006) '2001 debris flow and debris flood in Nam Ko area, Phetchabun province, central Thailand', *Environmental Geology*, vol 51, pp545–564

Ziegler, A., Giambelluca, T., Sutherland, R., Nullet, M., Yarnasarn, S., Pinthong, J., Preechapanya, P. and Jaiaree, S. (2004) 'Towards understanding the cumulative impacts of roads in upland agricultural watersheds of northern Thailand', *Agriculture, Ecosystems and Environment*, vol 104, pp145–158

Part V

Where Do We Go from Here?

17

Issues that Need to be Addressed: Assessing Experience

Uno Svedin

In addressing the issue of how science can relate to policy – with special regard to issues of vulnerability and resilience in a global environmental change context – the earlier chapters in this book have moved between these two poles, on the one side the vulnerability/resilience thematics, and on the other, the science/policy interface in general. In the many chapters these discussions have, in addition, also moved between the poles of generalities and specific examples.

You could in this situation – among the mass of assembled insights and elaborations – ask yourself if there is much more to say! This is especially the case if the task now is to say something about 'Where do we go from here?' However, it still might be of interest to rerun the central points of crossover lines in the theme of the book (i.e. the thematic issues about vulnerabilities/ resilience in their own right and on the other the various forms of connectivities that link to the science/policy nexus). You could, in fact, say that the relationship between science and policy is a much broader one than when it 'just' deals with the topic of vulnerability and resilience.

However, you could also argue that the very nature of the vulnerability/resilience challenges is of such a kind that the 'fit' or 'match' between the knowledge production system (e.g. in the spirit of Gibbons et al, 1994) and the system dealing with initiatives and enacting these in the policy and politics arena are especially questionable in this case.

Let us, thus, for the sake of checking such a hypothesis go through a small list of specific highlights in the particular case of vulnerability/resilience to see if it can tell something about the challenges for the science/policy interface.

Many of the following items have, indeed, already been discussed earlier in this book; but let us now look at them again with that special kind of watchfulness that may give an extra illumination. So, which are the test points of concern chosen here?

'Systemicness'

It is already clear that vulnerabilities (and resilience) are deeply systemic factors. It would even be very difficult to talk about these issues in a more one-dimensional way or more specifically to use a 'one cause/one effect' approach. No, the theme of vulnerability/resilience comprises topics deeply ingrained in the multi-causality of systemic-ness and its multidimensionality. The feedbacks and feed-forwards are intrinsic aspects of the two topics; thus, their closeness to non-linearity features is not something that comes over and above some simpler 'normal state', but is the very fabric by which these topics are made. Thus, with the non-linearities also follow the 'uncertainties' and the 'surprises'.

Does this indicate something for the science/policy interface? Probably, yes. On the one hand, the traditional process of science in providing a scientifically 'soundest possible' and 'stable' input to the policy process is shifting. The 'speaking truth to policy' posture may have to be modified. If we consider the current and recent Intergovernmental Panel on Climate Change (IPCC) scenarios as a case to illuminate this point we can see that the provision from the scenarios is no longer a strict prediction of how things will be, or most probably will be. Rather, it is to provide a wide set of, as far as possible, consolidated views on possible outcomes given a sequence of varying conditions. The more 'strict prognostic' approach can here be seen to have weakened. On the other side, the elaboration around various forms of 'what if' approaches has increased. And so has the challenge to become more precise on what the uncertainties are in any of the statements developed ('How much can be said under the current conditions of understanding?'). This has also called for increased sophistication in the way in which the science side has had to differentiate between the varying forms of uncertainties and those forms where it 'just needs some more work to settle the issue' to very profoundly unsettled uncertainties of intrinsic nature.

But moving into this type of question in the analytical domain may not be so bad for the science/policy dialogue, as these features are already well recognized in the policy arena. Nevertheless, there is a call for enhanced didactics from both sides about the interpretation of what this 'new' situation really means: a science that cannot deliver 'certainties' any longer, but which in a changed frame of mind also has become more relevant to the concerns of the political arenas.

It may well also be that the constant elaboration by the policy stratas in their 'stakeholder' shapes in defining what seems more or less important from a 'practice perspective' will now promote new interests in the science domain, encouraged to absorb new types of overriding questions. Here the specific knowledge, even if tacit, among the practitioners may blend its knowledges and insights with those coming from the scientific domains, providing its accumulated procedural and methodological considerations to consolidate the knowledge base.

The multi-scale features

Vulnerability and resilience do operate in a multi-scale 'universe'. You could even argue that the vulnerability features emerge exactly through these phenomena, operating simultaneously at multi-levels. The 'multi-levelness' may in this perspective be seen even as a source of the vulnerabilities. This holds not only for the threats, but also for the counteracting potential capacities that in terms of increased resilience also have their roots in the multi-level capacities. It is sufficient for us to remind ourselves of the increased food security that during the last century has been gained through global trade. This is not to draw that statement too far, as globalization also has impacts upon the vulnerability side of the balance. The point is that the scale from micro to meso to macro provides several clues for driving functions of positive and negative kinds for resilience. Scale matters!

The emerging width of understanding of such multi-layered features (not the least in its institutional forms) may also increasingly fit better the challenges that policy-makers face every day. But the science (including a broader analytical) side of the science/policy junction can probably now feed in very much more interesting insights about 'real life' phenomena than was ever done before. This may prove especially important in the global environmental types of domains where probably the very core of solutions that need to be designed lies in finding the appropriate and also 'politically possible' solutions using the multi-layered 'realities'.

So this feature, about the scales, is improving on the science side in terms of being more thoroughly understood. It remains to be seen if the policy side at the high level globally may provide the sufficient design inputs in shaping the new governance systems that are badly needed for planetary survival. Some of the recent progress in the interest in the climate change domain – in some patchy way spread among parts of the world and among a growing number of types of actors – shows that there are interesting signs for such emergent activism. But will it be sufficient and will the actions provide solutions in time? That remains to be seen.

The connectedness between the different domains of knowledge

We are not talking here only about the important interface between natural science and social sciences/the humanities, but also how technological knowledge will be brought into the overall developmental picture. Quite often the knowledge domains of different character are served by quite different knowledge producer constituencies. This situation is gradually improving, but probably not sufficiently in terms of connectedness and links of understandings. We can just take the case of 'sustainability-oriented' knowledge-producing domains and those (others) dealing with the dramatically developing 'information society'. The connectivity between these two camps is not very great. The knowledge backgrounds between the two camps are very different and the core of concerns devoted to topics is very disconnected. And this is just one case.

If we are now looking at the way in which the vulnerability/resilience issues in their 'science community' arena is materializing in relation to a 'policy' arena, it is quite obvious that the connectivities in the knowledge-producing system are still too weak to face the grand issues at the planetary scale.

The normative/ethics dimensions

Vulnerabilities and resilience topics almost immediately cry out 'Vulnerable for whom?'; 'Resilient for whom?'; 'On whose account?'; or 'Who wins and who loses?' (as is often expressed in the climate change debate, taking into account not only geographical but also social differentiations). The earlier mentioned aspects of 'uncertainty' (emerging from the 'systemicness' character) will thus have to join hands with the ethics aspects in relationships (e.g. to risk assessment and in the analysis of distributions of gains). This is relevant both within our generation and between us and future generations.

The intense discussion that has emerged in the 'economics constituency' about the report by Nicolas Stern (2006) in the climate change economics field points at exactly this challenge. How are we to insert judgements on distributional ethics (including the weight of thousands of dead in gradually heating up areas of the world) and on related uncertainties in the modelling systems providing policy action maps for decision-makers globally?

This could be broadened from 'climate change' to a row of associated challenges (e.g. around 'security' concerns in the food, water and energy sectors). This holds true for a wide number of actors to whom the relevance of these concerns for the policy side of the science/policy nexus are demonstrated.

The normative aspects are usually a domain of high interest in the political and policy sphere. The interesting thing is now that the analytical side – as in the climate change domain – will have to increase its probing activities into these normative and ethical aspects. This may change the context of gravity in the science domain and may change the way in which science will be conducted.

Governance

We have so far said a lot about the analytical 'scientific side' of the science/policy interface. But we should also take note that the governance and institutional aspects are under pressure for reform. The new challenges, as exemplified by the resilience and vulnerability questions, will call for – yes, even demand – new styles of leadership and the expression of broader concerns. We are probably already seeing the beginning of a bottom-up pressure aimed at higher political levels of political leadership to live up to the requests from a broad constituency of voters. We also have a timing issue at hand. Will the reforms be sufficient in terms of vigorous imagination and daring activism?

Rounding up reflections

These few examples have been highlighted to show that the vulnerability/resilience challenges probably have the nature of pushing the scientific and the policy sides more strongly together by the mere momentum of the tasks involved. This will probably also profoundly alter the ways in which both sides will perform their 'traditional activities'. So we will not only see a change in the crossover interface between science and policy (which would only concern the changes to some sort of procedural advancements only, as important as this may be); the changes will also be deeper in terms of roles, initiatives to be taken and the style of the common work. Here the idea of mobilizing, similar to 'the first man on the moon mission', is now directed towards the alleviation of some of the most burning planetary environmental challenges (sometimes also referred to by John Shellenhuber as a new 'Manhattan project'). This is just the type of move that would need the involvement of a new alliance between science and policy and at a much grander scale. The situation calls for grand innovations!

In closing, then, the question returns: 'What issues need to be addressed?' Well, some of the reflections above point not only to the current points of sensitivities, but also to what they may mean for the future. It means that much that has had its forms given for long time (maybe as much as more than half a century after the end of World War II) will now have to be renewed. This has to do, for example, with grand issues such as our developmental models and forms of consumer societies. It has to do with the way in which our monitoring systems are designed – and not only the one dealing with economics indicators – so that sufficient images are provided about the landscape of our quickly changing world. This, in turn, has to do with the need to scrutinize current value systems, including the involved issues of distributional ethics, as well as the ethics of managing a planet that has already moved into the historical phase of 'The Anthropocene' (suggested as an appropriate label by Nobel

Laureate Paul Crutzen). In this era, mankind has grown to become the dominant evolutionary force. This calls for reflection, consideration and profound action. Or, to rephrase the challenge: how will we deal with our responsibility when we find ourselves the guardians of the planet – and this having happened in just a very limited number of generations! This is the grand challenge of the crossroads between science and policy in its application to vulnerability and resilience!

References

Gibbons, M., Limoges, C., Nowotny, C., Shwartzman, S., Scott, P. and Trow, M. (1994) *The New Production of Knowledge: The Dynamics of Science and Research in Contemporary Societies*, Sage, London

Stern, N. (2006) *The Economics of Climate Change: The Stern Review*, Cambridge University Press, Cambridge, UK

18

Directions for Closing the Science/Practice Gap

Roger E. Kasperson

In the end, what can be said about paths forward that help to close, or perhaps (more modestly) just to narrow, the gap between science and decision-making? In addressing this question, it is important to note at the outset that the gap should probably not be fully eliminated. The gap often exists for good reason. The findings of science are never the complete story. Even where many of the problems noted in this volume have been ameliorated and narrowed, what the decision-maker needs to know almost always extends beyond the pale of science.

Knowledge and communication systems

The scope of what the decision-maker needs to know for sound decision extends to matters well beyond the purview of science. Scientific knowledge, to begin, is only one of the elements of a knowledge base necessary for sound social and policy decisions. A host of other considerations is typically involved, such as budget implications, effects on industry or consumers, public support or opposition to interventions, or other challenges needing attention. Also, at root, such decisions almost always involve public values, whether related to the acceptability of the risk or attitudes towards the particular mode of action contemplated. Of course, the decision-maker is typically juggling multiple issues that require attention, so that even the decision of what to work on and what to leave for another day is not an easy one.

A recurring deficiency in the scientific knowledge base relates to what is taken to be 'science' in most corporate and governmental venues. The finding in the US National Research Council (2009, p7) review of the US Climate Change Science Program (CCSP) that 'the underlying human dimensions research needed to understand and develop sound adaptation strategies is a major gap in the CCSP' appears repeatedly in some form in report after report of the National Research Council. The time is well past to recognize this gap as a systematic inadequacy in governmental-sponsored research on environmental and health risk problems. It also needs to be recognized that this is an inadequacy unlikely to be resolved any time soon. Indeed, until a grasp of these issues exists at the highest level, government agencies and corporations do not yet know what they need to know, much less be able to judge the quality of analyses they commission.

Yet another systematic limitation in the knowledge base required for responses to environmental changes and disasters is the lack of unity between expert and local knowledge. This comes through strongly in the cases on the Taiwan earthquakes (see Chapter 10) and the use of indigenous knowledge in the Canadian Arctic (Chapter 12). In Taiwan, the recovery from the 1999 Chi-Chi Earthquake involved the creation of a new emergency response system that drew upon expert knowledge at the national scale, and the use of local knowledge in the Chi-Chi township in central Taiwan, to build a stronger resilience and adaptive capacity to future potential disasters. The Taiwan response to the Chi-Chi Earthquake, in theoretical terms, involved the co-production of a knowledge system on different sides of a boundary using boundary objects. The overall goal was a risk assessment and emerging plan in which scientists at the national level assessed earthquake risk while local officials evaluated who and what would be vulnerable. An earthquake reconstruction council was established to manage cross-scale cooperation and coordination. Local university scientists and local government officials took the lead in improving the capacities of local government in disaster management. The result was a cross-scale boundary organization and a knowledge system that integrated scientific risk assessment at the national level with local vulnerability expertise at the local level.

Besides local knowledge, the need to create knowledge bases incorporating other types of knowledge is now becoming more appreciated, but is still at an early stage of development. Wandel et al in Chapter 12 note that inputs from indigenous knowledge to environmental assessments have been commonplace for several decades; but the actual *integration* of indigenous knowledge with formal scientific knowledge is still relatively new. Using examples drawn from the Canadian Arctic, they show how indigenous knowledge is evolving from 'data inputs' to a new emphasis upon co-management and community-based resource management. They argue that this evolution is reaping new benefits, such as greater relevance and appropriateness, more holistic understanding, enlarged participation and consensus, and, finally, more community acceptance and compliance.

Finally, Roberto Sánchez-Rodríguez (see Chapter 8) lays bare a continuing need across many societies: the lack of reliable data related to differential vulnerability. Vulnerability and resilience raise highly charged political issues. In his analysis of Tijuana, Mexico, Sánchez-Rodríguez finds additional evidence of a recurring failure to understand the interactions between social and biophysical systems. Beyond this, however, his study of Tijuana shows how, in the wake of rapid urbanization and industrialization, new patterns of social inequality and marginalization have appeared. But appraisals of the implications of such changes for vulnerability and adaptive capacity remain scarce and opaque, and public officials and experts have little incentive to assess the negative consequences and victimization that such changes leave in their wake. The Tijuana case, Sánchez-Rodríguez concludes, 'highlights the need to present the knowledge obtained within the broader context of development'. The message is clear: the knowledge system to support decisions requires a thorough analysis of ongoing changes in social equity and how they interact with the outcomes of changing risk.

Knowledge and communication systems, as various chapters in this book underscore, are fields for struggles over assembling power, controlling how problems are framed, and what knowledge is legitimate and what is not. The case of climate change is instructive. Riley Dunlap has tracked, over recent years, the efforts of 'climate change sceptics' to challenge the Intergovernmental Panel on Climate Change (IPCC) and mainstream scientific assessments of climate change. Their strategy, in effect, is consistent with the findings of this volume and earlier previous work by the Harvard University project on social learning. The Harvard project argues that the influence of scientific assessments depends principally upon potential users seeing them as *salient*, *credible* and *legitimate* (see also Mitchell et al, 2006, p314ff). Dunlap argues that, among these, the sceptics have relied principally upon seeking to discredit the credibility of the science underlying the scientific assessments of climate change. A noteworthy event occurred in December 2009 with the hacking into the emails of scientists at the Climate Research Unit at the University of East Anglia, intended to cast doubt on the integrity of scientific work conducted on climate change (*Science*, 2009, p1329; *Science*, 2010, p510). What influence this event and a following discrediting of an IPCC prediction that Himalayan glaciers would experience by 2035 have had on attitudes to climate change in Europe and North America is unclear, but recent polls suggest that while scientists appear to be moving into clearer consensus on the threat of climate change, members of the public appear to be moving to greater dissensus (Leiserowitz et al, 2009; Leiserowitz, 2010).

Conflicting definitions and conceptions

The problems compounding issues in the science/decision-making gap do not all lie with the decision-makers. Despite significant progress over the past decade, science still speaks with many voices when addressing environmental risk, vulnerability and resilience. Susan Cutter (1996), for example, has assembled a host of differing definitions of vulnerability that have appeared over several decades in the professional literature. More fundamental are that significantly different underlying conceptions of risk and vulnerability are often involved, leaving the decision-maker to cope with a quagmire of conflicting interpretations and metrics to adjudicate. Accordingly, to stay with vulnerability for the moment, different analysts focus on 'unsafe conditions' emanating from basic political and economic structures (Watts and Bohle, 1993; Bohle, 2001); some emphasize and position their analysis as primarily 'social vulnerability' (Adger, 1999); others use a more integrative concept of the 'social–ecological system' (Kasperson et al, 2009); and still others emphasize the risk to livelihood systems (Leach et al, 1999). It is not surprising that decision-makers often find these analyses confusing rather than illuminating, as our case (see Chapter 14) on the European Programme for Food Aid and Food Security illustrates, pointing to the pressing need for an overarching, agreed upon framework for analysis.

Two-way communications

Other problems arise from outside the domain of science or assessment. A consensus has long existed that risk communication is necessarily a two-way process, and government and corporate officials readily endorse that proposition. But practice and implementation are another matter. Programmes dealing with communities and publics invariably describe themselves as 'outreach' programmes, and outreach is what is invariably involved. The assumption remains that the primary responsibility of the decision-maker is to share with stakeholders information and analyses developed in supporting scientific research programmes. This is what 'transparency' has come to mean. The need for 'in-reach' programmes to supplement 'outreach' remains unmet. So, even when meetings or other interactions with stakeholders are arranged, the purpose and content of the meetings are invariably to distribute information and analysis possessed by the decision-maker. The need to 'listen', as opposed to 'talking', is not generally an objective of 'outreach' programmes. It is not surprising, then, that such programmes are rarely tailored to the differing needs or wants of various stakeholders. Despite all the writing and experience with risk communication and stakeholder involvement, 'sharing' expert information is still the norm and unlikely to change anytime soon (Moser and Dilling, 2007).

Power and contention

The spider webs separating science and decision-makers are fertile grounds for contestation and the assembling and exercise of political power (Ribot, 1995; Ribot 2010). At stake is the question of how the scientific results or events will be framed, what support or opposition can be assembled, and what legitimacy and credibility will be attached to the science. This is not simply a matter of the credentials of the scientists who did the work or the rigor of their analyses, contrary to the assumptions of science. Rather, as we have pointed out elsewhere (Pidgeon et al, 2003), the societal processing of the events and risk results typically determines the signal value of what emanates from science and assessment, leading to the amplification or attenuation and the entrance of the issue onto the decision-maker agenda.

At stake in this set of processes permeating the spider web network of actors and advocates is control over the message or signal. This is partly about information processing, particularly access to the media and internet bloggers, and partly about mobilizing power among the intermediaries populating the spider web. So the spider web is typically a domain of contestation and exercising power, with a multitude of actors pursuing conflicting objectives and seeking to mobilize the influence needed to shape the message or signal going to the decision-maker. Mobilization of power and influence is a central activity in the spider web (Ribot, 1995). Furthermore, we note below that the nature, or what we term the 'architecture', of the spider web extensively shapes the ways in which these activities proceed. Spider webs with highly diffuse and transitory actors pose major differences to assembling power than those spider webs that are simple and relatively stable.

Trust and confidence

Finally, permeating these dynamics at work in the spider webs is the degree of trust and confidence that exist pertaining to science and governance. Social trust, it has been widely noted (Kasperson et al, 1999), has declined over the long term in the US. Significant differences exist among different societies, even in the West, regarding the levels of trust in both science and governmental decision-makers. The Harvard group attaches particular importance to the credibility of scientific assessments in shaping the role that they eventually play in decision-making (Social Learning Group, 2001). Where trust is very low, both in science and the governance system, moving forward entails heavier burdens upon deliberation. It must be expected that distrust has its price – management options that rest on high levels of trust become less available. Where managers and decision-makers are held on a short leash, there is greater pressure upon lengthy and often time-consuming deliberations with those that bear the risks. In addition, greater sharing of decision-making may be required. This has been a lesson from experience with nuclear and toxic chemical waste facilities where provisions have been built in for local communities to participate in the performance of such facilities in meeting regulatory standards and the ability of local communities to participate in the shut down of facilities where standards are breached (Kasperson, 2005; Linnerooth-Bayer, 2005).

Compounding this issue is the illusion that exists among many managers and decision-makers that, whatever has existed in the past, they will quickly win over the needed trust required to operate quickly and efficiently. 'We are the good guys' after all, it is felt, so local people will quickly know that we can be trusted. However worthy these assumptions and motivations, it is probably unrealistically optimistic, at least in the timeframe in which decisions must be made. Substantial evidence exists in the social sciences that distrust is quite deep seated and the result of long-term evolution (Kasperson et al, 1999). The loss of trust in the US and many other Western societies goes across major social and economic institutions and is not limited to, or simply the product of, particular bad experiences with the representatives of those institutions. So realism and experimental studies (Slovic, 1993) suggest that trust, once lost, is very difficult to regain. The reality of spider web dynamics is that they will often occur in a climate of distrust, a climate that is unlikely to change despite the well-intentioned behaviour of scientists and decision-makers and within the time span in which decisions must be made.

Core questions revisited

In Chapter 1 of this volume, we identified a series of core questions that we asked each study to consider (though not necessarily take up in each case). Here we return to these questions and state succinctly what we believe we have learned about each.

1 *How adequate is the knowledge base to support efforts for vulnerability reduction and adaptation? How is that knowledge distributed among actors?* The 'adequacy' of the knowledge base is inherently difficult to assess. The cases suggest both examples where this has done better and other cases where poorer. It is clear that cross-scale assessment and multiple ways of knowing are required for a robust knowledge base. The European Programme for Food Aid and Food Security (see Chapter 14) and the Tijuana case (see Chapter 8) indicate how easily cross-scale interactions can go wrong. The Arctic and Taiwan cases (Chapters 12 and 16) suggest how other forms of knowledge can inform, and deepen, the knowledge base needed for decisions.

   The cases in the volume have largely not examined carefully the distribution of knowledge among actors, although the Philippines case (Chapter 13) demarcates bottlenecks in the flow of needed information throughout the network of actors. In this case, it was obvious that farmers did not possess the needed information to make effective adaptations.

2 *How may the science/practice interface best be structured and characterized? Who are the principal actors? What are their roles and interests?* Although it is typically assumed that this is a generic issue, the argument in this volume is that the answer is highly pluralistic. The principal actors vary according to the nature of the spider web and the decision process. Ton Dietz, for example, in Chapter 11, indicates how important the churches became in the local institutional culture in northwestern Kenya in outside interventions of international aid in reducing vulnerability to environmental change. Turning to a developed country case, Chapter 6

shows how Californian scientists and non-governmental groups played a major role in shaping the California Climate Scenarios Project and the aggressive climate change actions that eventually resulted. Obviously, there is not a single answer to this question.

3 *To what extent do actors make use of the knowledge available to them? How do the actors structure the vulnerability and resilience discussion in the transfer of knowledge? How relevant and pertinent is the knowledge to the needs of decision-makers and other actors?* This volume has uncovered limited new knowledge to bring to bear on these questions. It is apparent, as we have pointed out earlier, that conceptual confusion and conflicting frameworks for environmental risk and vulnerability analysis have been major barriers for more effective societal responses to reduce vulnerability and other related risks. The salience, or relevance, of scientific evidence has been underscored by both the cases in this volume, but also by the broad-based work of the Harvard University Project on Usable Assessments. If the knowledge rendered by science sheds only limited new understanding of the problems that decision-makers must deal with, this knowledge is likely to have little contribution to, or influence on, the decisions that are made.

4 *What barriers and failures limit the transfer of knowledge and feedbacks in the science/practice interface? Do the barriers and failures occur in the transfer of knowledge from science to practice or from practice to science? How important are the intermediaries between science and practice and who are they?* In the previous section we have identified and discussed four serious barriers that are found in each of the major types of spider webs and that contribute to the gap between science and decision-making. They are, respectively, knowledge base inadequacies; conflicting concepts, frameworks and nomenclature; lack of two-way communications; and social distrust and lack of credibility. We note, and argue, that these inadequacies do not lie solely in the domains of science and decision-making, but pertain to both. The case studies in the volume provide far-reaching testimony and evidence of these problems. Furthermore, it is unquestionably apparent that the intermediaries between science and decision-making are numerous and influential. We use the image of spider webs to convey differing architecture or networks that exist between science and decision-making. So the answer to the role of intermediaries is not single, but plural. The degree of importance of intermediaries is case specific but is also related to the three major types of spider webs. Where science and decision-making is more tightly linked, intermediaries play less of a role. Previous work at Harvard University groups the problems into three major issues – salience, credibility and legitimacy – and the cases in this volume find much supporting evidence for that trio of factors.

5 *How does the nature of institutions shape the science/practice interface? To what extent is institutional fragmentation a problem? How permeable are the boundaries of institutions to new information?* Obviously, institu-

tions play a major role in the extent to which they draw upon science in decision-making. The two US climate cases in this volume make that very clear: at the national level where the Bush administration set up a fragmented management process that supported research in the climate sciences but impeded action to mitigate carbon emissions, and at the state level where California drew heavily upon local scientists to formulate an aggressive state Climate Change Program. Political culture is important in the role of institutions and their functioning, and the US, for example, has a long tradition of holding expertise in check, including science, despite the prominence given to reports emanating from the National Research Council and the assurance from the Obama administration that the federal government would, in the future, not interfere with the scientific integrity of research and assessment reports.

6   *What major conflicts exist among actors and institutions in the interactions between science and practice? To what extent are the conflicts primarily about values or facts? Does social justice enter into the decision-making process?* These questions cannot be answered in the abstract for they are largely situation specific. As already noted, the type of spider web interface facilitates or impedes the entrance of conflicts into two decision-making processes. Science, of course, has its own values; but much of the conduct of science is oriented towards understanding and characterizing the nature of environmental risks. Indeed, the so-called 'red book' (US National Research Council, 1983) sought to establish a linear management process in which the final stage, after the science had been completed, involved the application of values to decisions to manage the risks. More recent works, such as the 'orange' (US National Research Council, 1996) and 'silver' (US National Research Council, 2009) reports, have adopted frameworks in which deliberation occurs in parallel with scientific assessments. In fact, the 'silver book' has engendered significant controversy by elevating the decision to be addressed as the stage of embarking on a risk assessment. It is not surprising, as the Tijuana case demonstrates, that social justice issues that challenge existing power relationships and established institutional structures are difficult to surface and incorporate within decision-making processes.

7   *What is the role of consensus in the science/practice interface? How is consensus built? How are conflicts resolved? Who are the consensus builders and mediators and what are the major processes that they use?* The issue of consensus is closely related to the types of interfaces between science and decision-making. First, there is the degree of consensus that exists in science in assessing environmental risks. If the consensus is limited and the interface complex and unstable, then value debates and conflicts may rapidly eclipse the findings of science. Similarly, where sharp dissensus exists over value and policy issues in the decision-making process, then value disputes are more likely to determine decision outcomes. The intermediaries can play an important role in mediating the disagreements that

exist among actors in the spider web and shaping consensual solutions.

8 *What factors contribute most to adaptive capacity? How large is the gap between the capacity to adapt and the adaptation that actually occurs? What causes this gap and how can it be reduced? What new elements of enlarged capacity would contribute most to greater resilience in the face of environmental change over the short run and the longer term?* Adaptation was not a major focus of the studies in this volume, so what can be said in response to these questions is limited. None of the cases attempts a careful assessment of the factors contributing to adaptive capacity or to the adaptation process that eventually unfolds. Generally, the limited scientific capability, especially in the social sciences, that typically exists in decision-making institutions necessarily limits actions geared to major adaptations. Planned adaptations in major sectors of society and economy also involve major transformations in institutions, values and behaviour, including major changes in established structures. Scientific knowledge is still quite limited as to how such large-scale planned adaptations may best proceed.

9 *To what extent has social learning evolved among the principal actors? To what extent has social learning ameliorated or exacerbated vulnerability? What most limits or facilitates the ability to learn from one's own experience and the experience of others?* None of the cases in this volume attempt a deep analysis of social learning and its constraints. As Kai Lee (1993) has reported in his *Compass and Gyroscope*, there are different types of social learning. In order to manage risks and reduce vulnerabilities by using markedly new strategies or approaches, a deeper kind of social learning (what Lee would call 'double-loop learning') is needed than commonly occurs. Learning from experience first requires that decision-makers can step outside the structures and assumptions that accompany 'lock-in' strategies. Invariably, societies and decision-makers proceed on basic management or technological courses that accumulate over time. Major shifts away from these well-established developmental paths generally require either major risk events or incremental decisions made over long periods of time. Rarely are such major shifts, such as major transformations in energy systems to combat climate change, the product of new scientific knowledge alone.

10 *Where is the science/practice interface vulnerable to failure? Where is the science/practice interface most vulnerable to failure to future risks?* Again, the answer to this question depends upon the interface under consideration. Simple, stable spider webs, for example, are prone to failure where science and decision-making remain strongly isolated from each other. Complex, unstable spider webs are prone to failure where means for mediating intermediaries are few and ineffective and where major value conflicts pervade the spider web. While the contributors to this volume did not assess society's ability to cope with different future risks, it is apparent that rare, catastrophic risks that involve large uncertainties pose particularly difficult challenges for all three types of interface architectures.

11 *How can the science/practice interface be improved?* This, of course, is a critical question and so the section to follow takes it up at length.

Navigating the spider webs: Closing the science and decision-making gap

So what does this volume conclude as to how the gap between science and decision-making can be closed, or at least narrowed? Contrary to widespread hopes and intuitions, there is not a single answer. The question is akin to the longstanding query in risk management: how do we best manage a particular technological or environmental risk? As with risk management, the question cannot be answered in generality, anymore than how can the acceptability of any particular risk best be judged? Despite some short-lived fantasy that, regardless of benefits, all risks could be reduced to a common metric and assigned an acceptability number (e.g. 1 in 1 million annual risk of loss of life), ensuing discussion quickly determined that, however much analysts and regulators might like it, complex problems cannot be reduced to simplistic solutions. So it is with strategies to narrow the science/decision-making gap.

The starting point is the nature of the interface or, in our terms, the type of spider web that occupies the domain between science and decision-making. Accordingly, we argue in the discussion that follows that the answer is pluralistic; different interface architectures pose different barriers and minefields for failure, and solutions must be geared towards the nature of the spider web in question. This is the beginning point for our strategies to narrowing, if not closing, the science/decision gap.

Simple, stable spider webs

This is the spider web, or interface architecture, that is most congruent with conventional 'pipeline' thinking about the interaction of science and decision-making. Communications between science and decision-makers can be more direct and less mediated by intermediaries than with other science/decision-making (or spider web) interfaces.

Accordingly, we suggest two strategies. First is the need for more direct interaction between the scientists who produced the report or assessment and the decision-makers who must act upon it. Here we think the notion of the 'co-production' of such assessments is particularly germane, right from the start. A recurring problem is that the science is typically addressed to 'curiosity science', which advances science in some arena, but may miss the issues of concern to the decision-maker.

These problems may not be fully resolved by discussion and negotiation. But for scientists who are concerned that their work contributes to better decision-making, enlarging the scope of analysis to include considerations relevant to decision-making needs has potentially valuable rewards to the downstream use of results. Furthermore, the decision-maker may see aspects to even the scientific issues that are not immediately apparent to the scientists or the analyst. In any event, some direct conversation and interaction may head off issues that may arise only when the work is completed.

Second, this is probably the best setting in which to explore further what boundary organizations have to offer. Since the network of actors is relatively simple and stable, a boundary organization can be established that has durability and encompasses the range of issues that may arise as society processes and debates what the primary signals and messages are. Such an organization should be established with a close eye to achieving mutual accountability and co-production of a number of boundary products. The interactions should also be mounted with a premium to building trust and mutual respect. Of course, the progress on these issues should be monitored and carefully evaluated, preferably by a third party of independent critical experts.

Complex, stable spider webs

This interface architecture poses substantially greater complexity for the integration of science and decision-making, particularly in the broader scope of actors and the range of particular issues and agenda. But at least there is considerable stability in such spider webs in the structure of involvement. Accordingly, the approach to better and closer interaction between science and decision-making must allocate particular attention to stakeholder participation. Clearly there is the opportunity for sustained interaction with stakeholders and for building a cadre of translators of the science for the complex array of actors in the spider web. This interaction with stakeholders will also provide opportunities for broadening the scope of the knowledge base to encompass issues not addressed in the scientific work. An ambitious programme of stakeholder involvement will also provide opportunities to negotiate conflicts and build greater consensus throughout the complex interface.

Complex, unstable spider webs

This is clearly the most challenging architecture for closing the gap between science and decision-making. There is a complex array of actors, as with the spider web above. But this architecture lacks the stable array of participants throughout the spider web. Some actors move out of the domain of discussion and deliberation, while other new actors enter. As a result, the spider web is composed of shifting and ephemeral coalitions organized around differing agenda. Of course, these coalitions compete with each other to frame and reframe the risk message or results. In this process, they shape the amplification or attenuation of the signal to society and decision-makers of the meaning of the scientific results. This is, in short, a very difficult setting in which to protect the integrity of the scientific work that has been accomplished.

So, an effective strategy for navigating this class of spider webs is challenging and opaque. Creating a stable cadre of actors to serve as effective intermediaries and translators of science and decision needs is difficult to achieve. The agenda of discussion and deliberation is likely to be a changing one in which the essence of the science can be quickly and likely lost. In such a setting, creating some coherence to deliberation seems essential. One possibility may be for some concerted work with the media, including internet and alternative media, to establish some common knowledge base for ongoing contention and debate. Some innovative approaches to public engagement, such as the use of focus groups or intensive work with the media, may be worth considering. In any event, it must be expected that the most difficult spider web context will carry its price and that a determined and sustained deliberation effort will be needed, one that requires substantive investment in time and resources and priority for 'in-reach' as well as 'outreach'.

References

Adger, W. N. (1999) 'Social vulnerability to climate change and extremes in coastal Vietnam', *World Development*, vol 27, pp249–269

Bohle, H. (2001) 'Vulnerability and criticality: Perspectives from social geography', *IHDP Update*, vol 2, pp3–5

Cutter, S. L. (1996) 'Vulnerability to environmental hazards', *Progress in Human Geography*, vol 20, no 4, pp529–539

Kasperson, R. E. (2005) 'Siting hazardous facilities: Searching for effective institutions and processes', in S. H. Lesbirel and D. Shaw (eds) *Managing Conflict in Facility Siting: An International Comparison*, Edward Elgar, Northampton, MA, pp13–35

Kasperson, R. E., Golding, D. and Kasperson, J. X. (1999) 'Trust, risk and democratic theory', in G. Cvetkovich and R. Löfstedt (eds) *Social Trust and the Management of Risk*, Earthscan, London, pp22–44

Kasperson, J. X., Kasperson, R. E. and Turner, B.L. II (2009) 'Vulnerability of coupled human-ecological systems to global environmental change', in E. A. Rosa, A. Diekmann, T. Dietz, and C. C. Jaeger (eds) *Human Footprints on the Global Environment: Threats to Sustainability*, MIT Press, Cambridge, MA, pp231–294

Leach, M., Mearns, R. and Scoones, I. (1999) 'Environmental entitlements: Dynamics and institutions in community-based natural resource management', *World Development*, vol 27, no 2, pp225–247

Lee, K. (1993) *Compass and Gyroscope: Integrating Science and Politics for the Environment*, Island Press, Washington, DC

Leiserowitz, A. (2010) 'Climate change risk perceptions and behavior in the United States', in S. Schneider, A. Rosencranz, and M. Mastrandea (eds) *Climate Change Science and Policy*, Island Press, Washington, DC

Leiserowitz, A., Maibach, E. and Roser-Renovt, E. (2009) *Climate Change in the American Mind: Americans' Climate Change Beliefs, Attitudes, Policy Preferences, and Actions*, Yale Project on Climate Change, Yale University, New Haven, CT

Linnerooth-Bayer, J. (2005) 'Fair strategies for siting hazardous waste facilities', in S. H. Lesberil and D. Shaw (eds) *Managing Conflict in Facility Siting: An International Comparison*, Edward Elgar, Northampton, MA, pp36–62

Mitchell, R. B., Clark, W. C., Cash, D. W. and Dickson, N. M. (eds) (2006) *Global Environmental Assessments: Information and Influence*, MIT Press, Cambridge, MA

Moser, S. and Dilling, L. (eds) (2007) *Creating a Climate for Change: Communicating Climate Change and Facilitating Social Change*, Cambridge University Press, Cambridge, UK

Pidgeon, N., Kasperson, R. and Slovic, P. (eds) (2003) *The Social Amplification of Risk*, Cambridge University Press, Cambridge, UK

Ribot, J. (1995) 'The causal structure of vulnerability: Its applications to climate impact analysis', *Geojournal*, vol 35, no 2, pp119–122

Ribot, J. (2010) 'Vulnerability does not fall from the sky: Toward multiscale, pro-poor climate policy', in R. Mearns and A. Norton (eds) *Social Dimensions of Climate Change: Equity and Vulnerability in a Warming World*, World Bank, Washington, DC, pp47–74

Science (2009) 'Stolen e-mails turn up heat on climate change rhetoric', *Science*, vol 326 (4 December), p1329

Science (2010) 'Climate science leader Rajendra K. Pachauri confronts the critics', *Science*, vol 327 (29 January), pp510-511

Slovic, P. (1993) 'Perceived risk, trust and democracy', *Risk Analysis*, vol 13, no 6, pp675–682

Social Learning Group (2001) *Leaning to Manage Global Environmental Risks, Volume 2: A Functional Analysis of Social Responses to Climate Change, Ozone Depletions and Acid Rain*, MIT Press, Cambridge, MA

US National Research Council (1983) *Risk Assessment in the Federal Government*, National Academy Press, Washington, DC

US National Research Council (1996) *Understanding Risk: Informing Decisions in a Democratic Society*, National Academy Press, Washington, DC

US National Research Council (2009) *Science and Decisions: Advancing Risk Assessment*, National Academy Press, Washington, DC

Watts, M. and Bohle, H. (1993) 'The space of vulnerability: The causal structure of hunger and famine', *Progress in Human Geography*, vol 17, no 1, pp43–67

Index

Accelerated Climate Prediction Initiative (ACPI) 155–156, 158
access to knowledge 36–37, 70, 215, 219–220, 247, 349
accountability 11, 34, 101–102, 217, 228, 258
action implementation 34–35
action plans 43
action research 320, 321, 323, 325, 331
'acts of God' 108
adaptation 24–25, 26, 43–44, 97–109, 177–179, 197, 234
  approaches to 98–100
  challenges/opportunities of 98–109, 113–114
  communication and 103–107, 109
  and connectivity 39–40
  defined/use of concept 25, 26, 29, 99, 322
  propositional inventory/literature review of 26
  and rural vulnerability case study *see* Tanauan City
  and social/urban vulnerability 197–201, 207, 208
  and stakeholder involvement 108–109, 111, 112–113
  three dimensions of 104
adaptive capacity 99, 104, 105, 107, 108, 109, 298, 429, 435
adaptive governance 234–235, 237–238, 237–240, 243–245, 254–256, 410
  and boundary organizations 256
  and communication/understanding 258–259
  five benefits of 254–255
  funding/project continuity of 261–262
  and information flow 256, 258, 259, 260, 263–264
  involvement of scientists in 254, 255
  and leadership 257–258, 263
  obstacles to 261–263
  and partnerships/networks 257, 258
  pilot study in *see* Shan-an village
  schematic display of 255
  and social learning/co-production 260
  success of 264
adaptive learning 43
adaptive management 12–16, 28, 39, 82–83, 236–238, 254
Adger, W. N. 195, 198, 199, 207, 255
AEC (Atomic Energy Commission) 31
agency 199–200, 208, 410–411
agenda setting 32, 42, 43, 74, 80, 85, 101, 102
agent-based models 308, 313–314, 327–330, 331
Agrawala, S. 35
agriculture 174–175, 178, 219, 238, 279, 286, 288, 310
  and environmental degradation 365–367
  farmer adaptation case study *see* Tanauan City
  and global market 327–328
  and privatization 364–365, 368
  rice *see* rice farming
  see also under Ping River Basin
anthropology 29
Arctic 18, 143, 147, 183
Armitage, D. R. 297

ASAL (Arid and Semi-Arid Land)
    programme (Kenya) 274–275, 284
  civil servants in 274, 275
  closure of 275
  corruption and 275
  impact evaluation study of 275–276
  local involvement in 274, 275
  perceptions of 282–283, 284, 285, 287
  projects 283
assessments 3, 4–5, 28–40, 53–54, 156,
    158–159
  iterative 27, 33
  as process/product 54
  and specificity 27, 33–34
  tailored 27, 32–33
Australia 60, 61, 97

Bammer, G. 201–202
biodiversity 108, 147, 389–390
Blann, K. 40
Boudreau, T. 99
boundary objects 216–217, 223, 239,
    247–248, 260, 428
boundary organizations 10–11, 18, 27,
    34–35, 43, 81, 82
  and adaptive governance 256–257
  and food security 347–348
  obstacles to 261–263
  roles of 102, 107, 114–115, 116,
    216–218, 238–239, 254–255
  and social learning/co-production
    217, 234–235, 260
  three features of 239
Brewer, G. D. 6
Brown, A. L. 201–202
Burton, I. 29
Bush, George, Sr 132
Bush, George W. 23–24, 134–135, 434

California (US) 152–159
  Choosing our Future report (PNAS,
    2004) 159, 162–163
  Climate Action Team (CAT) 155, 159,
    162
  Climate Change Center 152, 157, 159
  climate change reports in 153
  climate laws/policies in 152–153,
    154–155, 158, 160
  climate science in 155–159

Environmental Protection Agency
    (CalEPA) 154–155, 159, 161, 162,
    165
  government/academic institutions in
    155–158
  NGOs in 158–159
  PIER programme 156–157, 158,
    160–161, 163, 165
  political inactivity in 153–154
  water resources of 156, 157–158
California Climate Change Scenarios
    project 18, 151, 157, 159–165
  actors/institutions in 159, 162, 433,
    434
  challenges of 160–161
  distinguishing features of 160
  establishment of 152, 155, 159
  factors in success of 161–162
  lessons learned from 162–164
  presentation/communication by 162,
    163
  scientific credibility of 161, 163
  and scientific networks 158, 162, 163
  stakeholder involvement in 161
Canada 60, 61
Canadian Arctic 18, 291–304, 432
  caribou management in 294
  decision-making for 292
  environmental change in 291,
    299–300
  indigenous people/knowledge in *see*
    Inuit people
  land claims/management agreements in
    291, 292, 295–298
  multiple stresses in 291–292
  see also Inuvialuit Settlement Region;
    Nunavut; Ulukhaktok
Cane, M. A. 37
capability domains 277, 279, 283–285
capacity 82, 97, 99, 108, 110, 408
capacity-building 103, 104, 107, 176,
    188, 223, 228, 342
carbon emissions 38, 154, 172
carbon sequestration 149, 156, 172
Caribbean 60, 61
caribou 294, 300
case studies overview 19
Cash, D. W. 32, 237, 238, 239, 262
cattle *see* livestock farming

CBDM (Community-Based Disaster
Management) project (Taiwan) 248,
261
  pilot study *see* Shan-an village
  processes/procedures 251–254
  public participation in 248, 251, 260,
    262–263
CCSP (Climate Change Science Program,
US) 131–145, 151, 434
  budget/agencies of 136–139
  criticisms of 136, 137, 144, 145, 428
  feedback through workshops in
    142–143
  five core goals 135, 137, 140,
    147–150
  information products/exchange in
    132, 140–142, 143
  initial mandate/development of
    132–135
  and NRC 136, 137–138, 140–141,
    142, 144–145
  Our Changing Planet reports 133,
    143, 145
  postponement of action advocated by
    136
  recommendations for 145
  reorganization of 132, 141
  Strategic Plan 135–136, 140, 142
  Synthesis and Assessment Product
    reports *see* SAP
  three research foci of 135–136
  and uncertainty in metrics 140–141
  and VARIP core questions 143–145
  vulnerability/resilience in 135–136
Centeno, H. G. S. 328
CGIAR (Consultative Group on
International Agricultural Research)
341, 342, 347
char, Arctic (*Salvelinus alpinus*)
300–304
Chiang Mai (Thailand) 381, 386, 387,
392, 394, 396, 399–400
  floods in 401
  mapping 404
Chi Chi earthquake (Taiwan, 1999)
234–235, 240–264
  and boundary organization (NCDR)
    244, 245–247, 248, 251, 252, 253,
    256, 428

  collaborative approach after 243, 246,
    247, 254, 428
  extent/impact of 240–242
  institutional response to 242–248,
    254–256
  and legislation 243–245, 256–257
  Post Earthquake Reconstruction
    Council 242–245, 249
  role of science/scientists after
    245–247, 254
cholera 272, 277, 278
Choosing our Future (PNAS, 2004) 159,
  162–163
churches 272, 273, 274, 276, 278, 281,
  282, 287, 432
Cicerone, Ralph 116
CIG (Climate Impacts Group, US) 152
CLIMAS (Climate Assessment for the
  Southwest) 32–33
climate change 4, 18, 99, 108, 307, 308,
  310, 315, 316
  and boundary organizations 35
  cross-scale approach to 40
  and knowledge legitimacy 429
  normative/ethical aspects of 424
  perceived benefits of 172–173, 174
  and rural vulnerability 307, 308, 310,
    315, 316
  and social/urban vulnerability *see*
    social/urban vulnerability
  and uncertainty 134
climate change scepticism 133–134,
  136, 429
Climate Change Science Program *see*
  CCSP
climate models 147, 156, 161
coastal regions 60, 143, 174, 179, 291
collaboration 40, 43, 67, 82–83, 217,
  238, 239, 243, 246, 247
  and legislation 244
  pilot study in *see* Shan-an village
  see also adaptive management
co-management 237–238, 257, 259–260
  indigenous people and 292, 298,
    303–304
command and control management 13,
  14, 76
communication 28, 37, 43, 98, 100,
  103–109, 253, 427–429

and boundary organizations 116, 256
challenges/opportunities of 103–104, 106–107, 108–109
and lay public 107–109
participatory 108–109, 117
and policy-making/management 104–107
risk 15, 107–108, 109, 206, 430
and temporal/spatial scales 104–105
and three dimensions of adaptation 104
and trust 105–106
community-based projects 248
pilot study of *see* Shan-an village
see also CBDM
community diversity 75–76
community groups/organizations 204–205, 206, 243, 295, 324, 398, 408
conflicts of interest 24
Conforti, P. 328
connectivity 28, 39–40
see also knowledge transfer
consensus 17, 82, 428, 429, 434–435
Consultative Group on International Agricultural Research (CGIAR) 341, 342, 347
coping capacity 97, 99, 104
co-production 11, 53, 56, 85, 217, 224, 239, 255, 260, 428
barriers to 74–76, 80–81
and creating meaning 261
exclusion from 262
corporations 5, 15, 31, 194
corruption 275, 279
cost-benefit analysis 75
credibility 27, 31, 32, 100, 161, 163, 238, 433
see also under vulnerability/resilience assessments
cross-cutting 5, 26, 34, 350
Cross, R. 312–313
Cuba 97
cultural landscape 372–375
Cutter, Susan 430

dams 278, 286, 384, 387, 389, 392, 405
data collection 56–59, 112, 191, 192–193, 202, 205–206, 223, 411

debris flow disaster exercise 254
decentralization 181, 183, 198
de-contextualization 83
desertification 364–365, 367
DFID (Department for International Development, UK) 111
diet 36
Dietz, T. 237
Dilley, M. 99
disaster mitigation/risk reduction 52–53, 99, 233
see also Chi Chi earthquake
disease 174, 178, 183, 272, 277, 278, 279
displacement of people 272, 278–280
donor agencies 111, 114–115
double-loop learning 40, 435
DPR (Disaster Prevention and Response) Act (Taiwan, 2000) 244–245, 247, 252, 254, 256, 257–258, 263
droughts 16, 60, 61, 97, 110–111, 178, 183, 272, 273, 278
Dunlap, Riley 429
du Toit, A. 226, 229

early warning systems 97, 111, 114, 376, 400–402
earthquakes 60, 61, 97, 128, 338
see also Chi Chi earthquake
ECHO (European Commission Humanitarian Office) 341, 344, 354
ecological resilience 29, 30, 234
economic growth 170, 171, 173, 180
ecosystems/ecology 29, 38, 42, 234, 237
ecosystem services 108
Edison, Thomas 23
education/training 65, 103, 104, 158, 179, 253–254, 279, 408
Egypt 60, 61
EKV (Ecomuseum Kristianstad Vattenrike) 256–257
Ellis, J. 367
El Niño 190
embeddedness 101, 111–112
employment 108, 180, 194, 279, 365
energy efficiency 171, 183, 283
engineering resilience 29–30

ENSO (El Niño/Southern Oscillation) models 37
entitlement theory 191, 200
environmental change *see* GEC
environmental decision-making 4, 5, 6, 11
   and political economy 35–36
environmental degradation 190, 233, 251, 338, 365–367
EOCs (Emergency Operation Centres, Taiwan) 247–248
EPA (Environmental Protection Agency, US) 132
EP-FAFS (European Programme for Food Aid and Food Security) 18, 335–359
   actors in 337, 344–345, 348, 353–354
   assessment/monitoring of 351–352, 359
   direct/indirect operating modes of 345–347
   documentary/policy analysis of 337
   establishment/goal of 339–340
   evaluating, challenges in 344, 348
   institutional context of 336, 340, 356
   knowledge/flows/use in 335–336, 344–347, 348, 352–354, 358, 432
   and local contexts/needs 352, 357
   and NGOs 341, 342, 343, 346–347, 348, 351, 354–355
   Nicaraguan context 337–339
   and partners/intermediaries 341, 346–347, 351, 355–356
   political interests in 355–356
   and poverty reduction 349–350, 351
   programme operation modes 343
   range of operations of 342
   recommendations for improving 357–358, 430
   reform of 343–344
   research approach/objective for 336–337
   and science/practice interface 336–337, 344–348, 347–348, 354–357
   targeting criteria of 342–344, 345–346, 350–351, 353, 358–359
   two pillars of 340
   and VARIP core questions 336–337, 348–357

vulnerability knowledge base in 336, 348–349, 352–353
equity *see* social justice
EuronAid 341, 342, 344, 346, 348, 354
European Commission (EC) food security policy 337, 340, 341–342, 345
   as boundary organization 347–348, 358
   fund allocations of 341
   partnerships in 341
   Programme for Food Aid *see* EP-FAFS
   reform of 343–344
   Regional Programme on Food Security 357–358
European Union (EU) 152
expertise 5–6, 15, 24, 26, 76, 100, 102, 428
   and credibility 31
   cross-boundary 44

famine 98, 233, 272, 273, 278
FANR (Food, Agriculture and Natural Resources Directorate, SADC) 111
FAO (Food and Agriculture Organization, UN) 111, 224–225, 340, 341, 342
farmer associations 319, 321, 324, 331
farming collectives 365, 367–368
federal agencies 5, 15, 132, 155, 164, 240, 274
feedback 5, 7, 17, 80–81, 110, 223, 323
FEWSNET (Famine and Early Warning System Network) 111, 114
Field, Chris 158
fishing industry 9, 36, 381, 382
fish resource management 300–304
fish stocks 300
FIVIMS (Food Insecurity and Vulnerability Information Management System) 111, 113, 114
   see also under South Africa food insecurity study
floodplains 381, 382, 385, 387, 389, 390
floods 16, 18, 60, 61, 97, 178, 183, 234
flash 385, 401–402
   management *see* Ping River Basin
   and poverty 192–193

and runoff 193
and urban growth 189, 190–192, 193
Florida Everglades (US) 39
Folke, C. 257
food nutrition information 222
food security/insecurity 18, 191
　access issue 215, 219
　children and 215, 220–221
　and EC *see* European Commission food security policy
　European Programme for *see* EP-FAFS
　and income 219–220
　integrated strategy for 226
　monitoring 218, 223, 225
　and poverty reduction 349–350
　and scientific knowledge 218–219
　three key elements of 349
　see also South Africa; Southern Africa
forecasting 97, 178, 190, 239
forestry 183, 283, 390
framing/reframing 4, 6, 8, 9, 25–26, 54, 100, 104, 430–431
free market 365, 367–368

Gadgil, M. 40
Gallopin, G. C. 30, 63
GEC (global environmental change) 97, 98–99, 108, 117, 133, 198, 233, 235, 425–426
Germany 60, 61
GHG (greenhouse gas) emissions 38, 145, 153, 154–155, 157, 171–172
GIS (geographic information system) 192, 202, 223, 229, 245, 308, 330
Global Change Research Program (US) 131
global environmental change *see* GEC
globalization 24, 60, 61, 197
　and vulnerability 307–308, 327–328, 423
Google 56, 58, 60
governance 101, 204–205, 297, 385, 410, 425
　adaptive *see* adaptive governance
　polycentric 244–245, 255–256, 261
Gunderson, L. H. 38, 39, 236, 256–257
Guston, D. H. 216, 217

Haas, Peter 11
Hahn, T. 237
Harvard University (US) 54, 56, 84, 429, 431, 433
　Knowledge Systems for Sustainable Development 32, 76–77
hazardous waste 13–14, 385, 431
hazard simulation models 247–248, 260
heatwaves 97
Hilgartner, S. 31, 36
HIV/AIDS 110, 112, 226, 279
Holling, C. S. 13, 25, 29, 30, 38, 39, 236
Holman Char Working Group (HCWG) 301–302, 304
housing 191, 193, 194
Hull, B. 200
Human Dimensions of Global Change 23
hunters/trappers 292, 296, 297, 298, 299, 301, 302
hurricanes *see* storms

ICRAF (International Centre for Research in Agroforestry) 396, 403–404, 406
ICRC (International Committee of the Red Cross) 81, 272, 273, 341
IFSNP (Integrated Food Security and Nutrition Programme) 221–222, 223, 224, 229
IHDP (International Human Dimensions Programme) 233
income 219–220, 310, 313, 314, 315, 338
India 60, 61
indigenous knowledge 18, 178–179, 274, 275
　benefits of using 294–295, 303–304
　defined 293
　and fish management 300–304
　and hunting/trapping 292, 296, 297, 298, 299, 301, 302
　institutionalization of 295–298
　and resource/wildlife management 291, 292–293, 294, 295–298, 299–303
　and science 293–295, 298–303
　and vulnerability 294

INDEX 447

industrialization 189, 190, 323, 381–382
inferences/experimental propositions 26, 41–42
information flow 76, 256, 258, 259, 260, 263–264, 313, 324–327, 408
 barriers to 325–326, 408
 improvements to 326–327
 see also knowledge transfer
information products 132, 140–141, 158, 162
information society 424
Informing Decisions in a Changing Climate (NRC) 6, 136
infrastructure 57, 62, 132, 136, 142, 174, 178, 190, 279
 scientific 163
institutional change 206–207, 234, 242–248
institutional design 113, 181–182
institutional fragmentation 17, 81, 336, 356, 408, 433–434
institutional memory 164
Integrated Food Security and Nutrition Programme *see* IFSNP
integrated water resources management *see* IWRM
Intergovernmental Panel on Climate Change *see* IPCC
International Atomic Energy Agency 9
International Centre for Research in Agroforestry *see* ICRAF
International Human Dimensions Programme (IHDP) 233
internet 43, 44, 56, 143, 431
Inuit people 292–293
 caribou management by 294
 fish management by 300–303
 and land claims agreements 295–298
 narwhal management by 297, 299
 and science 294, 295
Inuvialuit Settlement Region (Canada) 292, 295–296, 297–298, 303
 see also Ulukhaktok
IPCC (Intergovernmental Panel on Climate Change) 9, 29, 38, 99, 103, 142–143, 151, 161, 233, 422, 429
IRI (International Research Institute for Climate Prediction) 35

irrigation 251, 310, 316, 317–318, 384, 385, 389, 390
 infrastructure 381, 387, 392, 397, 408
issue domains 77
IWRM (integrated water resources management) 382, 392–393, 394, 410

Jackson, J. 115
Jäger, J. 35
Janssen, M. A. 26, 30, 44
Japan 248
Jasanoff, S. 35, 36, 101–102, 110, 113
Jordan, L. 204

Katrina, Hurricane 14, 52, 197
Kenya *see* West Pokot District
Kerkhoff, L. van 77
knowledge/knowledge base 5–6, 35, 59, 97–98, 100, 187–188, 201–203, 336
 access to *see* access to knowledge
 adequacy of 16, 80, 143–144, 336, 348–349, 352–353, 432
 career-orientated 77, 80
 and collaboration 43, 67, 256
 communication of *see* communication
 conflicts in 81–82
 co-production of *see* co-production
 dominant producer/user categories 348
 interaction/connectedness 109–110, 112, 114–115, 424
 and legitimacy 27, 116–117, 429
 limitations in 428
 local *see* local knowledge
 and losses 52–53
 and scales *see* scale issues
 socially robust 101–102
 uncertainties 37–38, 179, 385, 387
 see also data collection
Knowledge Systems for Sustainable Development (Harvard University) 32, 76–77
knowledge transfer 52, 54, 65, 67, 70, 80, 205–208, 344–347, 348
 barriers to 80–81, 205–207, 325–326, 408, 433

and direct/indirect operating modes 345–347
in idealized social network 328, 329
and partnership/dialogue 324
pipeline model 6, 73, 74, 75, 78, 238
and science/policy gap 335–336
spider web model *see* spider web model
see also information flow
Kyoto Protocol 18, 38, 134, 151
Russia and 169–171

land claims 291, 292, 338
and resource management 295–298
land degradation *see* soil erosion/degradation
land privatization 364–365
landscape connectivity 372, 373
landslides 189, 190, 191–192, 234, 243
land use 39, 193, 319–320, 321–322, 329–330
planning 264, 390
language issues 28, 37, 65, 69, 81, 116, 258–259, 397
leadership 114, 154, 203, 239–240, 257, 262, 263, 425
learning systems 14
Lebel, L. 77, 404
Lee, Kai 13, 435
legitimacy *see* credibility
Leichenko, R. 198
Lemos, M. C. 32–33
life histories 276
livelihoods approach 111, 199–201, 207, 223
livestock farming 270, 272, 273, 277, 278, 279
overgrazing by 372
see also Mongolian rangelands
local disaster models 247–248
local government 111, 189, 195, 319–320, 323–324, 327, 428
see also under Ping River Basin
local knowledge 12, 18, 40, 66, 83, 100, 227, 229, 260, 428
local scale of assessments 33–34, 35, 40
long-term planning 112, 264
Luers, A. L. 33

malnutrition 215, 219, 350
maps/mapping 224, 226, 239, 247, 260, 312, 313, 329–330
memory 314
poverty 351
maquiladora industry 194, 337
marginalization *see* social inequality
MDGs (Millennium Development Goals) 109, 218
media 7, 11, 64, 65, 68, 69, 71, 97, 107, 159, 162, 397
Medvedev, Dmitry 177, 178, 183
memory maps 314
meta-analysis 18
Mexico 18, 60, 61, 97, 194, 206
see also Tijuana
migration 190, 193, 194, 251, 274–275, 278, 316, 365
Mileti, Dennis 57, 60
Mitch, Hurricane 338, 352
Mock, N. 113
Moller, H. 298
Mongolian rangelands 363–376
climate change impacts in 367, 374, 375
cultural landscape restoration in 372–376
early warning systems in 376
ecosystem diversity in 363, 374
environmental degradation in 365–367, 375
free-market economy in 365, 367–368
goats/cashmere production in 368, 369, 370, 371
land-use/tenure changes in 364, 367–369, 372, 374
livestock numbers/incomes in 367, 368, 369–370
livestock privatization in 364–365, 368–369
modern technology in 375–376
negdels in 364, 365, 367
pastoral networks in 363, 364, 367, 368, 375–376
pastoral system in 363–365, 367
policy changes in 370–372
population dynamics in 365, 366
provinces/administrative units in 373–375

public services in 374
resilience in 375
storms/droughts in 365, 368, 372
sustainability in 372, 376–377
tragedy of the commons in 372, 375
vulnerabilities in 367–368, 375
water resources in 365, 366–367
monitoring 14, 15, 104, 178, 218, 223, 225, 228, 245, 352, 359, 401–402
wildlife 294, 297, 298, 299–300
monsoon 381
Morehouse, B. J. 32–33
Moser, S. C. 238
Mozambique 112, 227
multidisciplinary approach 12, 23, 41, 42, 98, 99–100, 188, 189, 196, 234, 238
to leadership 203
multiple stressors 79, 83–84, 197, 234

Naess, L. O. 207
NAFTA (North American Free Trade Agreement) 194
narwhal management 297, 299
NAS (National Academy of Sciences) 36
national security 178, 182
natural hazards 29, 38, 97–98, 108, 197, 338
and knowledge/losses 52–53
management 52, 53, 74
multiple 241–242, 244, 251, 263
and urbanization 188–191
NCDR (National Science and Technology Centre for Disaster Reduction, Taiwan) 244, 245–247, 248, 251, 252, 253, 255, 256, 257–258, 263
communication and 258–259
enabling legislation for *see* DPR Act
funding for 261
institutional flexibility of 256
knowledge/action links of 255–259
leadership and 257–258, 263
partnerships/social networks and 258
success of 264
trust-building/norms/value formation in 259

negdels 364, 365, 367
negotiations 85, 102, 105, 238, 239–240, 259, 261, 383
Netherlands 12
donor agencies from 269, 274–275, 281
networks *see* social networks
NGOs (non-governmental organizations) 9, 52, 79, 110, 113, 158–159, 179, 180, 242
and food security 341, 342, 343, 346–347
and governance 204–205
perceptions of 281–282, 284–285, 287
Nicaragua 327–329
food aid programme to *see* EP-FAFS
food insecurity in 338–339
international aid to 337
natural hazards in 338
poverty in 337, 338
PRODELSA programme in 337, 342, 357
NOAA (National Oceanic and Atmospheric Administration, US) 132, 134, 138, 152, 155
North American Free Trade Agreement (NAFTA) 194
Nowotny, H. 101, 111–112, 113
NRC (US National Research Council) 6, 136, 137–138, 140–141, 142, 144–145
Nuclear Non-Proliferation Treaty 9
Nunavut (Canada) 292, 295–297, 298, 303

Obama, Barack 4, 131, 136, 144, 145, 434
Obersteiner, M. 53
O'Brien, K. 198
one-size-fits-all approach 13, 16, 188, 195–196
Ostrom, E. 30
Our Changing Planet (CCSP) 133, 143, 145
outreach 4–5, 65, 67, 85, 100, 117, 159, 164, 308, 313, 430, 438
Owens, S. 110

Pakistan 60, 61, 97
participatory evaluation (West Pokot, Kenya) 269, 270, 276–287
  capability domains in 277, 279, 283–285
  impact assessments in 283–285, 287
  participants in 276
  perceived changes in 279–280
  perceptions of ASAL programme in 282–283, 284, 285, 287
  perceptions of best/worst projects in 285–287
  perceptions of government in 280–281, 287
  perceptions of NGOs/churches in 281–282, 284–285, 287, 288
  reconstruction of past events in 277–280
  workshop elements 276–277
partnerships 116, 150, 163, 183, 239–240, 258
  see also networks
paternalism 83
path dependency 14
Patt, A. 197
peace initiatives 273, 278, 282, 286, 287, 288, 289
peer review 36, 101, 161
permafrost 175, 183, 291
Peru 60, 61
Philippines 18, 59, 60, 61
  see also Tanauan City
PIER (Public Interest Energy Research) programme (California) 156–157, 158, 160–161, 163, 165
Pimm, S. L. 29
Ping River Basin (Thailand) 18, 381–411
  actors in 391–400, 410–411
  agriculture/irrigation in 297, 381, 384, 385, 388, 389, 390, 398, 405
  aquaculture in 382, 385, 387
  biodiversity/ecosystem loss in 389–390
  bureaucrats/bureaucracy in 391–394, 401, 408–409, 411
  community water management in 397–398
  construction projects in 388, 392, 406
  consultants in 394–395, 410

  dams in 384, 387, 389, 392, 405
  disaster prevention/mitigation in 393–394
  environmental degradation in 384, 397
  fishing in 381, 382
  floodplains in 381, 382, 385, 387, 389, 390
  flood protection debate in 382
  floods in 381, 382, 383, 384, 387, 399–400
  groundwater use in 318, 388, 390, 394
  ICRAF in 396, 403–404, 406
  indigenous groups in 388–389
  industry in 381–382, 392, 398–399
  institutional fragmentation in 408
  and Interior Ministry (MOI) 391, 393–394, 402, 403
  IWRM in 382, 392–393, 395, 410
  knowledge system/networks in 383, 400–411
  knowledge system, opportunities in 409–411
  knowledge system, tensions in 406–409
  land-use changes in 382, 385
  land-use diversity in 388–389, 405
  local government (TAO) in 391, 393, 403, 405, 409, 410
  marginalized people in 382, 383, 397, 411
  MONRE (Ministry of Natural Resources and the Environment) 391, 392, 393, 405
  planning in 404–406
  politics in 399–400, 405
  preparedness/early-warning system in 400–402
  public awareness campaigns in 400–401
  rainfall variation in 384–387
  researchers in 396–397
  resilience in 388–390
  river basin organizations (RBOs) in 392, 393, 394, 395, 397, 402–403, 404, 406, 409, 410
  Royal Irrigation Dept. (RID) 391–392, 398, 400, 402

INDEX | **451**

stakeholder engagement in 395, 397–399, 401, 402, 403, 405, 406, 410, 411
tourism in 382, 389–390, 399
upland/lowland challenges in 389–390, 393, 397
urbanization/industrialization in 381–382
vulnerabilities in 382, 383, 384–388, 397
water/land use conflicts in 382, 383, 398
water quality/pollution in 382–383, 384, 385, 387–388
Water Resources Dept. (DWR) 391, 392–393
water scarcity in 383, 384, 392, 399, 402–403
watershed networks in 403–404
water users in 388, 391, 397–399, 402
place-based analysis 12, 34, 234
policy communities 102
political agendas 24, 27, 35, 36, 42
  see also agenda setting
political ecology 233
political economy 27, 35–36
pollution 11, 60, 61, 171, 382–383
population growth 189, 194, 279
poverty 190, 191, 192–193, 194, 197, 198, 199, 219–220, 277, 338, 365
  alleviation 180, 218, 341, 346, 349–350
  see also food security/insecurity
power relations 102, 430–431
Preparing for an Uncertain Climate (USOTA) 133–134, 136
Pritchett, L. 188
PRODELSA (Food Security and Local Development Programme, Nicaragua) 337, 342, 357
propositional analysis 18
propositional inventory 26–28
public awareness 131, 155, 224, 400–401
public health 36, 277–278, 279, 286
  see also disease
public participation 15, 76
  and adaptive governance 235, 243, 244, 248, 251, 254, 257, 259–260, 262–263

  and indigenous people 295
public perceptions 28, 39, 179–180
Putin, Vladimir 170

radioactive waste storage 13–14
rainfall 190, 192
rapid rural appraisal technique 112
RBOs (river basin organizations) *see under* Ping River Basin
Red Cross *see* ICRC
Regional Hunger and Vulnerability Programme (RHVP) 99, 113, 115
regional-level projects 114, 115, 155–156, 163, 165
Regional Vulnerability Assessment Committee (RVAC) 111
Reid, P. 200
RESAL (European Food Security Network) 347, 348, 351, 354, 355, 356
research funding 4, 24, 36, 62, 79, 81, 164, 165
resilience 24–25, 26, 43–44, 51, 136, 234, 239
  and adaptive management 39
  approaches to 98–100
  assessment 51, 52, 53–54, 62–63
  challenges/opportunities of 98–109, 113–114
  communication and 103–107
  and connectivity 39–40
  defined/use of concept 25, 26, 29–30, 63, 99, 430
  ecological/engineering 29–30
  normative/ethical aspects of 424–425
  propositional inventory/literature review of 26
  and science/policy interface 421–423, 429
  and stakeholder involvement 108–109, 111, 112–113
  see also vulnerability/resilience assessments
Resilience Alliance 30
resource depletion 187
resource management 39, 291, 292–293
  diagrams of actors in 297, 299
  and fish stocks 300–304

and indigenous/scientific knowledge 298–304, 428
and land claims agreements 295–298
participatory approach to 295
RHVP (Regional Hunger and Vulnerability Programme) 99, 113, 115
rice farming 315, 319, 320, 381, 385, 389
　alternatives to 316, 317
RISA (Regional Integrated Sciences and Assessments) programme 152, 155
risk 9, 13, 38, 75, 136
　assessment 245–247, 424
　communication 15, 107–108, 109, 206, 430
　redistribution 397, 401
　social amplification of 6, 7
　see also uncertainty; vulnerability
Robertson, D. 200
Roux, D. 201
Rüdig, W. 30–31, 36
Russia 171–182
　adaptation options for 177–179
　climate change debate in 172–174
　climate change impacts in 173, 183
　Climate Doctrine (2009) 176–177, 178
　climate policy of 175–177
　climate science in 172
　economic growth in 170, 171, 173
　GHG emissions of 171–172
　national security of 178, 182
　NGOs in 180
　public awareness of climate change in 179–180
　regional participation in 180–182, 183
　and UNFCCC 173, 175
　vulnerability of 171–175, 179
　water resources/agriculture in 174–175, 178
Russia and Kyoto Protocol 18, 169–171
　and EU/WTO 170, 182
　impacts/challenges of 175–176
　and public perceptions 180
　rationale for ratifying 169–170
　and regional interests 181
　ten key messages of 182–183

and uncertainty 173–174, 176
Russian Federation, Climate Doctrine of (2009) 176–177, 178
　adaptation strategy in 178
RVAC (Regional Vulnerability Assessment Committee) 111

SADC (Southern African Development Community) 111, 114
　Regional Indicative Social Development Plan (RISDP) 218
Salvatici, L. 328
SAP (*Synthesis and Assessment Product*) reports (CCSP) 133, 140–141, 142–143, 145
　metrics in 141
　style of 140
　topics of 140
scale issues 33–34, 35, 40, 63, 78–79, 104–105, 196, 235–239, 255, 423, 432
Schrecker, T. 36
Schröter, D. 197
Schwarzenegger, Arnold 152, 154, 163
science/policy/practice interface *see* SPPI
science/practice gap 3–9, 23–24, 188, 289, 322–324
　and boundary organizations 10–11
　and institutions 433–434
　and knowledge transfer 335–336
　in poor countries 189
　and power 430–431
　reducing 197–201, 436–438
　reducing, obstacles to 195–197, 205–207, 208, 428–430, 433
　and sustainability science 11–12
science/practice interface 116–117, 201–207, 307, 326–327
　actors in 6, 8, 9, 16–17, 101, 432–434
　bottlenecks in 322–324, 331, 432
　communication links at *see* communication
　consensus in 17, 82, 428, 429, 434–435
　core issues of 16–17, 80–83, 143–145, 336–337, 348–357, 432–436
　feedback in 4–5, 98, 327
　and food insecurity 215–216, 221
　and governance 425

improvements to 356–357, 436–438
and institutional fragmentation 17, 81, 336, 356, 433–434
metaphors used in 6–9, 11, 100–103
normative/ethical aspects of 424
and other issues/actors 5
roles blurred in 101
and scale 423, 432
spider web model of *see* spider web model
and trust 431–432
and vulnerability/resilience 421–423
weak points in 435
see also under South Africa food insecurity study
scientific advisors 27, 35–37
scientific disciplines 27, 30–31, 62, 98, 99–100, 188, 202, 235
see also multidisciplinary approach
scientific illiteracy 3, 4, 15
sea-level rise 60, 61, 136, 147, 150, 174, 291
seed companies 312, 318, 325–326, 327, 332
Sen, A. 211
Shan-an village pilot study (Taiwan) 249–254, 259
background to 249–251
CBDM practices/procedures in 251–254
debris flow disaster exercise 254
disaster management map/database 252–253, 254
funding/project continuity for 261–262
public participation in 262–263
site/time specificity 27, 28, 33–34, 195
Smit, B. 24–25, 29
social capital 223, 240, 243, 257, 259–260, 262
social context 7, 112
social-ecological systems 24, 28, 30, 38, 39–40, 53–54, 78–79, 84, 196, 234, 237
social inequality 187, 189, 190, 192–193, 194–195, 199, 215, 429
and globalization 197–198, 307
social justice 17, 99, 198–199, 208, 434
social learning 14–15, 77, 82–83, 239, 256, 260, 264, 331, 429, 435

social memory 85, 240, 243, 257
social networks 102, 142, 158, 162, 163, 240, 247, 257, 258, 331–332
analysis 312–313, 324–327
pastoral 363, 364, 367, 368, 375–376
and vulnerability scenarios 327–330
Social Science Citation Index (SCCI) 56, 58, 60
social security 219, 220
social/urban vulnerability 188–209
and adaptation 197–201, 207, 208
external/internal sides of 192
and industrialization 190
and inequality/marginalization 187, 189, 190, 192–193, 194–195
and science/practice interface 201–207, 202–207
study, data collection in 191, 192–193, 202, 205–206
study, stakeholders in 203–205
and topography/landscape 191–192, 193
and urban growth 188–191, 194–195
soil erosion/degradation 60, 233, 251, 279, 365–366
South Africa food insecurity study 18, 215–230
and administrative levels 218
and boundary objects/organizations 216–218, 223, 224–228
challenges/limitations of 228–230
children in 215, 220–221
data collection/analysis in 219, 223, 226
DoA/Social Cluster and 219, 221, 222, 224, 225–226, 227, 228, 229
FIVIMS-ZA and 215–216, 217–219, 220, 221, 223–229
food insecurity drivers in 220
and household income 219–220
and IFSNP 221–222, 223, 224, 229
institutional challenges in 224–227
and knowledge creation/dissemination 218, 223–224
monitoring in 218, 223, 225, 228
objectives/argument 216
and science/policy interface 215–216, 221, 226–227, 228
and social inequality 215

Southern Africa food insecurity study 98, 100, 102, 109–115, 116
 boundary organization (RIACSO) in 114–115
 institutional design in 113
 multiple causes of risks in 112–113
 multi-sectoral approach in 112
 stakeholder engagement in 111, 112–113
 successes in 111–113
 trust-building in 106
 VACs in see VACs
 weaknesses in 113–114
Southern African Development Community see SADC
spider web model 6–9, 18, 101, 116, 144, 324, 344–345, 430–431
 simple/complex stable/complex unstable 9, 11, 431, 433, 434, 436–438
SPPI (science/policy/practice interface) 53–56, 58, 66–67, 68–69, 70, 83
SSCI (*Social Science Citation Index*) 56, 58, 60
stakeholder involvement 108–109, 111, 112–113, 161, 236, 238
START (SysTem for Analysis, Research and Training) 308
steppes see Mongolian rangelands
Stern, Nicholas 424
Stern, P. C. 6
Stone, D. A. 108
storms 60, 61, 97, 241–242, 253, 338
 see also typhoons
sustainability 11–12, 98–100, 115, 183, 403, 424
 and indigenous people 292
sustainable development 32, 35, 38, 53, 176, 187, 188, 198–199, 200–201, 207, 208
sustainable livelihoods 199–201, 207, 208
SWCB (Soil and Water Conservation Bureau, Taiwan) 241, 243, 244, 245, 253
Sweden 12, 237–238

Taiwan 18, 233–264, 432
 boundary organization in 245–247
 collaborative approach in 243, 246, 247, 249
 collaborative approach in, pilot study see Shan-an village
 Community-Based Disaster Management project see CBDM
 Disaster Management Decision Support (DMDS) project 247–248
 Disaster Prevention and Response Act see DPR Act
 earthquake in see Chi-Chi earthquake
 economic reform in 250
 EOCs (Emergency Operation Centres) in 247–248
 floods/landslides in 234, 240, 244
 institutional reform in 234
 Land Recovery Act in 264
 National Disaster Prevention and Protection Commission (NDPPC) 245, 246, 247, 249, 258, 261
 National Science and Technology Centre for Disaster Reduction see NCDR
 Post Earthquake Reconstruction Council (921 PERC) 242–245, 249, 251
 Soil and Water Conservation Bureau (SWCB) 241, 243, 244, 245, 253
 system/policy development in 244
 typhoons in 240, 241–242, 243, 244, 249, 253
Tanauan City (Philippines) 307–332
 action research in 320, 321, 323, 325, 331
 adaptation measures adopted in 317, 318
 adaptation to global change in 308, 315, 316, 327–328
 agent-based model for 313–314, 327–330, 331
 barriers to adaptation in 316, 322–324, 328, 331
 case study area 308–310
 case study methods 310–314
 crop diversification in 320, 328, 329
 ecotourism in 310, 315
 extension services in 319, 321, 327, 331

farmer associations in 319, 321, 324, 331
farmer information sources in 324–327, 331
farmer training in 318, 327
farmer typologies for 310, 314–316, 317, 325, 331
farm productivity in 310, 316, 317–318, 320, 322, 331
interviews/network analysis of 312–313, 317–324
land ownership/use in 312
land-use programme in 319–320, 321–322, 327
local government in 319–320, 323–324, 327
mapping 308, 312, 329–330
marketing in 310, 311, 315–316, 318–319, 320–321, 322, 331
science/policy bottlenecks in 322–324, 331, 432
science/policy support in 317–324, 327, 332
seed/fertilizer companies in 312, 318, 325–326, 327, 332
social networks in 312–313, 324–327, 328–329, 330, 331–332
survey/cluster analysis of 310–312, 313, 314–315
trade/climate scenarios for 326–327
vulnerability/network scenarios for 327–330
Thailand *see* Ping River Basin
Thomas, D. 198
Thomas, David 396, 403–404, 408
Tijuana (Mexico) 187–209, 432
data collection in 191, 192, 205–206
floods/landslides in 190–191
industrialization/urban growth in 188–191, 192, 193, 194, 429
inequality/marginalization in 189, 190, 192–193, 194–195, 429
local government in 189, 190, 191, 192, 193, 195, 202, 203, 206
maquiladora industry in 194
science/practice interface in 201–207, 208, 434
stakeholders in 204–205, 207–208
topography/landscape in 191–192, 193
timeframes *see* site/time specificity
Tompkins, E. 199, 207
Toraji, Typhoon 242, 249, 250, 252
tourism 310, 315, 382, 389–390, 399
trade liberalization 307, 326–327
tragedy of the commons 372, 375
transparency 140, 247, 259, 406, 408–409, 410, 430
trust/mistrust 9, 67, 75, 76, 81, 82, 105–106, 203, 237, 239, 240, 258, 288, 331, 431–432
tsunami 60, 61, 97
Turner, B. L. 25, 29, 30, 33, 34
Turnhout, E. 202
Twyman, C. 198
typhoons 240, 241–242, 243, 244, 245, 249, 253
community-disaster management for *see* Shan-an village

UCS (Union of Concerned Scientists) 23–24, 158, 159
Uganda 270, 279
Ulukhaktok (Canada) 298–303
fish resource management in 300–303
wildlife monitoring in 299–300
uncertainty 6, 13, 15–16, 28, 37–38, 39, 83, 173–174, 385, 387, 422
United Nations 110–111, 114, 337
Convention on Long-Range Transboundary Air Pollution 171
Food and Agriculture Organization *see* FAO
Framework Convention on Climate Change (UNFCCC) 38, 99, 133, 151, 233
regional climate change assessments in 152
World Food Programme (WFP) 111, 341, 342, 351
United States (US) 60, 61, 97
Agriculture Dept. (USDA) 132, 147, 149, 150, 238
climate change debate in 3–4, 5, 18, 133–134, 136, 434
Climate Change Science Program *see* CCSP

Energy Dept. (DOE) 132, 147, 148, 149, 150, 155–156
Global Change Research Act (1990) 18, 131, 132–133, 134, 140, 143
and Kyoto Treaty 134
National Oceanic and Atmospheric Administration *see* NOAA
see also California
urban fragmentation 190, 194
urbanization 189–191, 192, 323, 381–382
　illegal 192, 193, 195
　two paths of 193–194
　and vulnerability to floods 189, 190–192, 193
urban vulnerability *see* social/urban vulnerability
USAID (US Agency for International Development) 132, 138
usefulness of knowledge 27, 32
USGCRP (US Global Change Research Program) 132, 133, 134, 137, 139, 156
US National Research Council 5, 6, 11–12, 15, 428, 434
　decision support principles of 12

VACs (vulnerability assessment committees) 110, 111–115
　challenges of 115
　multi-sectoral approach in 112
　stakeholder involvement in 113
　strengths of 111–113
　weaknesses of 113–114
Van Tuijl, P. 204
variability 28, 38
VARIP (Vulnerability and Resilience in Practice) project 3, 56, 57, 58, 80, 335
Vietnam 60, 61
Vincent, K. 207
Vogel, C. 200
vulnerability 24–25, 43–44, 51, 136, 233–234, 429–430
　approaches to 98–100
　capacity 105
　challenges/opportunities with 98–109, 113–114
　communication and 103–107

　and connectivity 39–40
　defined/use of term 25, 26, 28–30, 63, 99, 113, 196, 349–350, 430
　empirical sources on 26, 28–40
　and government priorities 82, 307
　intuitive sources on 26, 41–42
　maps 252–253
　methodological guidelines for 197
　multilevel/interconnected 387
　and networks 327–330
　normative/ethical aspects of 424–425
　and participatory communication 109
　propositional inventory/literature review of 26
　reduction *see* adaptation
　and science/policy interface 421–423, 429
　and social processes 197
　and specificity 27, 33–34, 195
　and technical solutions 84, 196–197
　see also social/urban vulnerability
vulnerability assessment committees *see* VACs
vulnerability assessments 100, 111, 197, 246
vulnerability/resilience assessments 51, 52, 53–54, 62–63
　barriers to co-production in 74–77, 80–81
　case study aims 53–56
　case study characteristics 60–62
　case study locations 60
　case study publication details 88–89
　case study selection for 57–59
　core issues and 80–83
　and credibility/legitimacy 66, 71, 72, 74, 78
　data collection for 56–59, 70
　and date of publication 70, 78
　in developed/developing countries 69–70, 72, 78
　and disaster management domain 71–72
　gender factors in 67, 71, 78
　hazard types in 60
　information sources for 64, 66, 67, 70
　integrative 62–63
　knowledge producers in 63–65, 67–70, 73–74, 77–78, 79, 80, 84

and knowledge transfer *see* knowledge transfer
knowledge users in 58–59, 65–66, 71–74, 80
limitations of 76–77
questionnaire 58, 63–74, 93–96
questionnaire participants 90–92
questionnaire results 66–74
and recognition of publications 67–68, 73, 74
recommendations for 83–85
scientific disciplines/knowledge domains and 62–63
single-/multi-hazard 60, 69, 73, 79
and spatial scale 63, 78–79
SPPI and 53–56, 58, 66–67, 68–69, 70
Vulnerability and Resilience in Practice project *see* VARIP

Walters, C. 13
Wandel, J. 24–25
warning systems *see* early warning systems
water pollution 382–383
water resources/management 39, 105, 147, 156, 157–158, 174–175, 238, 279, 283
 see also dams; irrigation
watershed networks 403–404
Westley, F. 238
West Pokot District (Kenya) 269–289
ASAL programme in *see* ASAL programme
cattle farming in 270, 272, 273, 277, 288
chronology of events in 278
church activities in 272, 273, 274, 276, 278, 281, 282, 284, 287, 432
civil servants in 274, 275, 281
dam project in 281
disease in 272, 273, 277
drought/famine in 272, 273, 278
education in 270, 272–273, 275, 276, 279, 281, 282, 283, 286
gold rush in 273
government finance in 274, 275
history of region 270–273
indigenous participation in 274
local councils/district officers in 281
NGOs/donor agencies in 269, 274, 281–282, 284–285, 287, 288, 289
participatory evaluation in *see* participatory evaluation
peace initiatives in 273, 278, 282, 286, 287, 288, 289
violence/political conflict in 273, 274, 278, 281, 288
vulnerability of 270, 272–273, 288–289
water management in 281, 283, 286, 288
wetland landscape management 237–238, 256–257
WFP (UN World Food Programme) 111, 342, 351
White, G. F. 52, 53
women 71, 272, 277, 285, 397
women's groups 274, 276, 280, 282, 283
Woolcock, M. 188
World Bank 60, 204, 319, 340, 391, 395, 404
WTO (World Trade Organization) 170

Yaqui Valley (Mexico) 33
Yorque, R. 37, 38, 43
Young, O. R. 33, 40

Zimbabwe 112–113, 114